Cable Visions

Cable Visions

Television Beyond Broadcasting

EDITED BY

Sarah Banet-Weiser, Cynthia Chris, and Anthony Freitas

New York University Press

NEW YORK AND LONDON

NEW YORK UNIVERSITY PRESS
New York and London
www.nyupress.org

Library of Congress Cataloging-in-Publication Data
Cable visions : television beyond broadcasting / edited by Sarah
Banet-Weiser, Cynthia Chris, and Anthony Freitas.
p.   cm.
Includes bibliographical references and index.
ISBN-13: 978-0-8147-9949-9 (cloth : alk. paper)
ISBN-10: 0-8147-9949-3 (cloth : alk. paper)
ISBN-13: 978-0-8147-9950-5 (pbk. : alk. paper)
ISBN-10: 0-8147-9950-7 (pbk. : alk. paper)
1. Cable television—United States.   I. Banet-Weiser, Sarah, 1966–
II. Chris, Cynthia, 1961–   III. Freitas, Anthony.
HE8700.72.U6C355 2007
384.55'50973—dc22   2007009176

New York University Press books are printed on acid-free paper,
and their binding materials are chosen for strength and durability.

Manufactured in the United States of America
c 10 9 8 7 6 5 4 3 2 1
p 10 9 8 7 6 5 4 3 2 1

Contents

Acknowledgments

The idea for this volume emerged from a panel discussion about innovative practices in cable television in which we participated, along with Cheri Ketchum and Catherine Saulino, at the International Communication Association's annual meeting in San Diego in 2003. Despite the richness of the field of television studies, it seemed to us and to colleagues in attendance and with whom we discussed the matter that much of the work being done on subscription-based, multichannel television deserved to be collected in a form that would allow us, as scholars and as students, to examine the institutions, practices, and meanings that emerge from the hundreds of channels that are commonly called cable TV. The authors who contributed to this volume are among those established and emerging scholars working on issues pertaining to this kind of media. First and foremost, we extend our immeasurable appreciation to them. Each and every one brought enthusiasm and patience to this project, and the breadth of their collective knowledge never failed to impress us. At New York University Press, Eric Zinner and Emily Park offered confidence and guidance throughout the process; anonymous readers encouraged us with the depth and rigorousness of their critical engagement with the manuscript. Ronald Duculan, Josh Dunsby, Joy Van Fuqua, David Gerstner, Larry Gross, and Marita Sturken lent ideas and insights along the way. Cara Wallis at the Annenberg School of Communication at the University of Southern California was an invaluable research assistant. Colleagues at our various institutions provided untold support. Finally, we acknowledge the legions of scholars in various disciplines, many of whom are cited in this volume, whose work was foundational to Television Studies and which continues to inspire us.

Introduction

> Few technologies have been the subject of as many hopes
> and expectations as cable television. If one were to be-
> lieve its varied advocates, cable TV could carry almost
> every international sports and cultural event into the
> home, allow retail shopping from the living room, per-
> mit instantaneous referenda on public issues, and assure
> emergency monitoring of fire and burglar alarms. What
> more could anyone ask for?
> —Mitchell Moss, "Can Cable Keep Its Promises?"
> *New York Affairs* 6, no. 4 (1981)

Cable TV, on the brink of a boom in the 1970s, promised TV audiences a
new media frontier, an expansive new variety of entertainment and information choices. Cable seemed poised to provide access to a greater variety
of media forms and points of view than could be found on oligopolistic
broadcasting sources, as increasing channel capacity, a regulatory apparatus that had become amenable to the growth of the industry, and consumer demand for new services transformed how homes received television signals. The rapidly expanding industry appeared to lower barriers to
entry, allowing independent entities to launch cable networks and allowing new media forms and programming formats to emerge and prosper.
Music video, 24-hour news, 24-hour weather, movie channels, children's
channels, home shopping, and nostalgia channels devoted to programming long off-network shows were introduced.

Prior to the cable boom, most homes in the United States received only
a handful of broadcast television signals, which typically included, on VHF
channels, the local affiliates of the "Big Three" networks, ABC, NBC, and
CBS; in most markets, a PBS station; and, on the UHF spectrum, a few
scattered, independent stations. Now, the majority of American homes receive not only their local broadcast signals but also dozens, even hundreds,

of additional channels via cable or satellite dish.[1] In 1980, 17.6 million television-owning homes in the United States, or 23 percent of the market, subscribed to cable TV.[2] By June 2005, some 94.2 million households (86 percent of TV households) had come to subscribe to some form of multichannel video programming distribution (MVPD). Most (69.4 percent) were connected to cable; 86.3 percent of those cable subscribers received more than 36 channels. While the number of households subscribing to premium services (such as HBO, Showtime, Cinemax, Encore, or Starz!) has declined in recent years (to 28.1 million, in 2004), a growing number (26.3 million homes, by the end of June 2005) subscribes to digital cable tiers that have increased channel capacity and interactive capabilities such as video-on-demand. In 2005, subscribers paid an average $56.51 per month for cable-TV only (or $70.75 when bundled with high-speed Internet access via cable modem), and industry revenue for the year exceeded $66 billion. By June 2005, 26.1 million homes (27.7 percent) subscribed to direct-broadcast satellite (DBS) television services, which compete with cable to provide much the same service via a wireless alternative technology. Non-broadcast networks now draw a greater share of the television audience than broadcast networks (for the 2004–2005 season, non-broadcast audiences outpaced broadcasting's declining viewership, at 53 percent to 47 percent, respectively, in prime time, and an even greater share, 59 versus 41, for "all-day" viewing, over the 24-hour period).[3]

The distinctions and interconnections between broadcast and cable television have framed the way the cable television industry was both conceived and advertised. Indeed, the optimistic early "blue sky" era of cable television was accompanied by an ever-more cynical outlook on the overt commercialism of broadcast networks, leading to a simultaneous hopefulness and doubt about the symbolic power of television—especially as it concerned the empowerment of the viewer. While broadcast television was still held to the public interest obligations set out in the 1934 Communications Act, by the 1970s, TV seemed much more committed to a commercial system dependent on advertisers than to abstract obligations to the public's "interests."

Academic Attention: Cable Television Studies?

The topic of cable television within academic settings is situated firmly within Television Studies. However, despite the increasing interest and significance of cable television, there is still a striking absence of scholarly work on the various dimensions of the cable industry, including regulatory practices, the proliferation of channels, and actual programming. Previously, most television scholarship that has explored the implications for society of the shape of the television industry and the form and content of its programming have concentrated intensively on prime-time broadcasting and a few daytime genres such as talk shows, soap operas, and children's programming. A number of anthologies, edited volumes, and textbooks strive with much success toward comprehensive overviews of television history and approaches to TV studies; consider, for example, *Television: The Critical View,* edited by Horace Newcomb, which was first published in 1976 and has subsequently been expertly and frequently updated in new editions.[4] Yet volumes such as this one continue to draw most of their examples of industrial formations and programming from broadcasting's mainstream forms, underrepresenting the extent to which multichannel video programming distribution systems (cable and satellite) have usurped broadcasting as the means by which we now receive TV and captured the majority of the audience that once clustered en masse around a handful of broadcast channels and now fractures across dozens, even hundreds of channels.

There have been, certainly, a number of publications that focus on television content enabled by the rise of cable television. Several have taken the music video as their subject.[5] More recently, a few authors have turned their attention to analyses of cable and satellite TV's cultural and social alternatives.[6] Other scholars have focused narrowly on particular high-profile cable series.[7] This growing body of new research is a strong indication of the recent preeminence of cable programming and the beginnings of the scholarly attention it warrants.

This book examines the role of cable television in this proliferation of formats and channels and in the disaggregation of a once-mass television audience, as well as considering some of the implications of these trends. In this book, we look beyond broadcasting's mainstream, toward cable's burgeoning alternatives, to critically consider the capacity of commercial media to serve the public interest. The essays in the book represent new work by leading and emerging scholars in a variety of academic disciplines —communication and cultural studies, film and television studies, media economics, gender studies—as well as by industry professionals. The essays that follow focus primarily on cable television within the United States, but we also include discussions of globalization within television markets as well as examinations of cable's intersections with broadcasting, satellite television, and the Internet. The book is divided into three parts: Institutions and Audiences; Channels; and Programming. These categories are, of course, not mutually exclusive, and the chapters within each section necessarily touch on, draw from, and connect to the chapters in the other sections. For example, many essays in the Channels and the Programming sections explore the history of the networks and studios they examine. Likewise, essays in the Institutions and Audiences section address concerns about channel development and programming. In the remainder of this introduction we briefly outline some of the key elements and issues facing the cable industry.

How Cable Is Different: History and Regulations

The emergence and increased installation of cable technology in the American home during the 1970s was positioned and celebrated in terms of cable's difference from broadcast television. Cable television, it was claimed, could offer less "lowest-common-denominator" and crassly commercialized television, less intrusive advertising, and more interactivity on the part of the viewer, more viewer empowerment. In other words, cable TV would ostensibly "serve" the public interest in a way that broadcast television ignored. In fact, to some extent, the cable TV industry has brought new owners into the medium of television, upended longstanding programming traditions, and addressed audiences underserved by broadcasting. However, while offering many of these distinctions from broadcast television, cable television also shared a number of similarities with it. The cable market has been largely, but not exclusively controlled by the same powerful conglomerates that dominate other media markets; it has repeated and recycled broadcast programming concepts (and programming itself, in the form of repurposed and off-network syndicated series); and it has competed fiercely for those audiences already favored by broadcasters.

Cable's more relaxed regulatory environment and diverse patterns of ownership can be attributed directly to its distinctive origins. Broadcasting has, historically, been the most regulated arena of the U.S. media, reined in by a few not-so-little words first found in the Radio Act of 1927, which was written as powerful corporations positioned themselves to dominate the manufacture of radio equipment, the production of content, and its distribution throughout national networks. The act declared that broadcast stations would be licensed (and licenses would be renewed) only if it could be determined that these stations would serve "public interest, convenience, or necessity." This phrase was repeated in the Communications Act of 1934, in reference to both broadcasting (radio, and later, television) and telephony. The regulatory apparatus sought to shape broadcasting as a public trusteeship. Thus, a privately owned, for-profit industry was granted use of a public resource, the electromagnetic spectrum, through which its signals would be transmitted from station to listener and viewer. In exchange for access to this presumably scarce resource, commercial broadcasters would be obliged to *serve* the audience, even as they strived to profit by *selling* their audience to advertisers. That is, while the concept of public interest has never been precisely defined (beyond vague agreement that the broadcast industries should operate as a competitive marketplace and represent diverse viewpoints), broadcasting has been charged with the sometimes contradictory tasks of addressing not only the most potentially lucrative audiences but the nation as a whole, and addressing us not only as consumers but also as citizens.[8]

Thus it can be said that broadcasting emerged, in the form of radio and, following in the template established for radio, in the form of television, as an oligopolistic market whose players enjoyed significant market power and agreed to some constraints on the extent of that power. In contrast, as Megan Mullen shows later in this volume, cable TV, or as it was initially called, Community Antenna Television (CATV), originated in the late 1940s and early 1950s in the hands of individual, independent entrepreneurs without ties to established media corporations and with little scrutiny or regulation of their efforts. While cable television was spared the public-interest obligation and some of the specific ownership and content regulations pertaining to broadcasting, it would be misleading to characterize this market as entirely unfettered. Eventually, cable and direct-broadcast satellite television would be overseen by a combination of federal and local authorities. For example, the Federal Communications Commission (FCC) sets ground rules for the kinds of franchise agreements that can be negotiated by municipalities with cable systems. The FCC has also, at times, regulated cable pricing. And the television industry, like any other, can be subject to antitrust policies of not only the FCC but also the Department of Justice and the Federal Trade Commission.

However, most limitations on ownership and content placed on broadcasting are largely absent from cable and DBS regulations.[9] This absence is justified by the fact that broadcasters transmit their signals free-to-air over publicly owned spectra, to be captured by anyone with a receiving set and an antenna. In contrast, cable and satellite transmissions are received only by those audience members who choose to bring this kind of television into their homes by making monthly payments and who typically acquire or lease special equipment in the form of set-top boxes or roof-top dishes. Cable and DBS are, in the eyes of the courts and other authorities, "invited guests" (perhaps, more accurately, hired help) that a subscriber could banish simply by unsubscribing. Given the purposefulness that is said to characterize reception of cable and DBS signals, Congress has seen fit to let the market rule without some of the constraints placed on broadcasting by means of the "public interest" clause; moreover, the courts have ruled that, unlike broadcasting, cable and DBS channels hold the same First Amendment rights as print media.[10] In other words, the rules treat cable and DBS subscribers as citizen-consumers whose needs are best met by market-driven media.

Institutions and Audiences

Given its legacy from broadcast television, it is unsurprising that cable technology was characterized by both optimistic and pessimistic forecasts over the role cable would play in audience empowerment. Within the context of looser regulatory control and the installation of technologies that could receive programming delivered by satellite, the promise was that cable would address the problems that continued to haunt broadcast television. As Megan Mullen points out, early policy statements on the possibilities of cable (by, for example, the 1971 Sloan Commission on Cable Communications) associated the technology with a kind of "revolution" which could eventually "remedy all the perceived ills of broadcast television, including lowest-common-denominator programming, inability to serve the needs of local audiences, and failure to recognize the needs of cultural minorities."[11] The idealism of such a claim is quite similar to the language used at the emergence of other communication technologies: telegraphy, photography, radio, and television.[12] Like these technologies, cable promised to radically revise the viewers' relationship with media by encouraging a more active viewership and a newfound sense of political and cultural empowerment. Because cable did not use public airwaves, and thus was not regulated in the same ways as broadcast television, there were fewer constraints on its content. For instance, since cable was not (initially) conceived as being dependent on advertisers, it promised to cater more to the direct needs and tastes of specific audiences through niche channels. Capable of this kind of direct audience address, cable channels were seen as possible facilitators of viewer empowerment and a potential catalyst for citizenship.[13]

Importantly, this was not simply an idea that circulated in the world of telecommunications policy. As Thomas Streeter argues, the enthusiastic claims about the value of cable as the groundwork for an information highway were included in the policy reports and city documents produced by government agencies and others interested in the new technology and planning for its implementation. This discourse also found its way into public and industry discourse.[14] Moreover, since the burgeoning cable industry was not yet tied to major media conglomerates, it seemed to exist as a communication technology that encouraged independent entities to create networks and channels that allowed for alternative kinds of programming.

As was the case with broadcast television, the technology that enabled cable was not the essential factor invoked in the public and policy discourse concerned with the revolutionary qualities of cable. As Streeter articulates, although constantly labeled as "new," cable was not necessarily "new technology." Rather, it was the *use* of this technology that was understood as novel: the increase in channel capacity and the development of a market so that more channels were available were the potential catalysts for the "revolution." Indeed, it was precisely these "new" aspects of cable that made it "possible to speak of cable, not as an embodiment of social contradictions and dilemmas, but as a *solution* to them."[15] Cable was seen as a possible remedy to all sorts of social problems, such as racism and poverty, because it could "narrowcast" and thus more accurately represent an audience. And, as Streeter argues, "Cable, in other words, had the potential to rehumanize a dehumanized society, to eliminate the existing bureaucratic restrictions of government regulation common to the industrial world, and to empower the currently powerless public."[16] The chapters in Part I of this volume address the history and development of cable systems and audiences, and anticipate the convergence of digital media delivery to the home.

Channels

It is clear that the hyperbolic promise of cable was never quite fulfilled. In fact, the current structure of the cable industry resembles the familiar format of broadcast television in almost all ways. There was no televisually-inspired "revolution," and many cable channels and much cable programming are now owned and produced by the same transnational media corporations as the broadcast networks. However, the utopian discourse that framed the early debates on cable, especially in terms of the public interest, was essential for providing a context for channels such as CNN, C-SPAN, and even Viacom's Nickelodeon to emerge precisely because it focused attention on television's potential to empower viewers as particular kinds of citizens. To enforce public interest obligations would necessarily change the scope of the television industry, and it would certainly transform the dynamic among viewers, media owners, and advertisers.

In part, this dynamic changed through the proliferation of channels, whereby in a few short years the cable industry has added hundreds of different channels to the television landscape. A familiar question results from this kind of quantity: is a more diverse television system such as cable capable of accurately representing different audiences and publics, or does the proliferation of channels and the concomitant increasing centralization of media ownership indicate that the cable industry is a particularly profitable site for new niche markets and the commodification of identity? Joseph Turow identifies cable as one place for the cultivation by advertisers of "primary media communities," wherein communities of consumers are sought in order to nurture brand loyalty in a progressively more cluttered commercial landscape. Situating cable television alongside other communication technologies that are part of a new wave in marketing dedicated to narrow niche audiences and to the cultivation of "life-styles" in lucrative consumer groups, Turow understands cable as part of a new "fractured society." In fact, in seeing two cable channels, Nickelodeon and MTV, both owned by Viacom, as "pioneer attempts" to establish a kind of advertiser-shaped community, Turow argues that "While [Nickelodeon and MTV] started as cable channels, they have become something more. Owned by media giant Viacom, they are lifestyle parades that invite their target audiences (relatively upscale children and young adults, respectively) into a sense of belonging that goes far beyond the coaxial wire into books, magazines, videotapes, and outdoor events that Viacom controls or licenses."[17]

Cultivating this "sense of belonging"—the "something more" that constitutes a cable channel—has been a particularly successful way that cable channels have positioned themselves in the television industry, where it is not so much the programs on individual channels that are important, but the designs of the channels themselves: Nick is for kids, Animal Planet is for pet owners, Food Network is for foodies, etc. In Part II of this book, the authors look at specific cable channels as a way to parse out some of the contradictions in the regulatory history of cable—contradictions between the "blue skies" discourse that shaped the early cable industry and the intensely competitive nature of the current cable industry. As the authors in this section point out, some cable channels frame themselves according to a particular "theme," such as animals or food; others shape their identity around particular audiences, such as African Americans, children, or Latinos. These narrowcasting strategies are implemented with varying degrees of success in meeting both cultural and commercial desires.

Because cable channels often are designed to capture a part of the market share that is under-marketed, there are important ways in which cable channels "recognize" audiences that have been historically obscured by broadcast television. BET, HBO Latino, Bravo, Lifetime, Oxygen, and Logo, among others, have all tapped into consumer groups that have been underserved by broadcast channels, namely African Americans, Latinos, women, and gay men and lesbians. While this kind of recognition is crucial in a media society that often equates visibility with social value, it also works to commodify particular identities within boundaries established by cable companies. Additionally, niche channels have the tendency to marginalize distinct groups, identifying them as "different" from the mainstream (and thus deserving of their "own" channel), so that the result is not a harmonious multi-channel, multi-cultural televisual universe, but rather one in which broader audience channels continue to define the norms of representation.

Programs

The optimism that propelled the growth of the cable industry over regulatory hurdles and then throughout the consumer market for subscription television was not only based on promises of a more competitive television market and more channels within that industry. It was also the

programming that could be distributed via those new outlets that heralded a new era. But perhaps ironically, many of the most successful pioneering, nationally distributed cable networks (such as Home Box Office,
launched in 1972; The Movie Channel, 1973; and Showtime, 1976) focused
on feature films, most of which had already enjoyed theatrical releases.
Others, such as ESPN (1979), offered more of already familiar content: an
entire slate of sports programming, rather than the broadcasters' few time-
slots per week.[18] Tried-and-true genres would be, in many ways, cable's
bread-and-butter, and channels from Bravo to TV Land would fill endless hours with off-network series, interspersed with flagship original
programming.

Early advocates for cable touted the use of cable systems to produce
and air locally originated news in communities not served by broadcast
stations. They also envisioned telecourses and other educational programming that would be provided by schools and universities, as well as health-
care information offered by professionals.[19] While these kinds of programming would be part but hardly the eventual mainstay of cable, other content that had not found much of a home in broadcasting tried cable, too:
religious programming; in-depth coverage of governmental functions via
C-SPAN, launched in 1979, as a form of voluntary public service; and
high-end arts and cultural programming.[20] Innovative programming models such as these sought to serve audiences with interests and needs that
had been largely unmet by broadcasting; some flourished, others fizzled.

Still, continuing to be intent on attracting viewers with programming
that is at least somewhat different from what can be found on the broadcast networks, cable has forged new kinds of programming that is not
bound by broadcasting's decency regulations on profanity and sexual content. In addition, cable has also come to broach controversial subjects that
advertisers might shy away from in broadcasting but embrace in cable if
cable links them to a niche market. These conditions, which have freed
cable from broadcasting's "lowest-common-denominator" strategies, have
left it to devise tactical programming models which are examined in Part
III of this volume. One of these tactics, most closely but not exclusively associated with advertising-free, premium subscription cable, is the creation
of a "quality" genre, defined as adult-oriented drama with high production values like The Sopranos.[21] Another tactic involves developing programming that appeals to small but avid audiences which have demographic qualities coveted by advertisers and which can easily be lured away
from broadcasting with edgier programming such as Queer Eye for the

Straight Guy and *Chappelle's Show*. Elsewhere, cable repackages old genres in new ways—as in the form of single-genre channels providing 24-hour news (CNN, Fox News Channel), 24-hour sports (ESPN, ESPN2), or even 24-hour game shows (Game Show Network)—and operates as only one prong of a multimedia blitz organized around particular media brands, such as the Worldwide Wrestling Entertainment. Each of these tactics is addressed in Part III.

Cable's Contradictions

The utopian visions of the early cable industry promised a different, improved, more connected television landscape. Clearly, many of those initial goals have been reshaped within the competitive media-conglomerate context, and some of the same struggles over what constitutes the "public's interest" that plagued early broadcast television continue to have resonance in the more contemporary niche-channel cable universe. The essays in this book continue the exploration of the historical, regulatory, ownership, and programming parallels and disjunctures between broadcast and cable as well as other media. Additionally, these essays also address several fundamental questions that arise from cable's particular regulatory and ownership histories. Primary among them is the ability of cable and satellite television to serve the public: does cable, with its loosened FCC controls on content and its move toward niche channels, reflect a public's interest more accurately than broadcasting precisely because it is not bound by partisan regulatory policies? That is, how does cable's largely for-profit structure impact its ability to serve the needs of the public as citizens as well as consumers and to serve us as members of both distinct niches and as members of a common society?

Attempting to answer these questions definitively is a difficult task, primarily because the cable industry is thoroughly shaped by important contradictions regarding audiences, citizens, and the role of television as a kind of "public" service. Although it is this very *contradictory* nature of the cable industry that shapes the tone of this book, part of what we hope to do here is to recognize the symbolic power of television in terms of pleasure, of imagined community, and as a means of empowerment. At the same time, this symbolic power is always constrained by the political economy of the system, and this volume gives us a rich picture of the contradictions and complications of this media form. To this end, we interrogate the potential of cable to fulfill a "public interest" and take seriously the potential of the industry that was celebrated in the "blue sky" period.

NOTES

1. Throughout this volume the term "cable" is often used as shorthand to refer to channel line-ups that are available both from wired cable services and direct broadcast satellite (DBS) services. As several contributors illustrate, there are some important differences between the reach, costs, and offerings of cable versus satellite; however, the two provide a largely redundant channel line-up, albeit through different technologies.

2. Patrick R. Parsons and Robert M. Frieden, *The Cable and Satellite Television Industries* (Boston: Allyn and Bacon, 1998), 122.

3. These figures derive from Federal Communications Commission, Twelfth Annual Report: Annual Assessment of the Status of Competition in the Market for the Delivery of Video Programming (released March 3, 2006, MB Docket No. 05-255); see especially 4, 5, 11, 16, 19, 24, and 47–48. This series of reports is available via the Media Bureau at www.fcc.gov/.

4. Robert C. Allen, ed., *Channels of Discourse* (New York: Routledge, 1987); Robert C. Allen, ed., *Channels of Discourse, Reassembled,* 2nd ed. (Chapel Hill: University of North Carolina Press, 1992); Robert C. Allen and Annette Hill, eds., *The Television Studies Reader* (New York: Routledge, 2004); Glen Creeber, *The Television Genre Book* (London: British Film Institute, 2001); Christine Geraghty and David Lusted, eds., *The Television Studies Book* (London: Arnold, 1998); Horace Newcomb, ed., *Television: The Critical View* (New York: Oxford University Press, 1976, 1st ed.; 1979, 2nd ed.; 1982, 3rd ed.; 1987, 4th ed.; 1994, 5th ed.; 2000, 6th ed.; 2006, 7th ed.); and Lisa Parks and Shanti Kumar, eds., *Planet TV: A Global Television Reader* (New York: New York University Press, 2002). See also Lynn Spigel and Jan Olsson, eds., *Television after TV: Essays on a Medium in Transition* (Durham: Duke University Press, 2004), which examines TV's new forms of digital production and distribution, interactivity, and convergence with the Internet.

5. See, for example, Jack Banks, *Monopoly Television: MTV's Quest to Control the Music* (Boulder, CO: Westview, 1996); Simon Frith, Andrew Goodwin, and Lawrence Grossberg, eds., *Sound and Vision: The Music Video Reader* (New York: Routledge, 1993); E. Ann Kaplan, *Rocking Around the Clock: Music Television, Postmodernism and Consumer Culture* (New York: Routledge, 1987); and Kevin Williams, *Why I (Still) Want My MTV* (Cresskill, NJ: Hampton Press, 2003).

6. A comprehensive list is beyond our scope here, but would include: Sarah Banet-Weiser, *Kids Rule!: Nickelodeon and Consumer Citizenship* (Durham: Duke University Press, forthcoming); Heather Hendershot, ed., *Nickelodeon Nation: The History, Politics, and Economics of America's Only TV Channel for Kids* (New York:

New York University Press, 2004); Amanda D. Lotz, *Redesigning Women: Television after the Network Era* (Urbana: University of Illinois Press, 2006); Megan Mullen, *The Rise of Cable Programming in the United States: Revolution or Evolution?* (Austin: University of Texas Press, 2003); Lisa Parks, *Cultures in Orbit: Satellite Technologies and Visual Media* (Durham: Duke University Press, 2005); and Beretta E. Smith-Shomade, *Target Market Black: Selling Black Entertainment Television* (New York: Routledge, forthcoming), among others. See also Eileen Meehan and Jackie Byars, "Telefeminism: How Lifetime Got Its Groove, 1984–1997," *Television and New Media* 1, no. 1 (February 2000): 31–49; Marita Sturken, "Desiring the Weather: El Niño, the Media, and California Identity," *Public Culture* 13, no. 2 (Spring 2001): 161–89; and Toby Miller's writings on news, Food Network, and the Weather Channel, collected in *Cultural Citizenship: Cosmopolitanism, Consumerism and Television in a Neoliberal Age* (Philadelphia: Temple University Press, forthcoming). These are but a few examples of the dynamic range of writings on this subject.

7. See for example, the Reading Contemporary Television series published by I. B. Tauris, which includes volumes released and forthcoming on HBO's *The Sopranos, Six Feet Under, Sex in the City,* and *Deadwood*; on Showtime's *The L-Word*; as well as on the broadcast hits *24, Buffy the Vampire Slayer,* and *Desperate Housewives.* On academic responses to *The Sopranos,* in particular, see Dana Polan's chapter in this volume.

8. Regulations meant to ensure that broadcasting operates in the public interest by maintaining a competitive marketplace have included rules preventing any two of the major broadcast networks from merging; caps on the number of radio and/or TV stations that a company can own, which have been raised and in some instances removed over time; and the defunct Financial Interest and Syndication Rules, which forced the networks to buy some independently produced programming. In regard to content, the FCC has obliged broadcasters to serve the public interest through the Fairness Doctrine, which required representation of opposing views in regard to controversial news and public affairs programming (rescindment of this doctrine unleashed talk radio and talking-head punditry); limits on the number of commercials per hour and commercialism in children's programming; and rules that define and respond to complaints about broadcast "indecency" and "profanity."

9. There are, of course, some limits and obligations placed on subscription TV. For example, cable and satellite systems are not permitted to air material that may be deemed "obscene," but they are not at this time subject to the rules than regulate "indecent" material on broadcast airwaves. Since 1972, "must carry" rules (in continuously challenged and rewritten forms) have required cable systems to retransmit the broadcast signals of local stations, and since 1994, revisions to these rules have obliged cable systems to secure permission to retransmit local stations' signals, usually involving compensation. The general structure of the industry and

technological specifications—that is, rules pertaining to franchising, channel us-
age, pirated signals, consumer privacy, and other matters—have been developed
in the Cable Communications Policy Act of 1984, the Cable Television Consumer
Protection and Competition Act of 1992, and sections of the Telecommunications
Act of 1996 (which, among other things, rescinded the 1992 Act's rules regulating
fees charged to consumers). And, since 1972, the FCC has required cable systems in
at least the top 100 markets to set aside public access channels; see Felicia R. Lee,
"Proposed Legislation May Affect Future of Public-Access Television," *New York
Times,* November 8, 2005, E1, E7.

10. See *City of Los Angeles v. Preferred Communications, Inc.* (1986); *Turner v.
FCC I* (1994); *Denver Area Educational Telecommunications Consortium v. FCC*
(1996), *Turner v. FCC II* (1997); and *United States v. Playboy Entertainment Group,
Inc.* (2000).

11. Mullen, *The Rise of Cable Programming in the United States,* 1.

12. Daniel Czitrom, *Media and the American Mind: From Morse to McLuhan*
(Durham: University of North Carolina Press, 1982); Ithiel de Sola Pool, *Technologies of Freedom* (Boston: Harvard University Press, 1984).

13. Thomas Streeter, "Blue Skies and Strange Bedfellows: The Discourse of Cable Television," in Lynn Spigel and Michael Curtin, eds., *The Revolution Wasn't
Televised: Sixties Television and Social Conflict* (New York: Routledge, 1997), 221–36.

14. Ibid.

15. Ibid., 227.

16. Ibid., 228.

17. Joseph Turow, *Breaking Up America: Advertisers and the New Media World*
(Chicago: University of Chicago Press, 1997), 5.

18. HBO also offered sporting events—primarily boxing—from early in its
history.

19. Ralph Lee Smith, *The Wired Nation: Cable TV: The Electronic Communications Highway* (New York: Harper Colophon, 1972), esp. ch. 2 "New Kinds of Television," 10–21. Presciently, Smith notes that a great deal of programming on the
Professional Channel for physicians, which was promoted at the 1968 NCTA Convention, was provided by pharmaceutical companies; he also observes that educational programming could unfortunately be subject to "the dangers of commercial
packaging" (28–29).

20. See Mullen, *The Rise of Cable Programming in the United States,* on movie
channels, 105–10; on religious channels, 116–18; on C-SPAN, 123–25; and on ABC's
ARTS channel and CBS Cable, 156–58.

21. Distinctive television programming has been defined as "quality" long before the boom in the production of original cable series, in ways that both overlap
and diverge from the current model. For more on "quality" in broadcast-network
programming, see Jane Feuer, Paul Kerr, Tise Vahimagi, eds., *MTM: 'Quality Television'* (London: British Film Institute, 1985).

Part I

Institutions and Audiences

Introduction

At least since the 1970s and well into the 1990s, cable television has constituted a fast-growing new media market in which large, long-established
corporate media entities and feisty upstarts do battle and do business.
Over the last decade, cable has been joined by direct broadcast satellite
(DBS) systems as the primary means by which U.S. households receive television signals. This section explores aspects of the structure of this industry, which has both mimicked its predecessor and, in many ways, remade
the medium of television. What are the individual and corporate entities
that have sought to shape the structure of this industry? How did their
distinctive means of delivering—and regulating—already available TV
signals originate? What shape does the multichannel video programming
delivery (MVPD) market currently take? And how does a subscription-
based television market imagine and address its audiences differently from
advertisement-based broadcasting? To answer these questions, this introduction considers some of the ways that cable TV differs from broadcasting, as well as the development and structure of the cable industry and its
audiences.

The MVPD industry—more commonly known, and therefore referred
to here in most cases as the cable and satellite television industries—com-
prises a vast set of interacting institutions. It overlaps with—but is not
entirely coterminous with—those that comprise the broadcast industry.
Of course, within this volume, we cannot cover each and every aspect of
these institutions in depth, but we can indicate their scope and scale.
Companies, large and small, deliver video signals to our homes and other
locations via cable or satellite. Programming services (the networks or
channels discussed in Part II of this book) produce and distribute content
delivered via those cables and satellites.[1] Legislatures and regulatory agencies develop and carry out policies governing these industries; thus, for example, Congress writes rules that are enforced by the Federal Communi-

cations Commission (FCC), and local municipalities negotiate and grant franchise rights to cable systems.

These primary institutions interact with another set of institutions, some of whose involvement may be readily apparent; others, less so. Cable and satellite systems and networks depend on corporate entities, from globalized automobile manufacturers and fast-food chains to local used-car dealers and family-owned restaurants, that advertise on cable and satellite channels; ad revenue provides an income stream that supplements the fees charged by cable systems to their subscribers and returned to the networks. Another key corporate agent, Nielsen Media Research, counts and analyzes audiences, producing data that advertisers use to identify appropriate venues in which to expose their brands and products to viewers. Some 3,000 public-access channels serve their localities by providing production facilities and airtime to community groups.[2]

The cable and satellite television industries also encompass independent contractors who maintain the cable infrastructure and install the devices needed to receive signals in our homes. Customer service representatives at call centers, some local, some not, handle inquires about accounts. At factories in Mexico and elsewhere, workers manufacture and assemble converter boxes and remote controls. And another key industry player, the audience, surfs the channels, watches the programs, sits through the ads (or not), and pays the bills. In short, the cable- and satellite-TV industries comprise a complex array of technologies, corporate entities, regulatory agents, and laboring bodies, as well as those at leisure.

These interacting components of the MVPD industry have been deeply entangled throughout the history of cable and satellite TV. The shape of these entanglements has changed due partly to technological developments allowing for greater channel capacity, and especially to shifting market conditions, including the booming growth of media generally; trends toward globalization and concentration of ownership throughout media industries; and sometimes, actions by regulators that reconfigure the market.

In the early years of television's introduction as a commercial medium in the post–World War II period, the nascent cable industry was not so entangled with broadcasting. Rather, it consisted of bundles of wires in the hands of independent entrepreneurs who sought to hitch homes to new medium of television where geographical features interfered with broadcast signals. Geostationary satellites, which would eventually relay television content from the networks and to cable systems across the country, did not yet exist; they would come in to regular use by the communications industries in the United States in 1965.[3] But constraints on cable were not entirely technological. In the 1960s, broadcasters fought hard to stem its growth, and they convinced the FCC to ban CATV from the top one hundred markets.[4] By the end of the decade, however, various advocacy groups, think tanks, and journalists embraced cable as a technology that could diversify and even democratize the media. In this so-called "blue-sky" period, proponents touted cable as a potential solution to social ills, a conduit for all manner of financial transactions, and a site that would enhance political participation.

In the 1970s, when the FCC relaxed the rules that had contained cable's expansion, the industry began a growth spurt that continued through the end of the twentieth century. Cable system operators were assigned monopoly franchise rights, cables were strung (or buried), rules were written, prices were set, audiences were imagined and courted. Established media companies—including newspaper and other publishing interests, such as Newhouse and Providence Journal, the broadcast networks (which had initially opposed cable TV), and others—entered the market, speculating that it was an inevitable and potentially lucrative new medium. In the 1990s, many small- to mid-sized media companies got out of the business of delivering cable, in part because they may have lacked the deep pockets that would be required to rebuild the cable infrastructure to provide for new digital services, and in part because the largest multiple-system operators (MSOs) and other investors fueled a buying frenzy with capital, largely in the form of junk bonds, coming from Wall Street interests. This wave of consolidation left a handful of MSOs in control of most of the market. In 2005, there were an estimated 65.4 million cable-subscribing U.S. households. The largest MSO, Comcast, served over a third, with 21.5 million subscribers as of June 2005. The second-largest, Time Warner, served just under 11 million homes. Together, they served almost half of all subscribers. The six largest MSOs (a list rounded out by Cox, Charter, Adelphia, and Cablevision) serve well over three-quarters of all subscribers.[5]

Over the last decade, competition has developed as DBS systems, growing at a rapid rate as a result of new rules promoting competition in the Telecommunications Act of 1996, have enticed a significant number of viewers to give up cable—or to subscribe to a multichannel service for the first time. In many markets, DBS competes directly with cable systems that otherwise have enjoyed local monopolies. As of June 2005, just over 26 million households subscribed to DBS, which reached almost 28 percent of the market for multichannel video delivery (compared to only about 4 percent of the market in 1993). The market is shared among three providers, but dominated by two: DIRECTV, which is a subsidiary of News Corporation's Fox Entertainment Group, serves over half of all DBS subscribers (almost 15 million); EchoStar's DISH Network runs a close second, at 11.45 million. Dominion Video Satellite's SkyAngel serves an estimated 1 million households with Christian and family-oriented channels.[6]

In the chapters that follow, contributing authors explore aspects of the cable and satellite television (or MVPD) industry and that industry's attempts to organize viewers and consumers as audiences and as subscribers. They investigate the entrepreneurial innovations and deliberative negotiations that structured "pay" cable (and eventually satellite) TV as alternatives to broadcasting; the potential of a digital media environment to further diversify the television marketplace with vastly expanded channel capacity and interactivity; and the relationship of the cable channels and the programming to the TV audience, and how that relationship is necessarily reimagined in a subscription-driven system offering a diminished role to advertisers—or in the case of premium channels, in which advertising revenue is entirely absent.

To begin, Megan Mullen takes us back to where "the cable begins," to reverse the title chosen by Lisa Parks for an entry later in this section. Exploring the origins of Community Access Television (CATV), Mullen finds it in the hands of individual entrepreneurs in widely dispersed locations—from Arkansas to Oregon to Pennsylvania and New York State—as early as 1947. In each case, in which remoteness or geographical features prevented residents from receiving clear broadcast signals, CATV innovators typically solved the problem by relaying signals from an antenna placed on a mountaintop or tall building to homes via a wired network. Mullen introduces us to some of these early cable system operators, each one running a small, family-owned business outside of a major metropolitan area. Together, they laid literal groundwork for what has become a multi-billion dollar business passing approximately 108 million U.S. homes.[7] Mullen explores the motivations that drove their initial efforts and, in many cases, their ongoing development of the industry, even as it changed and grew.

Despite the efforts of these pioneers, cable grew slowly in the 1950s, 1960s, and even 1970s, when the means by which TV signals would be delivered to viewers was debated in terms John McMurria discusses. While the FCC dithered on whether to authorize or ban "pay-TV" systems, in the

1950s Congress passed a series of rules that put the brakes on the growth of pay-TV, at least temporarily; consequently, free-to-air, advertising-supported broadcast television was allowed to dominate the market, even though broadcasting was already being criticized in its earliest years for the banality of its content and its failure to provide educational programming. At the same time, however, the free-to-air system was hailed for its capacity to provide universal service, and it was supported in testimony at congressional hearings on the subject by representatives of groups advocating for veterans, senior citizens, women, civil rights, and labor unions, among others. Seeking to reform television as a more varied medium, some industry observers and players—ranging from electronics manufacturers and TV critics to educators and progressives in the federal government—pushed for adoption of pay-TV systems designed to provide additional services to subscribers. While hoping to enhance the capacity of the medium to serve the public interest, their efforts—which allowed for eventual authorization of subscription-based cable services—provided for pay-TV models. With access to these dependent on ability to pay, the principle of universal service was jettisoned in favor of market-driven solutions.

Reform-minded critics of broadcasting believed that cable-based pay TV would solve the "scarcity" problem that plagued broadcasting. If only a handful of over-the-air signals could be clearly transmitted within any given locality, then broadcasting would forever be intrinsically too limited to serve the public with adequate competition and diversity in form and viewpoint. According to François Bar and Jonathan Taplin's essay in this volume, not only did analog cable alter and expand the television market, but also its quantity and range of services would be further enhanced by a digital "revolution" which was engineered by a massive, nationwide upgrading of the cable infrastructure, as well as by an overhauling of regulations that allowed for technological convergence and cross-media ownership. Accordingly, the Telecommunications Act of 1996 permitted cable systems and telephone companies to compete in offering multiple services (television signals, telephony, and highspeed Internet). This development, in turn, gives rise to the question: What shape will the "digital future" take? On one hand, it might look very much like prior phases of the television industry's history in which a few powerful corporate entities exerted primary control. Alternatively, Bar and Taplin look toward the emergence of a multi-service Internet Television platform that would boast virtually unlimited channel capacity, as well as allowing for alternative distribution schemes and new producers to enter media markets. But, as the authors suggest, this very diversity could eventually destabilize the economic viability of a multichannel universe, as increasingly fragmented audiences might turn out to hold little appeal to advertisers. In sum, Bar and Taplin argue that present-day policymakers must consider the long-term implications of the directions in which they seek to reshape electronic media so as to fully protect the user's access to an open, interconnecting, and transparent market.

In the next essay in this section, Amanda Lotz turns her attention to a sector of the cable industry that differs in some dimensions from conventional ("basic") subscription service: the premium channels, such as HBO, Showtime, and Cinemax. These channels, which operate according to an alternative business plan, are able to provide programming that is, in some ways, distinct from that offered by other cable channels. Since, unlike basic channels, the premium tier is financed only by subscription fees—it sells no commercials and therefore earns no ad revenue—its success is measured by numbers of paying subscribers, rather than numbers of viewers which can be translated into ratings. Without advertising, premium channels are able to offer distinct narratives and scheduling routines, revising audience expectations about television generally. While premium channels depended on a heavy rotation of feature films and sporting events in their early years, they eventually added original programming, much of which would come to be known as the "best" quality television available by any means. What Lotz then explores are the cultural implications of a media environment in which the highest quality content may be available to only a relatively small percentage of the potential audience.

Lisa Parks locates audiences that are not effectively served either by over-the-air broadcast signals or by cable, and who, not unlike the early cable pioneers discussed by Mullen, sought out signals not readily delivered to them. She argues that an array of factors impacting the means by which homes (and other sites) receive television signals is frequently overlooked in TV scholarship. The shape of the cable industry, and, eventually, direct broadcast satellite (DBS) services are frequently regarded as the product of aggressively pursued industrial and regulatory agendas; meanwhile, the geography of these media is too often ignored. DBS has served not only to compete with other kinds of delivery systems, but also to function in so-called fringe areas where satellite hook-ups provide the only access to TV. Parks considers the spatial expression of this medium beyond "where the cable ends," finding, in remote coastal and desert communities, early and pervasive adoption of satellite technology. Her materialist analysis of the growth and use of this technology complements the political-economic analyses that occupy most of this section.

NOTES

1. Some original programming, of course, is produced by studios operated by broadcast networks and Hollywood film studios; and much cable programming consists of off-network TV shows and feature films in their post-theatrical release windows.

2. On public access, see DeeDee Halleck, *Hand-Held Visions: The Impossible Possibilities of Community Media* (New York: Fordham University Press, 2002); see also Eric Freedman, "Public Access/Private Confession: Home Video as (Queer) Community Television," in Robert C. Allen and Annette Hill, eds., *The Television Studies Reader* (New York: Routledge, 2003).

3. AT&T, RCA, and the Hughes Aircraft Company conducted early U.S.-based experiments with satellite technology, with launch support from NASA. Communications satellites were in use in the United Kingdom, Western Europe, Japan, and Brazil one to two years earlier. See David J. Whalen, "Communications Satellites: Making the Global Village Possible," National Aeronautics and Space Administration History Division, http://www.hq.nasa.gov/office/pao/History/satcomhistory.html.

4. Tom Streeter, "Blue Skies and Strange Bedfellows: The Discourse of Cable Television," in Lynn Spigel and Michael Curtin, eds., *The Revolution Wasn't Televised: Sixties Television and Social Conflict* (New York: Routledge, 1997), 223.

5. FCC, Twelfth Annual Report: Annual Assessment of Competition in the Market for the Delivery of Video Programming (released March 3, 2006, MB Docket No. 05-225), 15, 118.

6. A fourth DBS provider, VOOM, stopped operating early in 2005. The MSO Cablevision, through a subsidiary called Rainbow DBS, had launched VOOM as a High-Definition only service; its HD channel line-up was shifted to the DISH Network. FCC, Twelfth Annual Report, 37–38.

7. While past calculations have differed slightly, both the NCTA and Kagan World Media estimated that 99 percent of television-owning homes were passed by cable systems by the end of 2004, with the number of television homes in the United States set at the time at 109.6 million. See FCC, Twelfth Annual Report, 10; see also FCC, Eleventh Annual Report: Annual Assessment of Competition in the Market for the Delivery of Video Programming (released February 4, 2005, MB Docket No. 04-227), 12–13.

Chapter 1

The Moms 'n' Pops of CATV

Megan Mullen

Inasmuch as the early 1950s origins of the U.S. cable television industry have been documented at all, the story tends to be one of "mom 'n' pop" entrepreneurship. In other words, media historians—and others affiliated with the modern cable industry—readily herald the small-town inventors and businessmen who devised ways to bring the television signals of large metropolitan areas into the nation's hinterlands. Most people enjoy hearing an "American Dream" type of story, especially when contrasted with a broadcast television industry headquartered in major cities and dominated by established corporations.

There is merit in this approach to cable history. The early cable industry—better known as community antenna television (CATV)—began around 1950 in towns too small and remote for residents to receive broadcast signals using set-top ("rabbit ear") or rooftop antennas. People across the country were well aware of television at this point, even if they lacked access to it. CATV drew from an eclectic combination of makeshift technologies that captured broadcast signals from high places (such as mountaintops) and transmitted them to television sets in lower elevations. The pioneers of these technologies were loosely if at all affiliated with existing broadcast television interests. Most were appliance dealers wishing to sell television sets. Some were radio operators, familiar with broadcast technologies. A few were simply residents of the small, remote towns with a fascination for technology. Some had backgrounds in consumer electronics, and some had technical training from military service. Virtually all of the earliest cable television pioneers held strong business, civic, and family ties to their small communities.

Yet explaining their operations as simple "mom 'n' pop" ventures is an extreme oversimplification. They actually represent, in one way or another, a complex industry that was in formation almost from the day the first person thought to connect a tall receiving antenna by wire to a television set. Many CATV businesses—indeed the majority, if one is counting number of systems rather than subscribers—did continue to be run locally and for relatively little profit well into the 1990s (with a fair number existing even today). And in many cases the operators of these systems were married couples, with the husband managing day-to-day operations and interacting with other professionals and the wife running the office and keeping the books. Yet the most forward-looking and professionally run of the very early systems tended to differ from this model—a little in some cases and a great deal in others.

The leaders of the emerging CATV industry did more than simply sustain their own local businesses; they built (or helped build) other systems, developed new technologies and services, and worked through legal and policy issues at every level. These were individuals who perceived the potential of the infant industry and made the critical technological and business innovations that kept it from being a simple retransmission or utility service. Those systems that moved the industry forward represented a great deal more technical expertise and business savvy that the term "mom 'n' pop" alone would imply.

One goal of this chapter is to discuss how the earliest CATV systems in the United States were founded and developed. The other is to show how those CATV operators who chose to work together formed a cohesive industry that would grow and develop, rather than serve merely as a stopgap technology until more of the nation's residents had better broadcast television service. Let's begin with a look at five CATV systems that have been cited as "firsts."

The First Wave: Cable Pioneers

Davidson's Claim

Probably the least known CATV "first" is the story of James Y. ("Jimmy") Davidson of Tuckerman, Arkansas—a small town ninety miles northwest of Memphis. As a young adult in the 1930s, Davidson managed the local movie theater and in his spare time ran a radio repair business, using skills he had learned as a teenage hobbyist. Davidson served in the Navy's Signal Corps during World War II and, in a story strikingly similar to those of other community antenna pioneers, returned to Tuckerman to run an appliance store.

Shortly after his return, in late 1947, Davidson found out that Memphis television station WMCT would soon be starting operations. He was disappointed that Tuckerman was too far away to receive the station's signal, and so he and an associate set about building a 100-foot antenna tower near his store. During WMCT's start-up period Davidson eagerly awaited and watched test transmissions—at first simply test patterns, but eventually a live telecast of a college football game. At the time, Davidson's system had only one regular subscriber, the local telegraph operator, but he connected the American Legion hall to the antenna for this special event. A very large crowd was in attendance, and not surprisingly people clamored for CATV service a few months later (January 1, 1949) when the Memphis station began full-time broadcasting.

Davidson, who is eager to compare himself to other CATV pioneers, claims to have had the first CATV franchise in the nation.[1] He explains, "I knew that the utilities had to have a permit or franchise or whatever you want to call it to access the public right of ways. And I guess I just assumed that I had to as well."[2] Soon after starting the Tuckerman system, Davidson began to build community antenna systems for nearby towns. He and his associate would travel from one system to another in Davidson's own small airplane. He also started Davco, a CATV supply and consulting business serving those who wished to start systems of their own. Journalist Tom Southwick explains that "Davco specialized in building a complete cable headend and then flying it intact to the location where it was needed and where it could be installed in a day."[3]

Davidson went on to work with CATV/cable trade associations at both the national and regional levels. He remained active in the industry until the late 1970s and still follows its developments from his home near Little Rock. As of 2006, he and his wife were working on turning their home into a cable history museum and library.[4]

Parsons' Claim

A more commonly told story relates how L. E. (Ed) Parsons, a radio station operator in Astoria, Oregon, invented cable to please his wife, a former journalist familiar with broadcasting. The couple had traveled to a radio convention in Chicago during the mid 1940s and while there enjoyed the opportunity to watch broadcast television. At the time, there were no broadcast television stations operating in the vicinity of Astoria. In the summer of 1948, though, it was announced that KRSC, Channel 5– Seattle, was to go on the air—and so Parsons set about finding a way to make that channel available in Astoria (some two hundred miles from Seattle).

He found that KRSC's signal filtered through the mountains surrounding Astoria in bands or "fingers" of varying strengths. One of the fingers was accessible from the roof of the two-story apartment building where he lived, so he erected an antenna. He and his wife worked as a team to tune the signal, he on the rooftop adjusting the antenna and she in the living room reporting on signal quality via walkie-talkie. No clear signal was received, however, until Parsons relocated the antenna to the top of the nearby eight-story Astor Hotel and then connected it to his own building using twin-lead cable.[5] This was in November 1948, about a month after Davidson's successful retransmission of the football game. Not surprisingly, the hotel owners were receptive to Parsons' experiments, at least early on; television reception quickly became a draw for visitors and patrons.

Since residents of Astoria approached Parsons with requests to have the antenna service extended to their homes, he had to develop a relay system and increasingly sophisticated networking technology. What he came up with was a combination of a coaxial cable network to connect buildings and transmitting antennas for sending the signal across streets. The latter technology—actually a form of broadcasting—would later cause some in the industry to question his claim of being the true inventor of community antenna television. But this might simply be a way of calling attention to the fact that Parsons never became active in the CATV/cable industry.[6] In any case, Parsons *did* devise a workable means of distributing a broadcast television signal to residents of a community who would not have received television service otherwise. At first, Parsons charged subscribers directly for the needed equipment, most of which he had either invented or adapted himself. By the end of 1951, though, he had worked out a standard installation fee of $125 and a $3 monthly rental charge.[7] Following this, he was sought out by aspiring CATV operators throughout the Northwest. Traveling in his own airplane, like Davidson, he helped locate signals and design systems.

Publicity surrounding Parsons' Astoria system had also brought CATV to the attention of the Federal Communications Commission (FCC) as well as to the broadcast television industry. Systems in the East that launched around the same time raised concerns that the new community antennas might harm the growing broadcast television industry, and thereby jeopardize a future in which all communities might have their own local television stations.[8] Interestingly, for its part KRSC was receptive to Parsons' system, seeing the potential increase in audience size as a benefit (not all television stations would be this amenable in the future). For the FCC, this presented a confusing new regulatory issue, though. Communication between Parsons and the commission (reflecting considerably more concern with the technology than with the business aspects) contributed significantly to subsequent FCC inquiries into the nature of the emerging industry—especially once other systems were known to be operating closer to Washington, D.C.[9]

Among those with claims to the earliest CATV systems, Parsons is the only one who did not remain involved with the developing industry. In 1953, he moved to Fairbanks, Alaska, where he remained until his death in 1989. Although he did find ways to bring television programming to remote communities there, he considered his efforts in improving radio and telephone communication networks both in Alaska and in the Arctic to have been far greater accomplishments. He ranked the CATV system in Astoria "third" among his lifetime accomplishments.[10]

Walson's Claim

John Walson (formerly Walsonavich) of Mahanoy City, Pennsylvania, holds a fascinating, if not entirely convincing, claim to have built the first CATV system in 1948. Mahanoy City is a town in eastern Pennsylvania's anthracite coal region that lies in a narrow valley between two steep mountains. Most of its homes and businesses are row-style buildings on the sides of the mountains, structures little changed since they were built during the mining boom that began over a century ago.

During the late 1940s, Walson held two jobs: maintenance worker for the utility company and part owner of an appliance store. Known to be a tinkerer, he used his experience with electrical wiring to devise a way to improve television reception (and therefore sales of television sets) in his community. Walson's system at first relied on twin-lead cable, but quickly was converted to the more reliable coaxial cable. Within its first year, according to Walson's claim, his business, Service Electric, had close to one thousand subscribers (a good portion of the town at that time).

There is little dispute that Walson actually did connect a mountaintop receiving antenna by wire to homes in Mahanoy City, and at a very early date. And it is well worth noting that he went on to introduce a number of precedent-setting programming and technology innovations in his community, including increased channel capacity using, first, what he described as "stacked antennas" and, later, a microwave relay.[11] However, his claim to have invented CATV as early as 1948 is greatly disputed, as documented in detail by cable scholar Patrick Parsons. First, Parsons explains, it seems odd that Walson did not actually make any claim at all to having the first system until well into the 1960s. And since all of his business records were destroyed in a 1952 fire, there is no actual evidence to back the claim. Second, information in *Television Factbook,* an authoritative reference source that began in 1952, shows an inconsistency in data reported by Service Electric. Apparently Walson's business reported its founding date as 1950 until 1967, at which point it began listing the date as 1948. This ambiguity was also reflected by the recollections of cable pioneers Parsons interviewed. Third, there is a credible competing claim to the first CATV system in Mahanoy City.[12] Parsons provides additional information that seriously questions the likelihood that Walson had the first CATV system in the nation—or even in his own community.[13]

At the very least, though, Walson's story is worth consideration because it represents the degree to which early CATV lay outside the interests of the established entertainment industries, much less the general public. Even if he had launched the nation's first CATV system in 1948, it might have been a "tree falling in the forest" situation. Few if any news media could have perceived the significance of what had occurred because there was no context for it. Anyone living in a metropolitan area such as New York or even Philadelphia—the nearest hubs of "big media" reporting— would not have been aware that parts of rural America were unserved by broadcast television and thus had their lives changed when CATV service came to their communities. The most coverage Walson's story even *could* have received as early as 1948 would have been a local news article or perhaps a few lines "buried" in a larger newspaper or trade publication. But if even this much did not happen, it should not be surprising.

Tarlton's Claim

Panther Valley TV (PVTV), the system begun by Robert J. Tarlton in Lansford, Pennsylvania, in 1950, was apparently the first to receive any national press coverage; articles about it appeared in 1951 in *The Wall Street*

Journal, Radio and Television News, and *Television Digest.* Lansford is an anthracite valley town with geography similar to that of Mahanoy City (sixteen miles west of Lansford). Tarlton was the son of a department store radio salesman who had developed his own interest in electronics engineering. In the late 1940s he experimented with twin-lead cable prior to the launch of an actual CATV system, once arranging makeshift reception for a wealthy acquaintance. Tarlton did not wish to claim a "first" because he did not consider his initial experiments reliable means of television reception. But unlike those systems discussed above, PVTV was a professionally run CATV business from the start. As Tarlton explained, "Many times I've been accused of being too conservative, but I suppose it paid off because it forced me to look for something that would be dependable."[14]

When PVTV officially went into business in the spring of 1950, the system was built using coaxial cable and components manufactured by Jerrold Electronics of Philadelphia and designed specifically for CATV use. Tarlton had approached Jerrold's owner, Milton Jerrold Shapp, upon learning that Jerrold was in the business of supplying "master antenna" systems that served hotels and apartment houses in television-served urban areas. Shapp quickly perceived the potential for equipping CATV systems once Tarlton, and soon other CATV pioneers, began purchasing components for their systems.[15] In fact, Lansford's CATV system became the showcase as well as the testing ground for Jerrold components.

The role of Shapp and his company would only grow as the CATV industry evolved. Shapp had been in the electronics business since the 1940s, but it was meeting and working with Tarlton that got him interested in CATV. He first used the Lansford system to field-test components and subsequently hired Tarlton to help set up new systems during the years 1952–1958.[16] By this point, Jerrold was equipping most new CATV systems with only a "service agreement" in place, according to which Jerrold would receive a percentage of the profits for a specified period in exchange for maintenance and upgrades. Tarlton explains that working for such a business,

> I put in systems all over the country: Clarksburg, West Virginia; Fairmont, West Virginia; out in the mid-west and western part of the country; up in Berlin, New Hampshire; and Wenatchee, Washington. We found problems in all of these systems and we overcame them. I worked with the engineering people in a cooperative manner. We made field tests in these systems

and then the engineering department was able to overcome some of the problems. In the latter years with Jerrold, my primary thrust was in developing the systems, getting the systems installed and using my experience to set up the system, offices, and the entire gamut of operating the business.[17]

For his part, Shapp is credited with being the first in the CATV industry to successfully approach New York City banks for financial backing, surely a signal of the industry's future direction. As Shapp summed it up:

I got J. H. Whitney Company in New York . . . and I sold [them] on the idea of financing cable television systems. So our job at Jerrold was to get the franchise. First we sat down with the Whitney people and went over 40, 50 communities. They boiled it down. They would only go about 7, 8, 9, 10 systems. They boiled it down to the ones and getting franchises in those days was a heck of a lot easier than it is now. We had a perfect batting record of the first 8 or 10 systems. They put up the money for it and we formed an individual company in each community. Whitney, I think it was 50%, and Fox-Wells had about 10–12%, and I came out of that, Jerrold, we had somewhere around 40–35% of the system. So we not only got paid for all our efforts, paid for all of the equipment, but now also had a financial interest in the systems.[18]

Shapp's role makes it clear that by the mid-1950s CATV had already attained fiscal credibility.[19] It had also developed into a distinct industry, with a prominent trade organization, the National Community Television Association (NCTA), begun by Martin F. Malarkey of Pottsville, Pennsylvania.

Malarkey's Role

Malarkey, an appliance store owner, never claimed to have been the first CATV operator, even though his system, which launched in February 1951, was among the very earliest. Malarkey had a component-testing arrangement with equipment manufacturer RCA that was similar to the one Tarlton had with Jerrold; both electronics companies were active (and highly competitive) players in the rapid rise of CATV, and they had every reason to cultivate the symbiotic "laboratory" relationship with enterprising individuals such as these.

Malarkey is best remembered, though, for playing the key role in organizing the earliest CATV operators into an industry by founding the National Community Television Council (to be renamed the National Community Television Association).[20] The first annual convention of this important organization was held in June of 1952 at the Neccho Allen Hotel in Pottsville, with several smaller meetings having been held there over the previous year. It hardly seems coincidental that the NCTA, the most prominent trade organization in cable history, traces its origins back to Pottsville. The city's proximity to other early CATV communities—including Mahanoy City, Lansford, and several others—made it a logical meeting place. The NCTA will be discussed in more detail later in this chapter.

Malarkey's keen business sense surely was key to his leadership role in the CATV industry as a whole. Trans-Video's fees were comparable to other systems at the time: $135 for connection and $3.75 monthly. But shortly after the start of his system, Malarkey began thinking of ways to offer subscribers more than simply retransmitted broadcast signals for their money—an idea that was years ahead of its time. First, starting in 1951, Malarkey provided a channel of closed-circuit locally originated programs. This operation was described as follows:

> Trans-Video Corp., Pottsville, Pa., put on a half hour live show June 27th in cooperation with one of their local radio stations. This show was staged at the antenna site and included interviews of city officials, heads of civic organizations, system personnel, visiting dignitaries, professional artists and most touching, an interview of the winner of the Soap Box Derby and his family.[21]

Reports such as this were common during the late 1960s and early 1970s, once the federal government had begun to mandate local origination. Until that point, though, CATV systems originating programs were few and far between. That Pottsville was the first is confirmed by the recollections of E. Stratford Smith, who, while working in the FCC's Common Carrier Bureau, visited Trans-Video in 1951.[22] There is also documentation that Malarkey would switch among the various signals received at the headend (the number of which exceeded system capacity) to select certain programs he felt would be more popular with his subscribers.[23] This strategy, though somewhat more common in CATV's earliest years, was similarly forward-looking in that it predicted a time and place when cable service would need to be marketed as something distinct from broadcast television. Innovations such as these were taking place within only a few years of the very first experiments with any sort of community technology.

So even though historians are unlikely ever to verify that one of the systems discussed thus far actually preceded the others or was the most pioneering in its use of "modern" technology, these sorts of claims seem trivial in comparison with the larger issue of what these endeavors and many others like them represent in the history of television in the United States. Malarkey's situation should make this clear. Although he started a couple of years later than other pioneers, he built on their knowledge in ways that moved the industry forward. The community antenna systems discussed here, along with a handful of others, quickly developed into successful local businesses, setting important precedents for such things as initial installation charges and subscription fees.

The Second Wave

Marveling at the rapid growth of CATV in eastern Pennsylvania makes it easy to overlook the fact that it took less than a year for the industry to start spreading to other parts of the United States. By 1953, there were three hundred CATV systems operating in states including New York, Maryland, New Hampshire, Washington, Oregon, California, and Texas.[24] This second wave of early systems constitutes a category unto itself. It included a number of operators active in the increasingly cohesive industry and eager to learn from the successes and failures of the earliest CATV entrepreneurs. Among this group were business men and women who sought to do more than simply provide a community antenna service and were particularly active in exploring new technologies, marketing schemes, and emerging legal issues. Even at this very early stage, the industry's founders were becoming aware of both the potential and the complexity of their business.

Oneonta

Some of the eastern Pennsylvania operators had already perceived that their subscribers wanted to receive the full complement of network offerings; they also were eager to receive New York independent stations in addition to the network affiliates from Philadelphia because of the sports and movies carried by the independents. But the full marketing potential of the additional channels arguably was not fully realized until operators began to migrate from the region and use their expertise to launch new systems. One such businessman was William Calsam, who had begun the CATV system in Schuylkill Haven, Pennsylvania in 1951. Schuylkill Haven is at the southern edge of Schuylkill County (technically outside the anthracite region) and only six miles south of Pottsville. Calsam knew Malarkey and apparently had begun his system using Malarkey's advice or at least his example. In 1953, however, Calsam sold the Schuylkill Haven system and moved to Oneonta, New York, to begin a CATV business there.[25]

Calsam found a business partner in Oneonta, local attorney and real estate owner Albert Farone, and they, along with a handful of other investors, founded Oneonta Video. Calsam epitomized the professional values that guided what was by this point the close-knit and forward-looking NCTA. He became NCTA treasurer in 1954 and remained active in the organization until his 1965 retirement; he was also a charter member of the New York Community Television Association. Oneonta Video's owners were eager to make Oneonta a showcase for new CATV technologies that would allow subscribers to receive a range of desirable channels instead of simply the nearest available broadcast signals. The full complement of broadcast network programming was readily available via community antennas from affiliate stations in nearby Binghamton, Syracuse, and Schenectady. But Calsam and Farone wanted to offer more and thus pursued two new technologies.

First was closed-circuit local origination of the sort Malarkey had been using in Pottsville. Calsam's grandson Robert McMinn, who was very familiar with his grandfather's business, recalls that a moderately successful local programming operation existed in Oneonta briefly in the early 1960s.[26] A more lasting and prescient programming scheme was Eastern Microwave. Like Walson and Malarkey, Calsam perceived that independent stations from New York City would be popular with his subscribers; as they continue to be today, stations including WOR and WPIX were known for extensive schedules of professional sports and syndicated Hollywood movies. Yet Oneonta (like Mahanoy City) lies just a little too far from New York to receive those stations' signals using a community antenna. Calsam's solution—similar to one in use by Walson and Service Electric around the same time—was to build a microwave relay between Oneonta and New York. In the case of Oneonta Video, this involved two mountaintop towers or "hops" between Oneonta and New York. Though technically a separate business, Eastern Microwave had shared ownership with Oneonta Video.

Within a few years, Eastern Microwave had extended its operations to CATV systems throughout central New York, including systems owned by Oneonta Video as well as others. When Oneonta Video was sold to Newhouse Broadcasting in 1965, Eastern Microwave was an important part of the package. The business grew, eventually connecting with microwave networks in Pennsylvania (including Service Electric). It continued to deliver distant independent stations to local cable systems, setting a critical precedent for today's cable "superstations." Eastern Microwave also delivered the newly formed Home Box Office cable network to subscribers in New York and Pennsylvania starting in 1972, before the cable network began satellite transmission in 1975.[27]

Calsam and the Oneonta Video story demonstrate that, in its earliest years, CATV was already perceived at least as much as a lucrative investment opportunity as a service to a person's own community. Calsam had left his own community very soon after starting the CATV system there in order to start a system in a town he probably had never visited before, much less held personal connections with. Farone, on the other hand, had many connections to Oneonta, but no previous interest in CATV. A shrewd businessman, as Farone was known to be, might have perceived Oneonta Video as a community service to some extent, but for him it was primarily a good investment opportunity. Several individuals affiliated with Oneonta Video recall that Farone had very little to do with day-to-day operations.

Meadville

The story of Meadville Master Antenna (MMA) of Meadville, Pennsylvania contrasts sharply with that of Oneonta Video in that it is a story of entrepreneurship and community service, carried out by individuals with very strong family ties to their community. Yet Meadville nonetheless differs in many ways from the "mom 'n' pop" prototype as well. Meadville Master Antenna was owned jointly by George Barco and his daughter Yolanda Barco. The system was launched on June 16, 1953.

As their "day job," the Barcos ran an extremely successful law practice in Meadville. Their prominence in the community lent credibility to their CATV business. At the same time, their combined legal expertise helped

them navigate the uncharted terrain of CATV. The Barcos and James Duratz (husband of George Barco's other daughter, Helene), who served as
general manager for the system from the late 1950s through the 1980s, were
a powerful presence in the emerging industry. Yolanda Barco's story is
unique for its time; women had little voice in the CATV industry of the
1950s, and the "moms" of any genuine "mom 'n' pop" businesses did more
to prove this rule than to provide the exception to it. If wives were to accompany their husbands to professional meetings—as sometimes was the
case—there would be flower shows and shopping excursions to keep them
busy. By contrast, Duratz confirms that, indeed, his sister-in-law was virtually the only female presence at NCTA meetings for many years (with
the exception of various recording secretaries).

For his part, Duratz, educated in economics and self-taught in engineering, was adept at implementing the latest technical innovations (often
at the behest of the Barcos, following their attendance at industry meetings). Like Tarlton, the Barcos and Duratz were affiliated with Jerrold Electronics during the early decades of their operations. Since MMA started
after Jerrold had begun manufacturing CATV components, Shapp had approached the Barcos about having one of Jerrold's service agreements in
Meadville, but the Barcos wished to retain more control of the business.[28]
Nonetheless, as Duratz explains, the Barco family maintained a strong relationship with Shapp and his company which clearly benefited both parties. As he put it, "Well, it was interesting that most of the development
that Jerrold created was field tested by us. When we'd get something you
never knew if it was going to work. You'd just have to put it out in the field
and try it and make adjustments."[29]

In its later decades MMA would continue both to be technologically innovative and to serve the specific needs of its community; in this company, these two goals were always seen as intertwined, not competing. In
the early 1960s the system underwent a major rebuild, among other things,
allowing MMA to launch a local programming operation, CTV-13. A full-
time manager, James Strickler, was hired, and CTV-13 instituted a regular
program schedule in 1967. This was two years before the federal government announced that it would soon require local origination (i.e., locally
produced programs aired on an available cable channel), and CTV-13
continued long after that requirement was revoked. Reflecting on CTV-13,
both Duratz and Strickler point out that the local origination operation
was never a profit-making venture; it was always subsidized by both
MMA's other services and the Barcos' law practice.[30]

One can only speculate as to how other communities might have fared with local origination had their original owners not sold them before this practice began to be supported by the federal government. Certainly it is not a common practice today, with the exception of scattered public access operations. It is not financially feasible, and the multiple-system operators (MSOs) that now own most local cable systems clearly have no personal or philanthropic stakes in the communities they serve.

The Barcos, of course, had another source of revenue—the law practice —and thus had less reason to be concerned about maximizing the revenue potential of their CATV system. MMA was not sold until the late 1980s, and even today it is owned by a small, regional MSO, Armstrong Cable, whose owners were acquainted with the Barcos. Arguably this is more a testament to the Barcos' commitment to Meadville than it is a statement about cable television economics; indeed, in many ways it contradicts those economics because of the scale economies made possible by more extensive multiple-system ownership. But inasmuch as the law practice contributed to the Barcos' commitment to their community, this practice also proved extremely beneficial to the newly organized NCTA during the 1950s.

Early Frustrations of the CATV Industry

As discussed above, the first meetings of the NCTA actually had occurred in June 1952, fully a year before the Barcos even began their CATV system. Their absence at this early stage certainly did not diminish their role in the organization through many subsequent years, however.[31] Rather, it reminds us of how unprecedented the very earliest CATV systems were, as well as the extent to which these inventors and entrepreneurs had to employ stopgap measures to keep their industry from being defeated before it even got off the ground. The issue that led to the NCTA's first meetings seems minor by contemporary standards: CATV systems in Pottsville, as well as Lansford and Honesdale, Pennsylvania, had been selected by the Bureau of Internal Revenue (now the Internal Revenue Service) as test cases. The bureau felt that an 8 percent excise tax, then being applied to leased wire services (such as newswires and the services used by the broadcast television networks to distribute programs to their affiliate stations), should also be applied to community antenna systems. The few dozen operators gathered at the charter meetings agreed to contest the tax.

Although it would be a few years before the case went to trial, the emerging industry's defense would be based on the fact that, for many smaller communities, it was the only way of getting television service—which, of course, larger communities received over-the-air without any charge at all.[32]

By the time the case did reach the courts, many more CATV systems had been launched, including those in Meadville, Oneonta, and others whose founders were among the vanguard in industry leadership. Indeed it was the Barcos who both prosecuted and provided the plaintiff for the test case, a prototypical MMA subscriber (Pahoulis) who had paid $70.49 in taxes for his service. Initially, the U.S. district court in Pennsylvania held that the tax did apply in the *Pahoulis* case. But this decision was overturned (along with a similar West Virginia case that had been taken over by the cable organization) in the Fourth Circuit Court of Appeals in 1956.[33] Surely this helped the fortunes of the emerging industry.

The Barcos were joined in their legal efforts during the early 1950s by E. Stratford Smith, an attorney, who had been working for the FCC's Common Carrier Bureau at the time the first CATV systems were being built. In fact, he had been responsible for reporting on both the Astoria and Pottsville systems. Smith quickly became so fascinated with the CATV systems he had observed that he left the FCC and went to work for the NCTA. As he explains,

> My personal impressions were that it was a coming thing and that it was going to be much more important than it was then. I won't sit here and tell you that in the early 1950s I envisioned the cable television industry as it is today; I did not see it that far ahead. But I did see it as a much more important way of getting television service around the country than the FCC did at the time. I can't say that the FCC had a position. They did not even mention CATV in the Sixth Report and Order, the television allocation report in 1956. Nobody there seemed to see it as meaning anything.[34]

Smith went on to serve as secretary to the NCTA for several years during the 1950s, and throughout the 1950s and 1960s tried many cases involving CATV. One of these was the landmark 1968 case, *Fortnightly v. United Artists,* which challenged CATV's right to retransmit copyrighted television programs. He won the case for the industry, a temporary victory that led to more formal guidelines on copyright issues involving cable television.

Although Smith owned some small CATV systems on the West Coast for a while, his role in the emerging industry was not primarily one of investment or entrepreneurship. Instead, his early arrival on the CATV scene signals the rapid professionalization of a grassroots industry. Without an industry already cohesive enough to have identified common problems and a distinct position vis-à-vis various government bodies, there would have been no role for him. Clearly, though, there *was* a role for Smith— and soon other categories of professionals—due to the fact that several of CATV's earliest pioneers chose to move quickly beyond the basic local service stage of operations.

The Real Moms 'n' Pops and the Role They Have Played

At the beginning of this chapter, I mentioned that the vast majority of CATV systems continued for many years as small "mom 'n' pop" operations. This fact should not be dismissed lightly since these low-profile systems have provided the only source of television service to many of the nation's rural residents well into today's era of satellite channels and multiple-system operators (MSOs) such as Time Warner and Comcast. For years, many of these systems were too small and isolated to be of much interest to the MSOs.[35] And in many cases, local operators did not wish to give up their businesses.

A particularly interesting story in this vein is that of Bainbridge Cable, a system that is "typical" in spite of a rather atypical beginning. David Coe founded Bainbridge Cable in 1953 to serve the village of Bainbridge, New York, some thirty-two miles from Binghamton and the nearest television stations. Interestingly, Bainbridge native Coe began the CATV system while still attending college, as a project for one of his engineering courses at Cornell University. He and his father financed and ran the system. In August 1967, *Electronics World,* a popular publication for electronics aficionados, included Coe's story (immediately following that of Robert Tarlton) in an article entitled "CATV: Past, Present & Future." The thrust of the article was to herald individual "ingenuity and enterprise," and the story it tells of Coe parallels those told of his predecessors, Parsons, Walson, and Tarlton: "As news of the successful experiment spread, more and more of Coe's neighbors began to ask if they could be connected to his antenna. And the senior project became a business. Coe started by charging $90 for connection to the system and $2 per month."[36]

Bainbridge Cable actually remained independent until its acquisition in 1994—long after surrounding systems had begun to be acquired by MSOs.[37] Still, as *Electronics World* was eager to point out as early as 1967, while systems like Bainbridge Cable flourished, they were not industry leaders. Coe remained able to keep pace with the MSO-owned systems as new technologies and programming options came along. He began offering HBO in 1978, just two years after it became available via satellite, and was about to add data services in 1995. His business could not be characterized as "ahead of the curve," but neither did it lag far behind.[38] The smaller and more isolated a local cable system was, the less likely it was to be acquired by an MSO. And while MSOs grew in total subscribers, the independently owned systems, barely covering their expenses, remained fairly numerous.

Even as of 2007, Shen-Heights TV of Shenandoah, Pennsylvania (one of the early anthracite region systems, founded in 1951), remains independent. Shen-Heights owner Martin F. Brophy was a child when his father, Frank Brophy, founded the company. At the time Shenandoah, already on the decline from the anthracite heyday, had a population of around fifteen to twenty thousand residents. Today it has only around five thousand. Brophy took over the system as a young man, following his father's death in the early 1960s. He, along with several other independent system owners, belongs to the American Cable Association, a lobbying organization, and the National Cable Television Cooperative (NCTC), which allows independent owners to aggregate and receive the sorts of bulk programming discounts as the MSOs.[39]

From 1955 until around 1990, the number of CATV systems was at least doubling by the decade. By 1960 there were 640; 1,325 by 1965; 2,490 by 1970; 4,225 by 1980; and 9,575 by 1990.[40] Many local systems flourished on their own at least into the late 1960s. But few remained independent as long as the systems owned by either Coe or Brophy. Most were bought by MSOs during the 1970s or 1980s (and were subsequently acquired by even bigger MSOs), becoming part of a heavily consolidated industry dominated by the NCTA. That this major trade organization, founded at cable's very beginning, would come to be dominated not by "moms 'n' pops," but by enormous synergistic media corporations is no coincidence. The earliest and most formative efforts of the organization were geared toward avoiding taxes and other legal obstacles to growth. The technologies developed and shared within the organization served small communities well, but proved even more efficient and effective when serving multiple communities.

The earliest founders clearly recognized the potential of their growing industry—at least in the abstract—and organized to build and protect it. Individuals like Davidson, Parsons, Walson, and Tarlton established technological and business precedents. Others like Calsam and the Barco family joined them in expanding the technological options for CATV operators. And individuals such as Smith and the Barcos played equally important roles in upholding the position of free enterprise against government regulatory bodies. This is not to say that government controls on CATV/ cable have not been called for over the years. Rather, one must recognize that the perceived imperative to control new technology can easily overwhelm the realization of that technology's full potential. Just as radio might have remained a "wireless telegraph" and the Internet a tool for researchers and the military, cable might have continued as a community antenna service and nothing more.

The late 1950s and 1960s would prove to be heavy years of regulation for CATV—far more than the industry's pioneers could have anticipated. These years saw CATV gain recognition from more people than just those who relied on it for television reception. It went from being an interim technology in America's hinterlands, to a potential threat to broadcast television, and then to the promise for a multichannel television future—all in the span of just over one decade. Yet the legacy of the early 1950s, when CATV was little understood by regulators or the general public, is essential to understanding the development of modern cable television. Had the industry's inventors not also had among them individuals with astute business sense and forward-looking legal expertise, CATV might well have been shut down by one or more government bodies in well-known and continued efforts to promote the fortunes of "free" broadcast television.

NOTES

1. A franchise is formal permission from a municipality to operate a CATV or cable television system there.

2. James Y. Davidson: An Oral History. (Interviewed by James Keller, no date. Unpublished manuscript archived at the Cable Center, Denver.)

3. Tom Southwick, *Distant Signals: How Cable TV Changed the World of Telecommunications* (Overland Park, KS: Primedia Intertec, 1998), 9.

4. James Y. Davidson, personal communication, July 2005.

5. Twin-lead (or ladder-lead) cable consisted of two copper wires connected at intervals by plastic bridges. Early in CATV history this was replaced by the coaxial

cable that dominated the industry well into the 1990s—until the development of fiber optic cable.

6. Mary Alice Mayer Phillips, *CATV: A History of Community Antenna Television* (Evanston, IL: Northwestern University, 1972), 14; E. Stratford Smith, personal communication, July 29, 2004.

7. Mayer Phillips, *CATV*, 23.

8. It is unclear why the FCC was unaware of Davidson's system at the time. However, this might help explain why the existence of Parsons' system has been more widely documented.

9. Details on Parsons' system are drawn from Mayer Phillips, *CATV*, and L. E. (Ed) Parsons: An Oral History. (Interviewed by Richard Barton, 1986. Unpublished manuscript archived at the Cable Center, Denver.)

10. Parsons: An Oral History.

11. Microwave is a point-to-point, line-of-sight, over-the-air mode of signal transmission that relies on one or more dish-type antennas.

12. The competing claim was by local police chief, William McLaughlin, who did run a CATV system in another part Mahanoy City for several years. In 1970, that system became part of Service Electric. See Mayer Phillips, *CATV*, 9–10.

13. Patrick R. Parsons, "Two Tales of a City: John Walson, Sr., Mahanoy City, and the 'Founding' of Cable TV," *Journal of Broadcasting & Electronic Media* 40, no. 3 (June 1, 1996): 354–365.

14. Robert Tarlton: An Oral History. (Interviewed by Lorna Rasmussen, 1993. Unpublished manuscript archived at the Cable Center, Denver.)

15. For years, Jerrold remained the most prominent supplier of CATV components, especially in the Northeast. Shapp went on to become governor of Pennsylvania in the 1970s.

16. Tarlton: An Oral History.

17. Ibid.

18. Milton Jerrold Shapp: An Oral History. (Interviewed by Mary Alice Mayer Phillips, 1986. Unpublished manuscript archived at the Cable Center, Denver.) See also Ray Schneider: An Oral History. (Interviewed by Marlowe Froke, 1993. Unpublished manuscript archived at the Cable Center, Denver.)

19. After selling his interest in Jerrold in the late 1960s, Shapp went on to serve two terms as governor of Pennsylvania during the 1970s.

20. Subsequently the National Cable Television Association and now the National Cable & Telecommunications Association.

21. *NCTA Monthly Bulletin* 1 (August 15, 1953): 2. Cited in Mayer Phillips, *CATV*, 43.

22. E. Stratford Smith, personal communications, State College, PA, 1994–2004. Also discussed in Mayer Phillips, *CATV*, ch. 6.

23. *NCTA Monthly Bulletin* 1 (November 15, 1953): 4–5. Cited in Mayer Phillips, *CATV*, 44.

24. *Television Factbook* (Washington, DC: Warren Communications News, Annual).

25. Personal Communication with Jean Calsam McMinn and Robert McMinn (William Calsam's daughter and grandson), Schuylkill Haven, PA, August 4, 2004. The system was acquired at that point by Malarkey and named Schuylkill Haven Trans-Video.

26. Robert McMinn, personal communication, August 4, 2004.

27. Megan Mullen, "Distant Signals from the Future: A Case Study of Eastern Microwave and Its Operations in the Northeast During Cable's Pre-Satellite Era," *Quarterly Review of Film and Video* 16, no. 3/4 (Summer 1999): 391–404.

28. James Duratz: An Oral History. (Interviewed by James Keller, 2001. Unpublished manuscript archived at the Cable Center, Denver.) It is a testament to the Barcos' business acumen that they could refuse the service agreement. By this stage the service agreement was all-but-mandatory for systems wishing to use Jerrold equipment.

29. Ibid.

30. James Duratz and James Strickler, personal communications, 2004.

31. Several preliminary meetings had been held over the preceding year.

32. "Community TV Group Forms National Lines," *Broadcasting-Telecasting* (June 16, 1952), 84.

33. *Lilly v. United States*, 238 F. 2d 584 (CA 4, 1956), *Pahoulis v. United States*, 242 F. 2d 345 (CA 3, 1957), reversing 143 F. Supp. 917 (W.D. Pa., 1956). Discussed in Mayer Phillips, *CATV*, 32–33.

34. Ibid.

35. A few such systems remain even to this day. A few operators have clung tenaciously to their independence, but for most current non-affiliated systems, their fairly isolated locations and low subscriber numbers make them unpopular choices for acquisition by large media conglomerates.

36. Jerry E. Hastings, "CATV: Past, Present & Future," *Electronics World* (August 1967): 25.

37. It was first sold to regional MSO Gateway Communications and subsequently to Time-Warner Cable, the owner of virtually all the local systems surrounding Bainbridge.

38. David Coe, personal communication, Bainbridge, NY, July 1994.

39. Martin F. Brophy, personal communication, Shenandoah, PA, July, 28, 2004.

40. National Cable-Telecommunications Association. After 1990, existing systems began to consolidate regionally, reversing this growth pattern.

Chapter 2

A Taste of Class
Pay-TV and the Commodification of
Television in Postwar America

John McMurria

> The whole conception of pay TV is a natural result of
> dissatisfaction on the part of the American public. . . .
> Television has largely become a kind of soporific, a mild
> narcotic having the effect of preventing the majority of
> the viewers from exercising their thinking faculties.
> —Solomon Sagall, President of Teleglobe
> Pay-TV System, January 21, 1958

> This is a brainstorm that someone thought up to make
> life just a little more miserable for the low-income class.
> We are taxed to death now. Why more? I have $450 in-
> vested in my TV set already. I think it is time the public
> fought back.
> —Wilbur G. Kellis, letter to the editor of the
> *Cincinnati Post* opposing pay-TV, July 6, 1955

In 1950, long before HBO launched its pay-TV service in Wilkes-Barre, Pennsylvania on November 8, 1972 with a hockey game and a Hollywood movie, the Federal Communications Commission (FCC) authorized three pay-TV companies to conduct tests in New York City, Chicago, and Los Angeles using devices that required viewers to pay a fee to unscramble programs transmitted over the air.[1] The tests were a technical success, but as the differing perspectives above exemplify, the prospect of having to pay for television programs was not universally embraced.[2] Movie theater owners, having lost audiences to suburban flight and the rise of television, lobbied against this new form of television that presented yet another competitive challenge to their business. Many struggling UHF television stations favored pay-TV as a potentially new revenue source,[3] but most network-affiliated stations opposed pay-TV, and their opposition got the attention of elected officials who wanted to maintain good relations with local stations that had the power to impact their reelection campaigns. To consider the matter, Congress held four public hearings between 1956 and 1969.[4] As multi-day public forums that included industry participants, non-profit organizations, civil groups, and concerned citizens, these hearings exposed a broader set of contestations over the cultural value of television during its first two decades of rapid growth, as well as differing concerns over pay-TV's challenge to the universal service principle of free-to-air broadcasting.[5]

Pay-TV advocates, including electronics manufacturers, television critics, performing arts institutions, leading educators, and elected officials, shared the perspective that commercial broadcast television had catered to the low tastes of a mass audience with a barrage of sitcoms, soap operas, westerns, and variety and quiz shows. Seeking more middle- and high-brow alternatives to satisfy their cultural tastes, this professional class of reformers believed that, given a choice to pay for television, viewers would likely choose more "sophisticated" programs such as televised Broadway plays, symphonic music, opera, ballet, university lectures, and literary adaptations. As Barbara and John Ehrenreich have observed, this professional class, which emerged during the turn-of-the-century Progressive era and rapidly grew in the 1950s and 1960s, served an important mediating function between capital and labor. As experts in their fields, who required lengthy training and who were also committed to maintaining ethical standards and to performing public services, professionals defended their intellectual independence from the capitalist class but also demonstrated "contempt and paternalism" toward the working classes even as they championed the "public interest."[6] So while pay-TV enthusiasts criticized profit-seeking commercial broadcasters for lowering cultural standards, they also invoked the capitalist ideals of "consumer choice" and "deregulated free markets" in advocating for a pay-TV alternative.

The broadcast networks reinforced these professional-class standards of cultural taste in arguing that pay-TV was unnecessary because the networks were already airing plenty of Broadway plays, serious news, and literary adaptations. At the same time, a less hierarchal defense of commercial broadcasting came from a broad spectrum of civil society, whose representatives upheld the cultural value of popular genres and fought to preserve the universal service principles of broadcasting. Veteran's associations, women's groups, senior citizens, civil rights leaders, labor unions, and local officials thought pay-TV would siphon programming away from free-to-air broadcasting and provide alternatives for only those who could afford them. All but one of the more than a dozen opinion polls submitted to the hearings from industry groups, newspapers, and third-party organizations indicated that a strong majority of viewers opposed pay-TV.[7]

Television historians have represented the emergence of pay-TV as a technologically progressive alternative to free-to-air broadcasting that was stalled by a bureaucratically incompetent FCC which capitulated to the lobbying power of the broadcast networks and movie theater owners. Otherwise, according to these historical narratives, the Hollywood studios and other "ambitious entrepreneurs" would have developed pay-TV alternatives much earlier than the mid-1970s.[8] The hearings' statements from civic groups, educators, and individual citizens as well as industry participants and policymakers, which are presented here, suggest a different picture, however, one that makes visible the perspectives of those who are often ignored in policy decision-making and written histories while also taking seriously the taste hierarchies that informed the decision-making process. Congress convened these pay-TV hearings not to prod a foot-dragging FCC but to rein in a commission that was moving forward too quickly without due concern for the potential impact pay-TV might have on the free-to-air broadcasters whom members of Congress had political reasons to protect.[9] But revealing more than the influence of political patronage, the postwar pay-TV policy debates also reveal a negotiated transition from a liberal regime of television governance which emerged in the Progressive and New Deal periods and which justified commercial use of the "public spectrum" through what Thomas Streeter describes as "a faith in the broad liberal framework of property rights, the market, and minimal government, coupled to and qualified by a faith in expertise, administrative procedure, and a reified, paternalistic notion of the public good," to a neo-liberal regime which has increasingly relinquished universal public service principles and paternalistic fiduciary oversight to the supposed vigor of a deregulated market.[10] The taste-conscious professional class of reformers that spanned party lines was not unified in advocating exclusively for neo-liberal television reforms—indeed, many supported the

Public Broadcasting System in the late 1960s as a means to uplift television culture.[11] Nonetheless, their vocal support for a pay-TV system that called for television programs to be bought and sold as commodities in a "free market" revealed a particular cultural context for the establishment of a broader social, political, and economic neo-liberal consensus that solidified in the late 1970s. This consensus provided a foundation for the first federal cable legislation that was passed in 1984 and that largely defined cable systems as private companies rather than public utilities.[12]

A Brief History of Early Pay-TV

Following the successful technical tests of pay-TV devices in the early 1950s, the FCC held hearings in 1955 on whether to authorize pay-TV on a nationwide basis. Soon after, in 1956, the Senate Committee on Interstate and Foreign Commerce held pay-TV hearings as well.[13] Preoccupied with the slow growth of UHF stations, neither the Senate Committee nor the FCC took immediate action, but in October 1957 the FCC authorized more extensive pay-TV tests. Surprised that the FCC did so without legislative authority, the House Committee on Interstate and Foreign Commerce scheduled hearings on the matter in January 1958.[14] Although the Senate and House introduced a number of bills to restrict or prohibit pay-TV, none was passed. As a result, in 1961 the FCC authorized Zenith to conduct a three-year over-the-air pay-TV test in Hartford, Connecticut.[15]

The Hartford trial began on June 29, 1962, and attracted nearly 3,000 subscribers after its first year and close to 5,000 subscribers in its second.[16] Eighty-five percent of subscribers were middle class with reported annual household incomes between $4,000 and $10,000. Only 1.5 percent of subscribers earned less than $4,000 per year, an income group that comprised 29 percent of the total U.S. population.[17] In its first two years, the trial offered 30 hours of subscription programs per week comprised of feature films (86.5 percent), sports (4.5 percent), special entertainment that included plays, opera, ballet, variety, and nightclub programs (5.5 percent), and educational programs (3.5 percent). On average, subscribers paid between 75 cents and $1.50 for each program. Sports and motion pictures were not only the most frequently scheduled programs but also the most popular with viewers. However, only 3.1 percent of subscribers watched the cultural programming championed by pay-TV supporters.[18] The fifty educational features fared the worst as just 0.7 percent of subscribers tuned-in despite full cooperation from area educational institutions.[19]

Pay-TV over cable wires developed in the 1950s and 1960s as well. Experiments were conducted in Palm Springs in 1953, Bartlesville, Oklahoma in 1957, and Toronto in 1960.[20] A more controversial cable pay-TV system was launched in Los Angeles and San Francisco in the summer of 1964 by former NBC President Pat Weaver with support from motion picture distributor Matty Fox. They persuaded the Brooklyn Dodgers and New York Giants baseball franchises to move to California to participate in the pay-TV venture. The pay-TV operation offered exclusive coverage of home games as well as second-run motion pictures and cultural arts programs. But the California pay-TV systems angered baseball fans in the East and offered real proof that pay-TV could siphon popular programs away from free-to-air broadcast television. California theater owners joined with other civic groups to conduct a well-financed campaign to oppose the cable systems and collected enough signatures to place a referendum on the November ballot to prohibit pay-TV in the state. The referendum passed by a two-to-one margin. The California Supreme Court later ruled the referendum unconstitutional, but by then the cable systems had closed and filed for bankruptcy.[21]

In 1965, following the three-year test in Hartford, the FCC's Subscription Television Committee submitted a proposal to authorize pay-TV on a permanent, nationwide basis, albeit with significant constraints. Limiting over-the-air pay-TV to one station in large markets, the proposal restricted certain programs, including recent motion pictures, certain sports, and "series type programs with interconnected plots."[22] The House Subcommittee convened hearings in October 1967 and November and December of 1969 to consider the FCC's proposal and to deliberate over twenty-one bills to prohibit pay-TV. Because the Subcommittee took no action on these bills, the FCC began accepting over-the-air pay-TV applications, but over-the-air pay-TV was slow to develop. The first two stations did not begin broadcasting until 1977, and although the FCC received over 115 pay-TV applications by the end of the 1970s, only 6 pay-TV stations had begun operating.[23] In 1970, the FCC applied its over-the-air pay-TV content restrictions to cable pay-TV systems. However, in a 1977 landmark HBO case the Supreme Court ruled that the FCC had exceeded its jurisdiction and violated cable television's First Amendment rights.[24]

Pay-TV Advocates and Matters of Television
Taste in Postwar America

During the 1956 congressional hearings pay-TV manufacturer Zenith described broadcast television as a "continuous stream of 20-year-old horse operas and other hackneyed programs that now fill in most of the short spaces between the long commercials."[25] Similarly, in a letter to the FCC dated July 28, 1965, Pat Weaver remarked that the "promise of the [pay] cable service is an intelligent population, with elevated tastes, upgraded standards, and the opportunity to enjoy the pleasures of civilized and literate human beings."[26] These sentiments were indicative of broader perceptions among postwar political and cultural elites who thought the rise of commercial culture, and especially television, had created mediocrity, conformity, and passivity.[27] Even liberal democrats such as the Americans for Democratic Action (ADA) voiced these concerns. An organization founded in 1947 by liberal intellectuals and professionals,[28] the ADA, though generally committed to extending New Deal social welfare policies, developed increasingly intense anti-communist stances, which prompted the organization to defend free market capitalism against a communist threat and to champion civil liberties to protect citizens from an over-centralized government. Advocating for civil rights and labor issues, as well, the ADA, whose members were mostly college-educated, upper-middle-class white males, did not attract broad support among African Americans or rank-and-file union members.

In the 1950s, the tone of the ADA was set by one of its most prominent founders, Arthur Schlesinger, Jr., who wrote a series of articles that articulated a new course for liberals and for the Democratic Party. Since most Americans had already secured the "economic basis of life," Schlesinger argued, liberals should dedicate themselves to "bettering the quality of people's lives and opportunities" through focusing on the "more subtle and complicated problems of fighting for individual dignity, identity, and fulfillment in a mass society." The challenge for liberals was not to fight "a conspiracy of wealth seeking to grind the faces of the poor" but to curtail a "conspiracy of blandness, seeking to bury all tension and conflict in American life under a mess of platitude and piety—not the hard-faced men, but the faceless men."[29]

The ADA supported pay-TV at each of the four Congressional hearings. At the 1956 hearings the ADA stated that television had failed to produce programming for those with "less-than-majority tastes and interests," and it made clear whose interests and tastes were not being met:

> Many of our wisest and most objective men in public life, including educators, social scientists, creative writers, and artists, now believe that the past 30 years of experience in radio, and a few years of television, have proved by demonstration that the present economic system of broadcasting cannot satisfy the great and expanding needs for information, education, cultural, and entertainment services.

The ADA argued that if individual viewers could choose to pay for programs, editorial power would transfer from the audience-maximizing interests of network gatekeepers to the more diversified tastes of individual subscribers. Challenging the universal service principles of broadcasting, the ADA argued that television should be given the same regulatory freedoms as the press: "Without an optional, competitive system of audience payment, broadcasting can never be a truly free and diversified medium like print." Because there was "no precedent for assuming that freedom of the air requires free circulation, or no payment by the recipient," pay-TV was "completely consistent with traditional American freedom of the press." For the ADA, the public interest imperative for broadcasting was no longer tied to a free-to-air universal service over a public spectrum but to the logic of consumption wherein program makers responded to the direct interests of individual consumers in the marketplace. Indeed, in conceptualizing the pay-TV subscriber, the ADA fused the democratic citizen with the marketplace consumer: "A direct public payment for a product or service is a voluntary, selective ballot which not only measures the usefulness of that service but provides the economic incentive for invention, competition, and expansion."[30]

This conflation of individual consumption with democratic participation in the "citizen consumer" was echoed in the comments of other pay-TV supporters who invoked the box office as a metaphor for how individual choice could uplift television culture.[31] Consider prominent actor and producer Otto Preminger's support for pay-TV at the 1967 hearings: "In a capitalistic country like ours it is only the box office that will really prove what people want to see and not those free things which I personally don't believe." Appalled by broadcast network television, Preminger thought that pay-TV would bring "Shakespeare," "ballet," and "symphonies" to television and "raise the cultural level of the United States."[32] Other high-performing arts organizations supported pay-TV, including the Chicago Symphony, which argued in 1958 that TV broadcasters have irresponsibly "dinned Presley music into the ears of our impressionable youngsters."[33] In 1969, Blanche Thebom of the Southern Regional Opera suggested that while African Americans had benefited from civil rights legislation, the minority intellectual had no comparable benefits: "Democracy is dependent upon majority rule, but it must also recognize minority rights, and is doing so at this particular time in a most meaningful way in other fields and for other groups. The intellectually and or culturally sophisticated must also be served and currently, the mass media is doing less than a spectacularly outstanding job of it."[34]

While the "culturally sophisticated" supported pay-TV to defend their "minority rights," broadcasters defended their record of airing sophisticated programming. During the 1956 hearings, NBC, CBS, and the National Association of Radio and Television Broadcasters all cited NBC's recent airing of the Broadway play *Peter Pan* staring Mary Martin as evidence of network broadcast excellence.[35] Though cultural intellectuals disdained Broadway as a "middlebrow" amusement that pandered to the commercial impulses of mass culture, a postwar consumer-driven prosperity had produced a "status-panicked" professional class that turned to middlebrow culture in a more uncertain effort to acquire cultural prestige.[36] In their embrace of Broadway over sitcoms and soap operas, broadcasters shared a certain status-panic with other professionals, who were both for and against pay-TV. Together, they set the middle and highbrow cultural terms for the regulatory debate.

Leading educational institutions supported pay-TV in part to help finance struggling educational television stations. Although they had fought to win educational television (ETV) allocations in 1952, four years later, when the first pay-TV inquiry began, only fifteen of the 252 ETV allocated stations had gone on the air.[37] But in stating their educational goals, educators invoked the taste hierarchies that characterized their professional class positions. At the 1958 hearings the Adult Education Association of the U.S.A. commented that "many educationally negative consequences flow from a television system that is almost exclusively geared to the lowest common tastes of mass audiences."[38] The National Association of Educational Broadcasters invoked the citizen consumer: "It seems to this writer that the best traditions of the free-enterprise system suggest that pay TV be permitted to compete in the market place with the present commercial TV system. This should provide a desirable stimulus to both.

And the public, as it should in a democracy, will be the final judge."[39] The National Association for Better Radio and Television (NAFBRAT), whose forty-four member board included prominent educators and religious and community leaders, also conflated consumption with democratic participation: "'Let the people choose' has long been a tenet as fundamental in marketing economics as in electoral politics. Subscription television will offer the public an opportunity to pay for quality in the market place of programs, as it now pays for quality in the market place of television receivers."[40] Lawmakers shared the elite cultural biases of educators. In 1958 New York Congressman Isidore Dollinger said that "many people have spoken to me, and many Members of the House have, and with very few exceptions they have said that the 'Best pleasure I get is tuning the programs out.'"[41] Even some advertisers shared these taste sensibilities in voicing their support for pay-TV despite its potential threat to ad-supported broadcasting. For example, in 1955 the marketing industry magazine *Tide* polled a "leadership panel" whose members favored pay-TV by a wide margin.[42] As one advertising manager put it, pay-TV promised to "raise the intelligence value of television and increase its cultural possibilities for viewers."[43]

In summary, electronics manufacturers, advertisers, educators, lawmakers, liberal democrats, cultural critics, and the high performing arts supported pay-TV to liberate a minority of "discerning" viewers from the "lowest common tastes" of a mass commercial broadcast culture. There were certainly other "minority" groups who fell outside the middleclass white norms propagated by commercial broadcasters, including the civil rights movement which challenged racist broadcasters.[44] However, in framing the television problem as a matter of cultural taste and in articulating the distinctions between high and low culture as universal standards of judgment rather than as social categories produced through uneven access to education, capital, and leisure time, as sociologist Pierre Bourdieu has argued, this professional class of television reformers sought to legitimize its class position and cultural authority during the postwar period when television broadcasting emerged as America's most pervasive medium of popular culture.[45] In doing so, they also invoked a discourse of the citizen consumer which equated free markets and individual consumer choice with a more democratic culture, a discourse which came to dominate justifications for deregulating television and the telecommunications sector in the late 1970s and the Reagan-era 1980s. However, the cultural anxieties that motivated the pay-TV lobby to break from the universal service principles of broadcast television were met with broad resistance from organizations that viewed pay-TV as an affront to low-income citizens and an unjust move to privatize the public airwaves.

Civil Society and the "Freedom to Look and Listen"

The very idea of pay-as-you-see television violates the American concept of freedom to look and listen.

—James D. Johnson, letter to the editor of the
Cincinnati Post, July 6, 1955[46]

Because pay-TV potentially threatened to further impact the theatrical movie business, the National Association of Theatre Owners formed the Committee Against Pay-To-See TV in 1955 to coordinate signature petition drives in theater lobbies and finance advertising campaigns to "save free-TV." California exhibitors paid for most of the $750,000 that a coalition of opposition groups spent to win the California referendum against pay-TV in 1964.[47] Exhibitors, like the broadcast networks, argued that pay-TV would likely hurt the financial base of broadcasting by siphoning away quality programs, create more mass-produced rather than upscale programming, and benefit pay-TV manufacturers that were more interested in the profitable big cities than in helping struggling UHF stations in smaller markets.

While the theater owners had the lobbying power and financial resources to help stall pay-TV, they were joined by a broad spectrum of women's groups, veteran's associations, senior citizens, civil rights leaders, labor unions, and local officials who defended the cultural worth and universal service principles of commercial broadcasting. At the 1958 hearings, the spokesperson for the 2.5 million member General Federation of Women's Clubs said, "I am being asked by these groups of women before whom I appear, why seven men have the authority to sell our last free natural resource, the free air, the only thing seemingly that we have free."[48] At the 1956 hearings the General Federation's chairperson embraced television's "great influence on life in the home" and cited the programs that she and her members most valued:

I am a fan of Dave Garroway. I listen every morning for nearly 2 hours to his program. It is a wonderful contribution to family life. So, too, is the

Home program on the same network. So, too, are the spectaculars—Max Liebman's spectaculars—and what could be better than the productions of *Richard III* and the *Taming of the Shrew,* and the magnificent program that Edward R. Murrow put on discussing the problems of the Arabs and the Jews.[49]

While this embrace of classical theater revealed the white middle-class cultural sensibilities of this Progressive-era reform organization, in valuing morning talk shows and afternoon home improvement programs, these women defended the cultural value of the daytime television that white male opinion leaders often dismissed as trivial.[50]

Veteran organizations also strongly opposed pay-TV throughout the 1950s and 1960s. At their 1957 national convention, the AMVETS, "composed exclusively of World War II and Korean veterans," voted to oppose what they described as "one of the greatest giveaways of public property for private gain ever seen."[51] AMVETS defended broadcast television's universal service principles because most of their members were hospitalized and/or living on fixed incomes. They made the point that television belonged to the viewers because while the broadcast networks, stations, and advertisers had invested $3 billion in programming, facilities, and advertising costs, the nation's 42 million television viewers had invested $14 billion in television sets.[52] As 42 percent of American families made less than $4,000 and 20 percent made less than $2,000 according to 1950s census data, televisions sets that started at $200 were indeed expensive for many families.[53] Veterans also supported broadcasters because they aired free public service announcements promoting veteran rehabilitation and employment. The Armed Forces Reserve, for example, received $8 million-worth of free airtime during Military Reserve Week in 1956.[54] The Veterans of Foreign Wars, the Jewish War Veterans, and the Catholic War Veterans also opposed pay-TV.[55]

Voicing similar economic concerns, more than a dozen organizations representing senior citizens opposed pay-TV, as well. The National Association of Retired Civil Employees worried that their federal annuitants, most receiving less than $3,000 per year, could not afford the estimated $75 annual cost of pay-TV.[56] However, other senior groups supported advertising-free pay-TV because they were tired of being bombarded with commercial messages and seeing their favorite on-air personalities pitch products.[57] As President Max Friedson of the Congress of Senior Citizens in Miami explained, "we are distressed by the free-TV making merchandising peddlers out of such fine personalities as Hugh Downs, Arthur Godfrey, Barbara Walters, Perry Como, and many others whom the elder citizens admire."[58] Unlike other cultural critics who cited the variety shows and news magazines these personalities represented as evidence of commercial television's low cultural taste, these seniors embraced such programming and, instead, focused their critique on the commodifying effects of advertising sponsorship. Wealthy seniors and those who lived in middle-class communal environments tended to support pay-TV.[59]

While the interests of African Americans were marginalized throughout the pay-TV hearings, during the 1969 hearings the prominent African-American civil rights leader Terry A. François spoke passionately against pay-TV. François's testimony carried the weight of his public service record and extensive involvement in the civil rights movement, which included his leadership at the NAACP and Urban League and his status as the only African American at the time to be elected to a city and county board of supervisors in California.[60] As a board member of the Northern California ADA, however, François, broke from the ADA's official position supporting pay-TV to defend universal service principles because "television has become the principal source of recreation and of information for the Nation's poor, the racial minorities, and other disadvantaged—those Americans who cannot afford the more costly forms of entertainment outside the home."[61] François also doubted that African Americans would profit from pay-TV: "the possibilities for tremendous profits are awaiting the pay television system operators. I question what is being done to insure that black representation will be present to share in that gain. Of the broadcast stations around the country, I believe only seven radio stations are black owned."[62] Moreover, as the hearings followed two years of violent clashes between the police and protesting African Americans in a dozen cities across the country, François feared that pay-TV would create audiovisual segregation:

> I am equally concerned over the effect pay-TV may have on the limited advances which have been made in developing racial minority group programming. . . . Our society can ill afford this kind of divisiveness in mass media programming during these troublesome times. I do believe the effect will be that the middle class will be watching one type of program and the poor will be watching another. Many of the difficulties come about by failure to look to the needs and desires of the less affluent members of society.[63]

However, the few African Americans who spoke at the hearings were not unified on the issue. Juanita Handy, a social worker who was a member of the Providence Rhode Island branch of the NAACP and a former president of the Urban League of Rhode Island, invoked the citizen consumer in taking exception to François's testimony:

> Even as older Americans resent false propaganda campaigns which are blatantly aimed at what an ad agency conceives to be a "typical" old person, so, too, do most Negroes resent being lumped together to fit a stereotyped image. To be poor, uneducated, or living in ghettos, whether one is white or black, does not mean that one is also without the ability to prefer and appreciate good shows, good music, good movies, spectacular sports events and the like. . . . What is the greatest educational tool in shaping cultures and mores? TV. What makes TV networks improve their programs? Competition for the TV consumer's attention.[64]

Developing these themes in their own terms, most trade unions took a principled stance that pay-TV would discriminate against low-income viewers and create social stratification. Entertainment industry workers were split, however. "Above-the-line" actors and writers who worked mostly in theater and film supported pay-TV, while "below-the-line" technicians and rank-and-file performers in television and radio were opposed. For example, while Actors' Equity, representing theater actors, strongly supported pay-TV, believing that more pay-televised Broadway plays would benefit their members, the 20,000 performers with the American Federation of Television and Radio Artists (AFTRA) unanimously opposed pay-TV during their July 1955 convention on the grounds that few members made large salaries and that in 1955 only one quarter had made over $1,000 per year.[65] Believing that pay-TV revenues would increase star salaries and attract mass audiences, AFTRA members thought that pay-TV hurt staple network programs, including those aired during the day for "what is known as a woman-type audience." In other words, the pay-TV "million dollar spectacular" would ultimately "reduce the opportunities for rank-and-file members" of the entertainment industry.[66]

The American Federation of Labor and Congress of Industrial Organizations (AFL-CIO) also expressed unanimous opposition to pay-TV at its December 1957 constitutional convention. Stating that pay-TV would "break the pledge of the Federal Government, signed into law, that television would be free," the AFL-CIO argued in particular that pay-TV would also "work a hardship on the low-income part of the population."[67] Moreover, in addressing claims that pay-TV would help educational television stations, the Federation responded that "in view of our tax-supported public-education tradition and the fact that the TV waves are public property, there is serious objection to placing a price tag on educational broadcasting."[68] However, despite the 1957 vote, not all AFL-CIO members opposed pay-TV.[69] These included Hollywood guilds such as the Writers Guild of America, the Screen Extras Guild, and Screen Actors Guild (SAG), all of which supported pay-TV.[70] Since breaking with the more militant Conference of Studio Unions and in cooperating with the House Un-American Activities Committee in the late 1940s, SAG had espoused an anti-communist and pro-corporate stance on industry-labor relations.[71] The latter was especially evident during the 1969 hearings, when SAG president Charlton Heston embraced the "free enterprise society" and supported a pay-TV system that would "give the public the choice of better entertainment, culture, and education in the home."[72] Other spokespersons for the screen actors and extras told the 1969 House Committee that of the $112 million annually collected by union members, $54 million was for work in television commercials, $49 million for filmed television shows, and $20 million for motion pictures, thus sending a strong message that they did not believe pay-TV would hurt ad-supported television.[73] With the Hollywood guilds' strong support, on September 11, 1967, the AFL-CIO Executive Council voted unanimously to reverse their 1957 position and support the FCC's national pay-TV plan.[74]

Indicative of the influence that the performing arts guilds had on the AFL-CIO's new position was the number of unions that opposed pay-TV at the 1969 hearings. The East Coast Council of Motion Picture Production Unions, speaking on behalf of 30,000 workers, was "unalterably on record opposed to pay TV" even though roughly 10,000 of their members were also members of the Hollywood Film Council which supported pay-TV.[75] The spokesperson for the 15,000-member New York local of the International Union of Electrical Workers called pay-TV "the most horrible thing that has come upon us in many, many years."[76] The Greater Boston Labor Council and the New England Machinists and Aerospace Workers both passed resolutions opposing pay-TV; in the words of the council's executive secretary-treasurer, the council "did not think that it was in the public's interest to allow private corporations the privilege of using public airwaves for exploitation and profit."[77] The International Association of Machinists and Aerospace Workers, representing one million workers, the

280,000 workers with United Transportation Union, a United Mine Workers local, and a United Auto Workers local likewise urged Congress to ban pay-TV.[78]

Conclusion

As sites of public contestation over the cultural value of television during its initial years of rapid growth, the four Congressional pay-TV hearings reveal that the history of pay-TV in postwar America involved more than a regulatory agency protecting an entrenched broadcast industry against the competitive threats of technology pioneers. The hearings exposed class conflicts between professional educators, political liberals, lawmakers, above-the-line entertainment workers, and performing arts organizations who supported pay-TV in part to elevate programming tastes, on the one hand, and civil groups representing women, war veterans, senior citizens, civil rights leaders, and rank-and-file union workers who defended low-income citizens' right to access affordable quality television, on the other. In defining television as a box-office industry or an electronic press, rather than as a public utility, the professionals who shared a disdain for popular television propagated the idea that television viewers should no longer be thought of as citizens with equal rights to receive information and entertainment over the public airwaves but as individual consumers with rights to purchase television from competing providers within a market economy largely unencumbered by government constraint. Well before Ronald Reagan's FCC chairman Mark Fowler campaigned to treat television as a commodity bought and sold in a free market—a vision that was crystallized in his metaphor of television as a "toaster with pictures"—the pay-TV advocates of the 1950s and 1960s, which included political liberals as well as free-market conservatives who shared the tastes common to their professional class, laid a conceptual foundation that began to fundamentally shift the liberal Progressive and New Deal–era principles for broadcasting. Based on the idea of universal public service and a faith that government administrators could hold commercial broadcasters accountable for the paternalistic care of television viewers, these principles gave way to a neo-liberal framework guided by the ideals of consumer choice, free market competition, and privatization.[79] By the late 1970s these deregulatory frameworks had defined cable wires as privately owned communication conduits unlike broadcasting's public airwaves. And in 1996, when

Congress rewrote the Communications Act of 1934 "to promote competition and reduce regulation," the neo-liberal consensus had been solidified.[80] Eight years later, universal affordable access to television had become even more distant as cable subscription fees increased 45 percent, or three times the rate of inflation.[81]

In 2006, as broadband Internet emerged as a new communications infrastructure for distributing television and other video content, Congress supported further neo-liberal deregulations in bills that would allow phone companies to offer video without the consent of local municipalities and enable Internet service providers to charge fees for distributing content.[82] As cable operators have lobbied to oppose the efforts of citizen groups to keep broadband Internet access open and prevent municipal governments from offering universal broadband service, U.S. neo-liberal broadband policies have left the United States ranked sixteenth in worldwide broadband penetration in 2005, behind South Korea, Hong Kong, Netherlands, Demark and Canada.[83] Although in some cases population densities are important factors, these countries have taken a more active role in developing universal broadband access by implementing widespread digital literacy programs, investing in infrastructure, and promoting universal service through subsidies and grants.[84] In 2005, for example, Canada's broadband penetration was 77 percent compared to 57 percent in the United States, in part because the Canadian government initiated a National Broadband Taskforce in 2001 that supported broadband infrastructure development, subsidized pilot programs at the community level, and delivered electronic government services that stimulated demand for broadband capacity.[85] Similarly, the South Korean government invested $11 billion between 1998 and 2002 to support infrastructure build-outs, broadband in classrooms, digital literacy programs, and tax incentives for wiring particular areas for broadband.[86]

As we consider U.S. broadband policies in an effort to extend the "freedom to look and listen" to the more interactive right to comment and create, we might do well to remember how a status-conscious professional class helped dismantle the universal service principles established under Progressive and New Deal–era communications policies. In thinking about preserving these universal service principles while also finding alternatives to the liberal frameworks that put paternalistic regulators and profit-seeking advertisers and network executives in charge of the public trust, we might support more open distribution systems that do not reward the tastes of those professional classes that are able to pay for choices. But we should start by asking those who can't afford such services what *they* believe is worth watching and creating.

NOTES

The author thanks the following for their valuable feedback: William Boddy, Rick Maxwell, Anna McCarthy, Toby Miller, Dana Polan, Tom Streeter, Marion Wilson, and George Yúdice. The completion of this essay was assisted by a grant from the Faculty Research and Development Program, College of Liberal Arts and Sciences, DePaul University.

1. On the history of subscription television, see H. H. Howard and S. L. Carroll, *Subscription Television: History, Current Status, and Economic Projections* (Knoxville: University of Tennessee, 1980); Megan Mullen, "The Pre-History of Pay Cable Television: An Overview and Analysis," *Historical Journal of Film, Radio and Television* 19, no. 1 (1999); Richard A. Gershon, "Pay Cable Television: A Regulatory History," *Communications and the Law* 12, no. 2 (June 1990); Patrick R. Parsons and Robert M. Frieden, *The Cable and Satellite Television Industries* (Boston: Allyn and Bacon, 1998).

2. Sagall is quoted in The House Committee on Interstate and Foreign Commerce, Subscription Television, 85th Cong., 2nd sess., January 14–23, 1958, 356 (hereafter referred to as House Committee). Kellis is quoted in Senate Committee on Interstate and Foreign Commerce, Television Inquiry Part III: Subscription Television, 84th Cong., 2nd sess., April 23–27 and July 17, 1956, 1279–80 (hereafter referred to as Senate Committee).

3. Senate Committee, 1214–15; House Committee, 546, 550, 556–58, 560, 567; House Subcommittee on Communications and Power of the Committee on Interstate and Foreign Commerce, Subscription Television, 90th Cong., 1st sess., October 9–11 and 16, 1967 (hereafter referred to as House Subcommittee), 213, 619–622.

4. Senate Committee; House Committee; House Subcommittee; House Subcommittee on Communications and Power of the Committee on Interstate and Foreign Commerce, Subscription Television–1969, 91st Cong., 1st sess., November 18–21 and 24 and December 9–12, 1969 (hereafter referred to as House Subcommittee–1969).

5. The Communication Act of 1934 states that all residents of the United States are entitled to "equality of radio broadcasting service, both of transmission and reception." Communications Act of 1934, Public Law 416, 73rd Cong. (June 19, 1934), 22.

6. Barbara and John Ehrenreich, "The Professional-Managerial Class," in Pat Walker, ed., *Between Labor and Capital* (Boston: South End Press, 1979), 5–45. See also Jackson Lears, "A Matter of Taste: Corporate Cultural Hegemony in a Mass-

Consumption Society," in Lary May, ed., *Recasting America: Culture and Politics in the Age of Cold War* (Chicago: University of Chicago Press, 1989), 38–57.

7. Senate Committee, 1269, 1278–83; House Committee, 66, 170, 301–2, 519; House Subcommittee–1969, 261–83, 369.

8. For variations on this policy history, see Michelle Hilmes, *Hollywood and Broadcasting: From Radio to Cable* (Urbana: University of Illinois Press, 1990), 130; Michele Hilmes, "Cable, Satellite and Digital Technologies," in Dan Harries, ed., *The New Media Book* (London: British Film Institute, 2002), 4–5; Christopher H. Sterling and John M. Kittross, *Stay Tuned: A Concise History of American Broadcasting*, 2nd edition (Belmont, CA: Wadsworth, 1990), 520; Gershon, "Pay Cable Television"; Mullen, "The Pre-History of Pay Cable Television"; Robert Britt Horwitz, *The Irony of Regulatory Reform: The Deregulation of American Telecommunications* (Oxford: Oxford University Press, 1989), 192; Parsons and Frieden, *The Cable and Satellite Television Industries*.

9. See House Committee, 2; see also Horwitz, *The Irony of Regulatory Reform*, 163–64.

10. Thomas Streeter, *Selling the Air: A Critique of the Policy of Commercial Broadcasting in the United States* (Chicago: University of Chicago Press, 1996), 109–10.

11. Laurie Ouellette, *Viewers Like You: How Public TV Failed the People* (New York: Columbia University Press, 2002).

12. See David Harvey, *A Brief History of Neoliberalism* (Oxford: Oxford University Press, 2005); Cable Communications Policy Act, Public Law 98-549 (1984).

13. Federal Communications Commission, Notice of Proposed Rulemaking "In the Matter of Amendment of Part 3 of the Commission's Rules and Regulations (Radio Broadcast Services) To Provide for Subscription Television" (Washington, DC: GPO, 1955). Reprinted in House Committee, 8.

14. Senate Committee, 1473; Federal Communications Commission, Notice of Further Proceedings (Washington, DC: GPO, 1957), reprinted in House Committee, 14–17.

15. Federal Communications Commission, Joint Comments of Zenith Radio Corporation and Teco, Inc., in Support of Petition for Nation-Wide Authorization of Subscription Television (Washington, DC: GPO, 1965), 1–92, reprinted in House Subcommittee, 241–335.

16. Ibid.

17. Ibid., 265.

18. Federal Communications Commission, Joint Comments of Zenith Radio Corporation and Teco, Inc., 81, reprinted in House Subcommittee, 324.

19. Howard and Carroll, *Subscription Television*, 28.

20. Hilmes, *Hollywood and Broadcasting*, 126–27; Gershon, "Pay Cable Television," 12–13; Mullen, "The Pre-History of Pay Cable Television," 42–43; Howard and Carroll, *Subscription Television*, 28–29; House Subcommittee, 366–71.

21. David H. Ostroff, "A History of STV, Inc. and the 1964 California Vote Against Pay Television," *Journal of Broadcasting* 27, no. 4 (Fall 1983): 371–86; Mullen, "The Pre-History of Pay Cable Television," 46–47; Gershon, "Pay Cable Television," 14–15.

22. Federal Communications Commission, *Report to the Federal Communications Commission by Its Subscription Television Committee* (Washington, DC: GPO, 1967), 3–4, reprinted in House Subcommittee, 11, 12.

23. Howard and Carroll, *Subscription Television*, 37–38.

24. Gershon, "Pay Cable Television," 19–20. For a discussion of the First Amendment and cable television, see Cass R. Sunstein, *Free Markets and Social Justice* (New York: Oxford University Press, 1997).

25. Senate Committee, 1320.

26. House Subcommittee–1969, 411–12.

27. Warren Susman, with the assistance of Edward Griffin, "Did Success Spoil the United States? Dual Representations in Postwar America," in May, ed., *Recasting America*, 24; Dwight MacDonald, *Masscult and Midcult* (New York: Random House, 1961); Dwight MacDonald, "A Theory of Mass Culture," in B. Rosenberg and D. M. White, eds., *Mass Culture: The Popular Arts in America* (Glencoe: Free Press, 1957), 60; Tony Bennett, "Theories of the Media, Theories of Society," in Michael Gurevitch, Tony Bennett, James Curran, and Janet Woollacott, eds., *Culture, Society and the Media* (London: Methuen, 1982), 36; Ouellette, *Viewers Like You*, 30.

28. Steven M. Gillion, *Politics and Vision: The ADA and American Liberalism, 1947–1985* (New York: Oxford University Press, 1987), viii–ix.

29. Ibid., 21, 124, 125. Also see Andrew Ross, *No Respect: Intellectuals and Popular Culture* (New York: Routledge, 1989).

30. Senate Committee, 1049–51.

31. On the citizen consumer see Toby Miller, "Television and Citizenship: A New International Division of Cultural Labor?" in Andrew Calabrese and Jean-Claude Burgelman, eds., *Communication, Citizenship, and Social Policy: Rethinking the Limits of the Welfare State* (New York: Rowman and Littlefield, 1999), 279–92; and Lizabeth Cohen, *A Consumer's Republic: The Politics of Mass Consumption in Postwar America* (New York: Alfred A. Knopf, 2003).

32. House Subcommittee, 221–35.

33. House Committee 481–83; House Subcommittee–1969, 216–19.

34. House Subcommittee–1969, 234.

35. Senate Committee, 1231, 1264, 1316, 1320.

36. On cultural intellectuals and Broadway see MacDonald, "Theory of Mass Culture," 64–65, and Louis Kronenberger, "Highbrows and the Theater Today: Some Notes and Queries," *Partisan Review* 26 (Fall 1959): 560–72. On "status panic" see C. Wright Mills, *White Collar: The American Middle Classes* (New York:

Oxford University Press, 1951); Marianne Conroy, "'No Sin in Lookin' Prosperous': Gender, Race, and the Class Formations of Middlebrow Taste in Douglas Sirk's *Imitation of Life*," in David E. James and Rick Berg, eds., *The Hidden Foundation: Cinema and the Question of Class* (Minneapolis: University of Minnesota Press, 1996), 114–37; Vance Packard, *The Status Seekers: An Explanation of Class Behavior in America and the Hidden Barriers That Affect You, Your Community, Your Future* (New York: David McKay, 1959).

37. Senate Committee, 1177–85.

38. House Committee, 574.

39. Ibid., 572–73. For more on the NAEB see Sterling and Kittross, *Stay Tuned*, 157, 267.

40. House Committee, 578.

41. Ibid., 325–26.

42. Senate Committee, 1443–47.

43. Ibid.

44. For the civil rights struggle over broadcasting in these years see Steven D. Classen, *Watching Jim Crow: The Struggles Over Mississippi TV, 1955–1969* (Durham: Duke University Press, 2004).

45. Pierre Bourdieu, *Distinction: A Social Critique of the Judgement of Taste* (Cambridge: Harvard University Press, 1984).

46. Reprinted in Senate Committee, 1279–80.

47. Ostroff, 381.

48. Ibid., 475–6.

49. Senate Committee, 1424–25.

50. For example, the House Subcommittee chair of the 1967 hearings said, "[h]ave you ever had to be in bed or in a hospital for a day or two and watch daytime television?" (House Subcommittee, 598). The 600,000-member National Federation of Music Clubs also opposed pay-TV for discriminating against low-income viewers. See House Subcommittee, 479–81.

51. Ibid., 491.

52. Ibid., 490.

53. For census see House Committee, 169. For receiver prices see Sterling and Kittross, *Stay Tuned*, 290.

54. House Committee, 491–92.

55. House Committee, 467–68, 568–69.

56. Ibid., 219–22.

57. House Subcommittee–1969, 404.

58. Ibid., 292.

59. Ibid., 231, 290, 476–77.

60. Ibid., 172–73.

61. Ibid, 174.

62. Ibid., 174–75.

63. Ibid., 175.

64. Ibid., 193–94.

65. Senate Committee, 1201–3, 1357–68.

66. Ibid., 1357–68.

67. Senate Committee, 1354–56.

68. House Committee, 528–29.

69. Ibid., 530–36.

70. House Subcommittee–1969, 139–43, 353.

71. For more on the transformation of SAG see Lary May, "Movie Star Politics: The Screen Actors' Guild, Cultural Conversion, and the Hollywood Red Scare," in May, ed., *Recasting America*, 125–53. For a detailed account of the CSU strike and lockout see Gerald Horne, *Class Struggle in Hollywood, 1930–1950: Moguls, Mobsters, Stars, Reds and Trade Unionists* (Austin: University of Texas Press, 2001).

72. House Subcommittee–1969, 139–43.

73. Ibid., 143, 229. The Communications Workers of America also testified in support of pay-TV; ibid., 353.

74. Ibid., 200–201.

75. Ibid., 160–72.

76. Ibid., 256.

77. Ibid., 346–47.

78. Ibid., 462–64.

79. Mark Fowler and Daniel Brenner, "A Marketplace Approach to Broadcast Regulation," *Texas Law Review* 60 (1982): 207–57.

80. Quoted in Edwin C. Baker, *Media, Markets, and Democracy* (Cambridge: Cambridge University Press, 2002). Also see Patricia Aufderheide, *Communications Policy and the Public Interest: The Telecommunications Act of 1996* (New York: Guilford Press, 1999).

81. Mark Cooper, *Cable Mergers, Monopoly Power and Price Increases* (Washington, DC: Consumer's Union, 2003); http://www.consumersunion.org/pdf/CFA103 .pdf (accessed February 19, 2004).

82. Kim Hart and Sara Kehaulani Goo, " 'Net Neutrality' Amendment Rejected," *The Washington Post*, June 29, 2006, D5.

83. See Free Press, "Community Internet: Broadband as a Public Service," http://www.freepress.net/communityinternet/ (accessed February 21, 2005) and Save the Internet, "SavetheInternet.com Coalition Statement of Principles," http:// www.savetheinternet.com/=principles (accessed August 13, 2006).

84. Rob Frieden, "Lessons from Broadband Development in Canada, Japan, Korea and the United States," *Telecommunications Policy* 29 (2005): 595–613; Dal Young Jin, "Socioeconomic Implications of Broadband Services: Information Economy in Korea," *Information, Communication & Society* 8, no. 4 (December 2005): 503–23.

85. See Website Optimization, "U.S.-Canadian Broadband Penetration Gap at 20 Points," http://www.websiteoptimization.com/bw/0506/ (accessed August 13, 2006).

86. Frieden, "Lessons from Broadband Development in Canada, Japan, Korea and the United States," 595–613.

Chapter 3

Cable's Digital Future

François Bar and Jonathan Taplin

The deployment of digital technology throughout the U.S. communication networks over the last decade promised "the end of scarcity" for television.[1] Not only would fiber optics and electronics increase by orders of magnitude the capacity of each infrastructure—telephone, cable, or wireless—but once-specialized networks would all become able to transmit video streams, adding ever more channels for television distribution. By contrast with the old oligopolistic world of broadcasting and cable, various competing companies would own these new infrastructures, so there would be no single gatekeeper to program distribution. At the same time, digital technology would slash the cost of the equipment needed to produce video programming, bringing the ability to create video content within reach of an ever-growing number of sources. Finally, digital technology would spur the proliferation of devices able to display video streams—not just television sets in living rooms, but also digital theaters, computers, laptops, game consoles, mobile phones, personal digital assistants, and liquid crystal displays mounted in cars, airplanes, and shopping malls. Internet Protocol (IP) would become the lingua franca connecting all the components of this new production, distribution, and consumption system. While scarcity had been the organizing principle of the old television broadcasting business, abundance would be the new rule. Amid the abundance of distribution channels, anyone could become a creator and find a distributor for his or her work; any viewer could access any video program, anywhere, anytime.

As of 2006, we can get a glimpse at this abundant digital future on the broadband Internet. In July 2006, on-line video site YouTube.com crossed a milestone as it served over 100 million video clips per day.[2] Its success is followed closely by Google Video and Yahoo, along with about 180 new online video start-ups.[3] In the classic cable and satellite networks, however, distribution pathways remain tightly controlled by a few owners, steadfastly opposed to granting open access to their networks. While telephone companies were allowed to enter the video distribution business starting in 1992, in 2005 they provided video programming to only 322,700 homes out of 94.2 million, a 0.3 percent share of the video programming distribution market that barely registers on the competitive landscape.[4] Though new digital cameras and editing software have allowed many to create videos, getting these widely distributed still requires deep pockets and substantial marketing muscle. Thus, all those screens populating our everyday lives tend to display very similar programming.

This chapter describes the recent digital transformation of the cable television networks. It outlines the tremendous potential for the way we create, distribute, and access video programming. It examines the reasons why this potential has yet to be realized and suggests policy options that could help address these issues.

The Video Distribution Competitive Landscape

The recent evolution of U.S. mass media policy has seen a continuation of the trend toward deregulation—the increasing reliance upon market competition, rather than government rules, to influence the behavior of media organizations. Competition in the provision of video distribution, it was assumed, would stimulate innovation in the underlying infrastructure, in distribution services and in content, insuring low prices and consumer choice. The Telecommunications Act of 1996 laid out a new framework to foster broader competition in the multichannel video programming distribution (MVPD) market, 87.7 percent of which was then controlled by cable companies.[5]

Competition to the cable industry was expected to come from a variety of sources. Policy makers expected telephone companies ("telcos") to become the strongest competitors. Telcos had long lobbied to be allowed in the video distribution business, seeking to expand beyond the telephone market. With the spread of digital technology, they seemed the most likely source of robust competition for the cable operators. Further competition would come from direct-to-home satellite distribution systems, primarily direct broadcast satellite (DBS) providers DIRECTV and EchoStar (later renamed DISH Network), but also Home Satellite Dish (HSD) providers.[6]

Satellite was in fact the strongest competitor to cable in 1996, accounting for 9.1 percent of the MVPD market. Another competitor was "wireless cable" operators, also known as "multichannel multipoint distribution service" (MMDS), which used spectrum in the 2Ghz band for local distribution. Finally, emerging broadband network providers, including the growing number of competitive local exchange carriers (CLECs), were expected to deploy fiber optics throughout the nation, competing with the incumbent telcos. By 1998, Qwest Communications International was well on its way to laying out over 16,000 route miles of fiber optics throughout the United States, a network that would have given it at the time more capacity than AT&T, MCI, Sprint, and Worldcom combined.[7]

The Telecom Act of 1996 allowed telephone companies to offer video distribution services and, in return, permitted cable operators to provide local and long distance telephone service. This would allow both telephone and cable companies to offer full bundles of services to their customers, including video, local and long-distance telephony, and a range of "advanced services" including Internet access. The combination of the emerging alternatives for video distribution with the new regulatory framework introduced in the 1996 act was expected to result in a vibrant competitive marketplace, with a multiplicity of high-capacity delivery conduits for video programming.

Reality turned out to be markedly different, as illustrated by Table 1. Ten years after the passage of the 1996 act, we are still awaiting a direct confrontation between Telecom and cable. The two compete actively for the provision of broadband Internet access, and the cable industry has made some inroads into the telephony business, with 5.6 million telephone subscribers as of December 2005, although this figure overstates the extent of actual competition since the cable provider is providing voice-over IP through its own network for only 1.2 million of these.[8] Phone companies were only beginning in 2006 to deploy high-capacity, fiber-based local networks and no significant CLEC has emerged to do it in their place. By 2005, the broadband service providers (including the telcos) served only about 1.5 percent of the U.S. MVPD market. Fierce regulatory efforts on the part of both cable and telco incumbents to limit CLECs' access to their networks' "last mile" connections with end customers partly explain this surprising lack of competition. The *Wall Street Journal* noted that "the competitive local exchange carriers (CLECs) have been forced to sell their accounts or file for bankruptcy because of heavy network construction spending."[9] With the CLECs in bankruptcy,

TABLE 1
Competing Technologies' Shares of the Multichannel
Video Programming Market (1996–2004)

Access Technology	1996	1997	1998	1999	2000	2001	2002	2003	2004	2005
Cable	87.7%	87.1%	85.3%	82.5%	79.8%	77.5%	75.9%	73.6%	71.6%	69.4%
Satellite (HSD + DBS)	9.1%	9.8%	12.0%	14.7%	17.4%	19.8%	21.6%	23.2%	25.5%	27.9%
Wireless Cable (MMDS)	1.6%	1.5%	1.3%	1.0%	0.8%	0.8%	0.6%	0.2%	0.2%	0.1%
Private Cable Systems	1.6%	1.6%	1.2%	1.8%	1.8%	1.7%	1.8%	1.3%	1.2%	1.1%
Broadband (BSP + OVS)	0.0%	0.0%	0.1%	0.1%	0.1%	0.1%	0.1%	1.6%	1.5%	1.5%

Source: Compiled from FCC's Annual Assessment of the Status of Competition in the Market for the Delivery of Video Programming (1996–2006).

Wall Street pulled the money plug, and the incumbents had the space to themselves.

With one notable exception, none of the alternative video delivery distribution technologies really took off. By 2005, wireless cable had declined to 0.1 percent, and broadband service providers and private cable systems each barely passed the 1 percent mark. The important exception is satellite, whose market share of video distribution rose from 9.1 percent in 1996 to 27.9 percent in 2005. Within that segment, the true success story of the past decade has been the rise of DBS from a 5.1 percent share of the MVPD market in 1996 to 27.7 percent in 2005. (HSD, the older, larger dishes, declined from 3.1 percent to 0.2 percent during the same period.) Cable subscribership remained essentially flat during that period, resulting in a declining market share from 87.7 percent in 1996 to 69.4 percent in 2005. Among the top four U.S. MVPD companies in 2005, together serving 63 percent of U.S. subscribers, two are cable systems operators (Comcast and Time Warner, with 23 percent and 12 percent respective shares), and the other two are DBS operators (DirecTV and Echostar, with 16 percent and 12 percent respective shares).[10]

As a result, the competitive situation in the video distribution market is not radically different from that of ten years ago. Viewers do indeed have more choices now than they did then, since the vast majority of U.S. households can choose between one cable provider and two direct broadcast satellite providers, in addition to traditional over-the-air broadcasting. However, the programming packages they can obtain through these different providers are remarkably similar. Both cable and DBS operators currently offer essentially similar channel packages and compete primarily on price. One differentiating feature for cable operators, however, is their ability to deliver high-speed Internet and telephony over the same cable infrastructure. Such bundles remain technically impossible for DBS providers. This has led DBS providers and telcos to cooperate and compensate with marketing for the shortcomings of satellite technology. Thus, SBC teamed up with Echostar and Verizon with DIRECTV to bundle their respective services to offer "one-stop shopping" packages including telephone service, Internet access, and video programming to their customers.

Ultimately, this piecing together cannot compete fully with the integrated delivery of video, data, and voice services on a single network platform. The telco-DBS alliances (much like cable operators' initial resale of telephone service) are temporary steps. Cable operators can best compete with direct broadcast satellite by offering an integrated high-speed network that supports symmetrical video, data, and voice communication. Similarly, telcos are unlikely to survive in the long term unless they too develop an integrated network platform that can support combined services, such as those promised by AT&T's project Lightspeed and Verizon's FIOS.[11] We explore this vision of "IP-Television" in the final section of this chapter.

The Age of Digital Cable

In the mid-1980s, as the U.S. telephone companies were beginning to deploy fiber optics and as direct broadcast satellite promised to compete directly for video distribution, the cable television industry realized that its collective survival would require significant upgrade of its technology infrastructure. It would have made little sense for individual systems to pursue technology upgrades in isolation, so they decided to join forces and establish CableLabs in 1988. The brainchild of Dr. John Malone, CEO of TCI (then the largest cable multiple-systems operator, or MSO, in the country), CableLabs was meant to be a cable industry version of Bell Labs, the research-and-development subsidiary of AT&T. The initial project of CableLabs in 1990 was the specifications for the 750 MHz hybrid-fiber/coax plant that would become the foundation of the digital cable TV system. This was followed in 1995 by DOCSIS (Data Over Cable Service Interface Specification), which defined how cable modems would offer high-speed data over cable systems. The underlying notion was that the cable plant would need to become a complete two-way system capable of delivering interactive services in order to compete with the increased channel capacity of satellite TV systems. It made sense for the industry to cooperate in this R&D effort not only to spread its substantial costs, but also to insure interoperability and achieve scale economies. This insight into the competitive edge offered by a true two-way broadband system turned out to be prescient: although it was started well before the Internet had become a mass medium, the digitization of the cable's infrastructure would turn out to be essential to the industry's Internet strategy. The cable industry's early commitment to data over cable gave it the edge in the early years of broadband deployment.

The Telecommunications Act of 1996 transformed the competitive landscape. In response, the cable industry began a system-wide network upgrade to the hybrid fiber/coax architecture. This would not only boost the capacity of the cable networks to several hundred channels, but would also be key to the deployment of new digital services, ranging from interactive television to Internet access and telephony. Investment banks backed the upgrades, and over the next six years almost $85 billion was spent on capital improvements to the cable plant, the largest capital improvements expenditure in the industry's history.

The initial rationale behind this upgrade was simply to compete with the satellite digital plant. By the time the majority of the capital improvements were completed, the high-speed Internet business had become a major source of cash flow on its own. The ability to deploy a 1.5 MBPS high-speed data system gave the cable companies an initial advantage over the Regional Bell Operating Companies (RBOC) which were beginning to roll out a technology called DSL (Digital Subscriber Line) that had been on its shelves for ten years. The RBOC were slow to deploy DSL, in part for fear it would cannibalize their extremely profitable business data services. For the cable industry, by contrast, Internet access provision was a new market that could only add to existing revenues. Given the RBOC's slow rollout of DSL, the cable companies moved quickly to establish a dominant position in the home broadband market. Initially priced at $49 per month, the service had a gross profit margin of almost 50 percent, making it the most profitable cable service to the MSO.[12] By 2000, the RBOCs began to aggressively market their high-speed Internet alternative. They chose to compete with cable on price and offered their service for $35 per month. However, under typical circumstances, DSL service works for customers only within 18,000 feet from a central office, and so the cable companies were able to maintain the higher price and compete on service quality.

The deployment of digital broadcasting service has not gone quite as smoothly as the rollout of cable modem service. Most cable customers already had 100 analog channels and balked at paying an additional $10 per month to upgrade to a 200-channel digital service. According to Nielsen Media Research, of the over 400 channels available on digital cable and satellite systems, the average viewer watches 15 channels but considers only 8 to 10 of those "destinations channels."[13] Most operators found that customers were signing up for three months of free HBO to convert to digital but went back to the analog service at the end of their free trial. The resulting "churn" in digital cable subscription led Comcast to introduce free Video-on-Demand as incentive to switch to digital.[14]

By January 2004, the cable network upgrade was essentially complete. A digital cable plant of at least 550 MHz passed more than 90 percent of all cable homes, capable of carrying 450 traditional TV channels, or 90 High-Definition channels.[15] The remaining task for MSOs is to broaden the digital cable market beyond early-adopters. The sooner they convert all of their subscribers to digital, the sooner they can free up 450 MHz of capacity they currently must devote to distributing all broadcast channel under the "analog must-carry" regulatory requirement. At that point, the stage will be set for the true Interactive TV age to begin.

Policy Principles

The regulatory principle behind the 1996 Act was that facilities competition would suffice to drive diversity, without requiring an extension of common carriage to cable. If one video infrastructure refused to distribute some programming, the logic went, there would be other competing infrastructures eager to do so. Furthermore, one of these infrastructures would be owned by the telephone company, whose regulatory history promoted open access. However, since substantive facilities competition failed to materialize, an integrated policy framework for cable and telcos has become necessary, one that would match technological convergence with regulatory convergence. A first debate on these issues was ignited by AT&T's 1998 acquisition of TCI, then the largest cable operator in the United States. As part of the cable franchise transfers, a number of city governments attempted to impose "open access" conditions on the broadband Internet service offered by cable operators.[16] Ultimately, these efforts failed, but the principle behind them remains hotly debated: in the emerging broadband world, should the owners of the Internet infrastructure be allowed to constrain the range of applications, the pattern of communication, or the kind of content that can use their network?

Before stepping down as FCC chairman in 2005, Michael Powell spelled out his version of this principle as the "Four Freedoms of Broadband." First, the "freedom to access content" would guarantee anyone's right to access any Internet content. Network owners could not selectively block or hinder content delivery. Implicitly, this also recognizes everyone's right to publish (or broadcast) on the Internet. Second, Internet users should be guaranteed the "freedom to use applications," so that infrastructure owners do not predetermine patterns of interaction among Internet users, nor the kinds of experimentation they can engage in. The third would guarantee the "freedom to attach personal devices" to the network, as long as they do not harm the network. In particular, network owners should not restrict their consumers to connecting only the equipment they supply. Finally, Internet users should have "freedom to obtain service plan information" to guarantee market transparency.[17]

Powell articulated these four freedoms with Internet service provision in mind, but they would logically extend to other services, including video distribution provided by cable operators or telcos. Yet today's cable TV infrastructure is far from guaranteeing these freedoms. In particular, it can be very difficult for niche or controversial programming to get distribution. The applications cable consumers can run on their cable-operator provided Internet are limited (among other restrictions, they cannot operate web servers); they are free only to attach set-top boxes supplied by their network provider and must comply with the provisions spelled out in restrictive "acceptable use policies" (AUPs). Cable companies argue that such restrictions are necessary to prevent network congestion, because of the limited uplink capacity of their Internet offering. However, emerging technical approaches, including, for example, the peer-to-peer mechanisms built within BitTorrent, could alleviate these concerns.[18] Conveniently, however, the AUP's restrictions prevent cable Internet users from competing in any way with cable providers for content distribution. More broadly, the natural economic incentives of the network owners, cable and telcos alike, will lead them to explore only the network architectures and services that promote their financial interests, preserving their traditional role as content gatekeepers and allowing them to leverage new income streams out of that control. They are less likely to encourage broad exploration of peer-to-peer architectures and applications that would encourage bottom-up content creation and horizontal exchange. Their economic incentives are more closely aligned with extending one-to-many, centrally controlled communication patterns than with fostering many-to-many interaction and horizontal discourse.

Laying the Foundation for the IPTV Revolution

The combination of the 1996 Telecom Act and the dot-com boom of the late 1990s led to massive deployment of fiber optics throughout the U.S. long-distance networks. More than 15 new Competitive Local Exchange Carriers (CLECs) and cable over-builders, including Qwest, Global Crossing, 360 Networks, and Level 3, received Wall Street financing to lay vast amounts of fiber optic cable, creating enormous potential backbone capacity. Qwest, one of the companies that built out the backbone, ran an ad in 2003 wherein a tired salesman pulls into a motel and asks the clerk if they have movies in the rooms, to which the clerk replies "every movie ever made." This is not an idle boast. Qwest's 34 strands of fiber could technically serve up every movie ever made on demand to every hotel room in the country.

Strategic planners at equipment companies like Cisco, Nortel, and Lucent, as well as many of their competitors and suppliers, looked at the amount of fiber optic cable being delivered in 1999 and 2000, projected the number of routers, switches, lasers, and other gear that would be needed to enable that fiber, and geared up production capacity accordingly. But this proved to be a miscalculation: the orders never came, partly because wave division multiplexing allowed carriers to increase throughput by two orders of magnitude for each strand of fiber and partly because local broadband connectivity did not grow as expected. The backbone providers simply left the "dark fiber" in the ground. The boom of the late 1990s' telecom financing was followed by a bust of overcapacity. The ensuing telecom market crash hit both the suppliers and the carriers.

The crash, it appears, might have a silver lining: as a result of over-investment, there is a considerable amount of cheap, unused transmission capacity. With that infrastructure, the conversion to an IPTV platform would seem within reach. One problem, however, is the lack of last-mile broadband connectivity. It was as if we had constructed the interstate highway system in the 1950s but neglected to build the on-ramps. Further, while the telecom industry had deployed completely new ways to distribute content, the content owners have remained deeply concerned with digital piracy and have refused to release content to the new IP distribution system.

Although the existing build-out of broadband to the home has been progressing well, with Merrill Lynch estimating 50 million home broadband subscribers by 2007,[19] the United States has recently slipped from thirteenth to sixteenth in per capita broadband diffusion in the world. On the surface, it is easy to see that densely populated countries like South Korea and Japan are relatively easy to provision for last-mile fiber connectivity. The American suburban sprawl makes this task quite expensive and time-consuming. However, the recent moves by the RBOCs to roll out fiber to the home may introduce a new force into the system to compete alongside the cable industry's nearly complete hybrid fiber/coax build out. This could help speed up the transition to a new system of IP-Television.

Let's assume that every home had access to universal broadband. A browser-based IP media terminal with a TV display connects to the Internet at two Mb/s minimum, capable of receiving streaming DVD-quality video on demand. This system would use the Internet's existing open standards (IP, HTML, MPEG) to avoid choosing a winner from the existing competitive technology and media companies. Obviously, simply using Internet standards does not guarantee an open distribution system free of gatekeepers. In fact, two starkly contrasting views of how that infrastructure could be used are currently emerging. Following industry watcher Robin Good, we call these "IPTV" and "Internet Television."[20]

IPTV is the vision of the future promoted by the established media groups, cable operators, telecommunication providers, along with Microsoft. It represents a substantial upgrade of the current access infrastructure, but one that is led by established carriers to set up a controlled platform. Control would reside in the network's architecture, the operating system of the media terminals (the new set-top boxes), and the digital rights management system. Within that environment, content distributors remain gatekeepers and, thanks to detailed information they gather about customers' viewing habits, leverage this new video distribution system into a highly targeted e-commerce platform.[21] By contrast, Internet Television proponents envision a world much like the world-wide-web, where anyone can set up a server that publishes his or her video content globally. In this view, by contrast with IPTV, publishers establish direct relationships with their viewers. These relationships can take a multitude of forms, supporting a wide variety of business models. The next section focuses on that latter vision and explores how such a video environment might function.

Prospects for an Open Media Ecology

A ubiquitous Internet Television platform would enable a very different media ecology—a world in which television technology would no longer require gatekeepers. In this world anyone who wanted to publish media would have no more trouble doing so than putting up a web site. Anyone could choose to sell programming by subscription, pay-per-view, or give it away for free with targeted advertising. There would be no need for gatekeepers assembling program grids and determining who could reach what audience, under what conditions. Anyone who wanted to access media could point the browser built into display screens and sound systems to that content, much as he or she does to access a web site. Traditional worries about media concentration would recede in this new world of abundance. While it is clear that the marketing power of major media conglomerates like AOL Time Warner or Viacom/CBS would have huge power in the marketplace, it would be the power to persuade, not the power to control.

Why would the current media distribution powers, whose enormous market capitalizations have been built on a world of scarcity, ever allow such a world of abundance to come into being? One answer could be that they may have no choice: Internet technology could make it very difficult for them to retain control. A better answer might be quite simply that they could make more money in that new world. The dominant music companies first tried to control the online distribution of their music through branded portals (Sony, Universal, etc.), but it was only when Apple's neutral iTunes service entered the market that this distribution method took off. To understand how an Internet Television system might benefit all parties, we must look at the five constituents that control the current media universe: talent, producers, advertisers, distributors, and telecom suppliers.

Talent

It is one of the great ironies of the age of media consolidation that giants like Fox, Time Warner, and Universal promote themselves as "brands."

In the world of entertainment, however, the artist is the brand. Nobody buys a Sony Music CD or goes to see a Miramax movie—they want to listen to a Springsteen song or to watch Scorsese's *The Aviator* with DiCaprio. New affordable digital tools for both music and video production further empower artists. For the many artists who do not have the mass appeal of Springsteen or DiCaprio, inexpensive production tools mean they no longer need the deep pockets of traditional labels and studios to create songs and movies. Furthermore, many artists would trade large up-front salaries for a real stake in the gross earning power of their work.

Beyond the professional creative world, millions of amateurs now use cheap production tools such as Apple's GarageBand and iMovie to create their own songs and movies, resulting in a profusion of "content" of all types. One only needs to look at the tremendous growth in decentralized production of low-bandwidth creative content, such as text blogs and pictures on flickr.com, each aimed at incredibly diverse and overlapping audiences, to get a glimpse of what could happen once similarly straightforward production and distribution mechanisms came to exist for high-bandwidth media like video and films. Recent developments provide an indication of what is to come: new video distribution tools have been introduced by YouTube, Google, and Yahoo that allow users to upload videos for free on to the network. YouTube's business model appears centered on generating page views and selling advertising, but other platforms are experimenting with alternative models. For example, on Google Video, the content provider can choose to charge a fee for streaming or downloading, with Google taking a small commission. In fact, the success of these Internet video distribution platforms suggests that the boundary between "professionals" and "amateurs" may be blurring. With these platforms, musicians and filmmakers can begin to produce and distribute their content to very small audiences, and support themselves in the process.

So how would the arrival of universal broadband support this content proliferation? The world of channel scarcity could not, because its production, marketing, and distribution cost structures required mass audiences. Further, while 500 cable TV channels certainly represented a quantum leap from a dozen broadcast channels, Internet Television provides an infinite number of channels, constantly reconfigurable to address a multitude of changing niche audiences. This world of abundant, cheap digital technology and distribution should help artists escape the traditional media "hit" economics. If only content with mass audience appeal is financed, then artists with a different perspective have a hard time getting distribution. As a result, the bulk of the industry's revenue traditionally comes from a very small number of artists. But the Internet's reach and the precision of its search technologies make it economical to produce and distribute very large numbers of cultural goods that individually appeal to niche audiences, a phenomenon Chris Anderson calls the "long tail."[22] In his example, while a record store might stock a total of 40,000 individual songs, the digital music service Rhapsody can afford to store digital copies of over 500,000 songs, and the least popular still sells enough to cover marginal storage and transmission costs. With universal broadband, this model would extend directly to video content. This allows distribution of a quasi-infinite variety of cultural, political, or educational video content, even though each item may only appeal to a very small audience.

Producers

Producers develop, create, and finance programming. Though many producers are also distributors (Time Warner, Viacom, Disney, etc.), it is important to separate the two roles in order to understand the Internet Television transformation. As an example, let's take Discovery Networks. Originally begun as The Discovery Channel, it bought existing nature, science, history, travel, and exploration programming from around the world as cheaply as possible and packaged it for distribution under The Discovery Channel brand. This proved quite lucrative as the demographic of educated, affluent customers attracted to this programming was attractive to higher-end advertisers (Mercedes, Merrill Lynch, etc.) who were just beginning to move their ads from high-end print publications (*The Wall Street Journal, The New Yorker, Vanity Fair,* etc.) into television. For Mercedes to advertise on a network sit-com was a waste of money since most of that audience couldn't afford their product; the lower pricing of The Discovery Channel, on the other hand, was a relatively efficient buy. However, two developments changed the economics for Discovery. First, they began to run out of programming they could acquire cheaply and therefore had to begin producing their own shows at a much higher cost per hour. Second, as the number of cable distribution channels began to grow (and then explode with satellite and digital cable), Discovery grew niche networks to defend its brand against imitators (Animal Planet, Discovery Health), each of which required programming 24 hours a day, 7 days a week, 52 weeks per year.

While the mainstay of the Discovery Networks (The Discovery Channel) averages $0.23 per subscriber home passed per month, its more narrowly niched networks like The Travel Channel get less than $0.03 per subscriber.[23] More importantly, the ratings of the new niche channels continue to fall so that by 2008 they may be bringing in 20 percent less advertising.[24] Extrapolating to a universe of 300 "Programming Services" on cable or satellite, the economics become increasingly tenuous. Discovery alone is responsible for programming 5,000 hours of original content per year.[25] Given 14 networks programming 365 days a year, the remaining 40,000 hours of daily programming would be reruns. For Discovery to keep a fresh look, the programming will have to get cheaper each year in order for it to reach break-even on the new networks, as there is no way the advertiser will continue to pay higher rates for an increasingly fractured audience. For instance, Discovery Health, the most popular of the niche networks, garnered an average rating of 0.2.[26] Increasingly, Discovery will shrink its broadcast production and put more content on demand.

In contrast, with Internet Television, Discovery could cut by half its programming budget and produce twenty great hours of new "on demand" programming a week with extraordinary production values. Their most fanatic viewer probably does not have more than ten hours per week to spend watching this type of programming. But if such a viewer did, Discovery could cheaply archive every single episode of programming it owns and make those accessible on a pay-per-view or subscription basis. The viewer could watch the programs on-demand, with full VCR-like controls. Discovery could offer a "My Discovery" option that would push pet shows to the pet lovers and alligator wrestling to the fans of that genre. Since the object of Discovery's business is to sell advertising, it could offer pet-food advertisers very targeted opportunities not only to advertise to the specific audience they wanted, but also to sell their product through interactive ads with e-commerce capability. All of the technology to enable this vision currently is in place.

Advertisers

It is clear from the recent network "upfront" ad-selling declines that advertisers are aware that the 30–second broadcast spot is an endangered species. As *The Wall Street Journal* recently reported, "Procter & Gamble Co. is sharply cutting its advance purchases of television commercials for the upcoming season . . . the latest change in the way companies reach consumers and how broadcast and cable networks draw revenue."[27] The movement of dollars away from the broadcast and cable networks continues as advertisers seek the more targeted opportunities presented by Internet advertising. The famous maxim by department store mogul John Wanamaker that "50 percent of my advertising expenditures are wasted; I just don't know which 50 percent" is truer than ever. This problem has been exacerbated by the introduction of the Personal Video Recorder (PVR), originally under the brand name TiVo and later bundled with set top boxes, which let viewers fast-forward through the commercials. However, the evidence that both the major cable companies Comcast and Warner are embracing the new ad paradigm of on-demand TV points toward a quicker adoption of the Internet Television paradigm.

A video-quality broadband network affords advertisers the Holy Grail: the ability to target like the web, combined with the ability to run full-screen 30-second commercials that allow interested users to click-through to the e-commerce page of the advertiser. If you are moved by the Gap ad, you can immediately buy the clothes. Furthermore, the ad buyer can specify a demographic target (females, 14–18, in specific zip codes) and pay only for that target.

Distributors

In an Internet Television world, the role of distributor would change. Today, there are five basic conduits for video media: theaters, video purchase and rental, cable TV, satellite TV, and broadcast TV. The classic distributor seeks to market its product sequentially through every one of the existing channels, hoping to extract as much profit as possible from each of the corresponding video windows. Unless they own a player in one of these channels, distributors must deal with a third party who can demand a share of the revenue from the transaction. Four of these channels are one-to-many "broadcast" channels, and while viewers have a choice, it is a limited choice, constrained by those who program the channels. Video purchase and rental form the exception, the only "addressable" channel that allows the distribution of hundreds of thousands of programs to very small audiences, on-demand. So far, however, this "on-demand" character is slow and inconvenient.

The emergence of broadband IP networks as a sixth conduit and the transformation of the cable TV system into an addressable network together introduce a fundamentally new distinction. In the new world of Internet Television it will be important to differentiate between broadband carriers and broadcasters. Broadband carriers would be comprised of all DSL providers (the Baby Bells), all cable providers with upgraded hybrid fiber/coax plants (90 percent of the nations MSOs), all ISP's offering broadband service (AOL, Earthlink, MSN), and all fixed wireless providers. Broadcasters would consist of all over-the-air TV networks and all satellite networks. In an Internet Television world the broadband carriers would make their money by providing metered service much like your cellular or utility service. Heavy users of streaming media would pay more than light users. Distributors of content could then sell to the carrier's customer base on an open-access basis and use the three basic models for payment: monthly subscription, pay-per-view, or ad-supported content.

Clearly the broadcasting model would not be able to compete because of lack of a two-way network. However, this transition to IPTV would be gradual, and the "event" type of programming, such as sports or award shows which demand a specific mass audience to be present at a specific time, would continue to be a staple of the broadcasting universe for a long time.

Conclusion

With the transition to Internet Television, TV is entering the third phase of its evolution. Each of these three phases corresponds to a distinct architecture of the medium, leading to a specific allocation of roles between content providers and consumers and resulting in a unique control structure. This is why Internet Television holds so much excitement. It promises to transform the current structure of control over the most important mass medium and thus revolutionize the patterns of communications which have been articulated around it. The implications for the social, cultural, political, and economic activities organized around the corresponding communication activities are equally far reaching.

Industry and policy choices at each preceding phase have led to the construction of television systems around a particular network technology, and these technological choices have had important implications for the medium's architecture and its economic characteristics. Radio transmission of video programming was central to the first phase of broadcast television, and the limited frequencies available resulted in channel scarcity, giving central control over the television system to a handful of broadcasters granted licenses to use these frequencies. Since they could not exclude individual users from receiving their signal, the core economic model was to assemble mass audiences and sell advertising.

The second-phase technology—co-axial cable—alleviated scarcity and ushered in a new era in which viewers could chose among hundreds of channels. Mass audiences then begun to fragment into niche audiences, with programming increasingly finely targeted to the tastes of particular demographic segments. While control over programming expanded from a handful of broadcast networks to several dozen cable networks and cable systems operators, it still remained largely centralized. With cable, however, individual viewers could be barred from receiving specific channels or programs, enabling new economic models such as subscription and pay-per-view that offered an alternative funding mechanism to advertising.

Internet technology will be the foundation of the next phase, Internet Television. More than a physical medium, like the radio waves and co-axial cables of the two first phases, the Internet is a software overlay that can tie any existing network into an "inter-network," thus potentially combining all the existing transmission capacity existing in not only the broadcast and cable but also the telecommunication networks into a single transmission system. However, unlike its broadcast and cable predecessors, it can be configured to support an infinite variety of communication patterns—not just one-to-many like broadcast or one-to-few like cable. As a result, the Internet potentially enables a far-reaching transformation of television, making possible the construction of a system in which anyone can offer programming for any type of audience, with any kind of access control—and pricing mechanism—he or she chooses. The resulting medium would look much like the web, for video.

However, the fact that such an open, symmetrical, flexible video communication system is technically possible doesn't mean it will necessarily emerge. In fact, the very programmability of the Internet that allows the creation of a system supporting a multitude of communication patterns also makes it possible to shape Internet Television merely as an expanded version of the centrally controlled systems that dominated the first two phases of television, what we have called IPTV. Which path television ends up following between these two opposed visions will largely depend on who is allowed access to the Internet distribution network and who has the ability to program its configuration. This is precisely why the policy decisions contemplated today are so critical, in particular those that will determine how we chose to implement the legacy of the "four freedoms of broadband."

NOTES

1. See, for example, Peter Huber, *Law and Disorder in Cyberspace: Abolish the FCC and Let Common Law Rule the Telecosm* (New York: Oxford University Press, 1997).

2. Reuters, "YouTube Serves Up 100 Million Videos a Day," July 16, 2006, http://go.reuters.com/news/.

3. "Fears of Dot-Com Crash, Version 2.0," *Los Angeles Times,* July 16, 2006, http://www.latimes.com/business/.

4. Federal Communications Commission, Twelfth Annual Report: Annual Assessment of the Status of Competition in the Market for the Delivery of Video Programming (released March 3, 2006, MB Docket No. 05-255), 4.

5. Data cited in this section, unless noted otherwise, was compiled primarily from Appendix B/Table B-1 ("Assessment of Competing Technologies") of the FCC's successive annual reports, filed from 1996–2006, "In the Matter of Annual Assessment of the Status of Competition in the Market for the Delivery of Video Programming," http://www.fcc.gov/mb/csrptpg.html.

6. HSD refers to the first generation satellite-to-home, using large antennae, while DBS refers to the newer version, using smaller antennae.

7. David Diamond, "Building the Future-Proof Telco," *Wired* 6, no. 5 (May 1998), http://www.wired.com/wired/archive/6.05/qwest_pr.html.

8. National Cable Television Association, "Statistics & Resources," at http://ncta.com/ContentView.aspx?contentId=54; FCC, Twelfth Annual Report, 33.

9. "Telecom Economic Slowdown," *Wall Street Journal,* July 31, 2001.

10. FCC, Twelfth Annual Report, 118.

11. See AT&T's and Verizon's webpages http://att.sbc.com/gen/press-room?pid =5838 and http://www22.verizon.com/content/consumerfios/about+fiostv/about+fios.htm.

12. Kagan World Media, *Broadband Cable Financial Databook—2003* (Carmel, CA: Kagan World Media, 2003).

13. Lynn Smith, "Channels Bloom, and Viewers Pick," *Los Angeles Times,* March 2, 2006, E15.

14. UBS Investment Research, February 7, 2003.

15. Kagan World Media, *Broadband Cable Financial Databook—2003.*

16. François Bar et al., "Access and Innovation Policy for the Third-Generation Internet," *Telecommunication Policy* 24, nos. 6/7 (July/August 2000).

17. "The Digital Broadband Migration: Toward a Regulatory Regime for the Internet Age," Remarks of Michael K. Powell FCC Chairman, at the Silicon Flat-

irons Symposium University of Colorado School of Law Boulder, Colorado, February 8, 2004, http://hraunfoss.fcc.gov/edocs_public/attachmatch/DOC-243556A1 .pdf.

18. Bram Cohen, "Incentives Build Robustness in BitTorrent," paper presented at "Workshop on Economics of Peer-to-Peer Systems," University of California, Berkeley, June 5–6, 2003 (available online at http://www.bittorrent.com/ bittorrentecon.pdf).

19. Merrill Lynch, Cable Television Report Card, October 19, 2004.

20. Robin Good, "Is IPTV the Future of Internet Television?," June 4, 2005, http://www.masternewmedia.org/news/2005/06/04/is_iptv_the_future_of.htm.

21. Mike Quigley, "The Real Meaning of IPTV," *Business Week,* May 20, 2005, http://www.businessweekasia.com/technology/content/may2005/tc20050520_4620 .htm.

22. Chris Anderson, "The Long Tail," *Wired* 12, no. 10 (October 2004), http:// www.wired.com/wired/archive/12.10/tail.html.

23. Bernstein Research, "Pipe Dreams" (May 2004), 28, retrieved from http:// www.bernsteinresearch.com/.

24. Ibid., 5.

25. Discovery Press Release, http://corporate.discovery.com/news/press/05q2/ 050411br.html.

26. Discovery Press Release, http://corporate.discovery.com/news/press/05q1/ 050201r.html.

27. Joe Flint and Brian Sternberg, "Ad Icon P&G Cuts Commitment to TV Commercials: Top U.S. Advertiser Explores New Ways to Reach Viewers; A Product-Placement Surge," *Wall Street Journal,* June 13, 2005.

Chapter 4

If It's Not TV, What Is It?
The Case of U.S. Subscription Television

Amanda D. Lotz

British television scholar Charlotte Brunsdon presciently anticipated the quandary that "premium" cable networks would introduce when she titled a 1998 essay, "What is the 'Television' of Television Studies?"[1] Brunsdon's essay attends to the diversity of approaches scholars have used to study television and identifies the multiple characteristics by which they have defined the medium. In the relatively brief history of critical television studies, scholars have focused primarily upon programs, audiences, and institutions as key sites of television research. Examinations of programming and audiences display exceptional breadth while most institutional studies begin from the distinction of public versus commercial systems. As television has multiplied into ever more varied forms during the "multi-channel transition" of the last two decades, this simple dichotomy has proven increasingly inadequate for understanding the complicated commercial distinctions among broadcast, basic cable, and premium or subscription networks and the significant variation within what was once simply "television."

The subscription cable networks common today originated in the 1970s and historically primarily provided a venue for screening theatrical films in an uninterrupted and unedited form—often well in advance of broadcast or basic cable release. In the last decade, however, HBO and, increasingly, Showtime have developed original series programming that has irreversibly changed audiences' perceptions of these networks, program norms, and the industrial practices of television programming. HBO became the darling of television critics and won numerous institutional awards in the late 1990s with its original series and films. To the further aggravation of broadcast network executives, HBO began using a strategic and contrary branding slogan that proclaimed, "It's Not TV, It's HBO."

This simple, yet complex slogan is loaded with both obvious and obscure comment about U.S. television at the dawn of the twenty-first century. As others have noted, it primarily marks the network's attempt to differentiate itself by distancing HBO from stereotypic notions of television as a "low art" form providing the "least objectionable programming" —assumptions that are heavily weighted with cultural capital and that allocate assessments of higher quality to forms with less accessibility. The slogan affirms that HBO isn't really like other television—the much derided "boob tube" or "idiot box"; its programming is sophisticated and smart—as its promotions suggest, "ground-breaking, critically-acclaimed, smash-hits." HBO became, like PBS, television that "high-minded" audiences admit to watching.

But the network imbeds much more in this slogan, and the distinction it asserts raises significant questions about how we understand and theorize relationships of media within society once we move beyond the red herring of cultural capital affirmation that the slogan is commonly believed to portend. If we change the emphasis from programs to institutions, the slogan suggests HBO's acknowledgement (although likely unintentional) of the shifting terrain of television as an industry, technology, and art form. The slogan "It's Not TV, It's HBO" means something quite different if "It" is the network's economic structure rather than its programs. In this reading, the slogan acknowledges the very different industrial practices and capabilities of subscription networks relative to those of advertiser-supported broadcast and basic cable.

The revered storytelling role subscription cable networks (and HBO in particular) began to offer U.S. culture, their ousting of broadcast network preeminence in programming distinction, and the discourses surrounding the competition among broadcast, basic cable, and subscription cable networks indicate the need for complex critical frameworks that acknowledge contemporary uncertainty about what is "television." The development of cable as a broadcast competitor differentiated by a variant economic model and regulatory status is central to the disruption of assumed understandings of television. Subscription networks are fundamentally distinct from what has been known as television and are, in that sense, "not TV." But HBO should not be granted the cultural capital it seeks in this distinction, as substantial adjustments throughout many industrial practices do make subscription networks very much a component of contemporary "television."

As subscription services, HBO, Cinemax, Showtime, The Movie Channel, and Starz use industrial practices that clearly differentiate their operations from commercial networks financed through advertising. Various components of the subscriber-supported economic model, particularly the function of programs and measures of program success, vastly differentiate subscription networks from their advertiser-supported competition, so much so as to raise the question of whether networks operating with such varied business models can be considered to be in competition with each other at all. Television criticism in an era of such multifaceted norms for the medium requires attention to more than its programs and distribution form; it must also consider the specific economic model that undergirds distinctions among broadcast networks, basic cable, and subscription channels—as each circumscribes a particular nexus of art and commerce. Such criticism must interrogate how institutional characteristics contribute to programming possibilities because the type of programming provider and specific institutional context yield particular constraints and abilities.

This chapter examines the fundamental attributes of subscription networks by exploring their distinct institutional and technological strategies. The analysis addresses both historical strategies as well as emergent competitive techniques enabled by "multiplexing," the DVD sell-through market, syndication, and video on-demand. The production of original series by subscription services introduces a variety of questions essential to the critical study of media because of the way these productions trouble previous understandings of industrially and culturally delimited program norms. Foundational theories of television studies formulated in the network era did not anticipate subscription networks, and critical television scholarship has not interrogated whether these networks are similar enough (because they are all transmitted through "the box") or too different (because of their variant financial models) to be considered "television." This analysis establishes key distinctions in the economic structure of subscription-supported services that contribute to, yet do not determine, their programming possibilities.

Distinctions of the Subscription Economic Model

The economic structure of subscription networks differs substantially from that of broadcast and basic cable networks and consequently alters their programming possibilities. These networks air no commercials and instead fund their programming through a monthly fee paid by those who subscribe to the network. Attempts to develop subscription broadcasting date to the radio era, but this delivery format did not materialize in an ongoing and comprehensive way until the 1970s.[2] Early attempts at subscription television tried to combine telephone-line and broadcast technology (Phonevision) or scrambled signals transmitted by broadcast that could be de-scrambled by inserting payment (Telemeter) or a decoder card (Subscriber-Vision) in a set top box.[3] Technological capability did not limit the initial development of subscriber television so much as the contradictory history of U.S. cable regulation impeded its progress. In an illustrative case of media policy being hijacked by particular commercial interests, the Theater Owners of America spent over one million dollars in 1964 to pass a ballot initiative in California that banned subscription television companies; the FCC later expanded on this legislation at the national level to prevent the "siphoning" of content away from broadcast television.[4] The threatened theater owners used hyperbolic claims about what the introduction of subscription television would mean to a society accustomed to "free TV" and effectively stifled this alternative system of commercial television until the Court of Appeals ruled in HBO's favor in 1977 and found that the FCC had exceeded its authority in regulating the content cable could provide.[5]

Once regulatory limits were eliminated, various entrepreneurial efforts sought a sustainable business model built on subscription service. Theatrical releases and sporting events provided the initial lifeblood of many of these networks, including HBO (1972); Showtime (1976); The Movie Channel (1979); and Cinemax (1980). However, the successful transition of HBO and Showtime into producing original series during the last decade particularly challenge established theoretical and practical understandings of the norms and possibilities of commercial television. The analysis here consequently focuses primarily on the version of "television" that is simultaneously "not television" by industry conventions in order to identify distinctions in the economic model that require specific attention when assessing subscriber-supported networks and their content.

Functioning as a subscription service radically alters the conventional financing and programming practices from the standard advertiser-supported system. Most obviously, it eliminates advertisers from the equation, which many point to as the determining factor in explaining the distinction of HBO's programming. Advertiser-supported television relies on the selling of audience members to advertisers, so that the programming serves as the lure to ensure the audience's presence for commercials. Subscription-supported services simplify the sales process by eliminating the advertisers so that viewers purchase the programming and pay directly for content (rather than paying indirectly through elevated product prices—there is no such thing as "free TV"). The purchase of a network service rather than a particular show (as in Pay-Per-View), however, adds complexity to this transaction, especially in the case of multifaceted subscription networks such as HBO and Showtime. These networks offer a broad range of programming that appeals to incongruous tastes. This range of targeted programming interests enables the networks to use the subscription fees of boxing fans to help finance original series, the fees of those subscribing for original series to buy movie rights, the fees of those subscribing to see original drama series help supplement documentary series' costs, and so on, thereby creating a radically different economic situation and, consequently, programming environment.

The business model of subscription services differs significantly from the model of advertiser-supported networks, and the discrepancies in these models contribute substantially to the different programming possibilities afforded to these different types of networks. Subscription networks' variation from the conventional advertiser-supported business model is particularly evident in their measures of success, their scheduling capabilities, and audiences' expectations of these networks. The different set of institutional rules and structural opportunities leads each of these factors to contribute to the program differentiation available to subscription networks.

Measures of Success

Although an advertiser-supported model has dominated U.S. television, the power of advertisers in the program creation and distribution process has shifted throughout history. In an advertiser-supported system, the industry determines success by the quantity of viewers that a network draws and often by the quantity of viewers with particular demographic characteristics. Networks are very cognizant that advertisers will pay more for certain types of audience members and place a premium on those ages eighteen-to-thirty-four, who are college-educated, and earn a yearly household income in excess of seventy-five thousand dollars. Consequently, much of a network's decision to develop certain series or scripts results from its perception of how content performs with this most valued demographic group.

Advertiser-network relationships were established in an era of industrial operation in which only three networks competed for the audience's attention. This competitive situation led networks to develop programming likely to reach a broad and heterogeneous audience, a strategy often pursued by creating programming least likely to offend any segment of the audience. The massive expansion in the number of broadcast and cable networks since the early 1980s has significantly altered any network's ability to provide a broad audience, yet advertisers remain wary of content that too narrowly circumscribes its target audience, and some withdraw from advertising in programs with content deemed controversial.

Advertisers now have much less day-to-day input regarding the content of the series in which their commercials appear than they did in television's first decade, but they remain present through the self-censorship invoked by the "standards and practices" and programming divisions of many networks. Advertisers do not need to verbalize or micromanage their relationships with networks to make known their wishes regarding audience composition and non-controversial content.[6] Daily network operations have internalized many of advertisers' perceived values and created great self-imposed censorial power.

Eliminating advertisers reduces institutional gatekeepers and passes increased importance to the preferences of the audience. Programming decision-makers at subscription networks are theoretically less likely to have external forces (advertising and management) quash their ideas and projects in an attempt to steer programs toward the mainstream because these networks derive value through their differentiation from what is available on advertiser-supported networks. Subscription networks can consequently take more significant programming risks and offer creative talent great leeway in following their vision and designing programming unlikely to appear on a broadcast or basic cable network (a reputation HBO particularly has cultivated).

Beyond these differences in process, the measure of success for a subscription network deviates significantly from the advertiser-supported norm and bears significant consequences for subscription networks' programming strategies and possibilities. Subscription networks succeed when they entice a substantial number of people to pay for the service. This goal requires a different product than if the goal is simply to encourage viewers to watch a program. How many or what type of viewers watch specific content takes on decreased importance for subscription networks; a willingness to subscribe affords viewers their value rather than the possession of particular demographic features important to advertisers. Subscription networks are not exceptionally concerned with how often or what the individual subscriber views, but rather that each subscriber finds enough value in some aspect of the programming to continue the subscription.[7]

This different measurement of "success" leads to fundamental differences in the function of programs because subscription networks need to sell the network instead of individual programs. Most basically, subscription networks must develop programming of such distinction that viewers are willing to pay for it. As HBO chairman Chris Albrecht explains, "Unlike broadcast and broad-based cable networks that are defined by their audiences, HBO is defined by the things that it puts on its air"—a position that industry journalist Allison Romano summarizes as, "It's not just about viewers tuning in. It's about their paying up."[8] Consumers base their subscription decision on the value of the programming, but there is a key and significant difference in the programming emphases of subscriber- and advertiser-supported networks.

The institutional characteristics of subscription networks allow them to create programs with distinctive voices and clearly demarcated "edges," by which I mean that they do not intend their programs to draw the most viewers or attract a broad audience.[9] They deliberately exclude audience members who will be offended by the normalization of particular stories and depictions, such as the polygamous family of HBO's *Big Love* or a suburban mom selling marijuana to pay the bills in Showtime's *Weeds*. This willingness to tell otherwise unavailable stories endears subscription networks to another audience that exhibits great loyalty and will pay monthly fees. These networks succeed precisely when they offer programs that hail the interests of a variety of audience members—and this often occurs through different programs. In fact, their strategy is most successful if programs have excessive edge so that audience members find the one show that addresses them more specifically than a broadcaster is likely to be able to do; that one program, and perhaps access to a few others, is enough to justify the subscription fees. Mark Edmiston observed this phenomenon in the magazine industry, noting that, "as you get narrower in interest, you tend to have more intensity of interest: the person is more likely to pay the extra money."[10] Likewise, in their text on the economics of the television industry, Bruce Owen and Steven Wildman describe a primary challenge for mass media industries such as U.S. television as "a trade-off between the savings from shared consumption of a common commodity and the loss of consumer satisfaction that occurs when messages are not tailored to individual or local tastes."[11] The subscriber-supported network approaches this dilemma much differently than a traditional broadcaster because viewers' interest is measured by their purchase of the network rather than their viewing of specific programs. Consequently, the subscriber-supported network is most successful if it provides a variety of programs, each tailored to what Owen and Wildman refer to as "individual or local tastes," while advertiser-supported networks most often seek to plane the edge of programs to make them acceptable to as many audience members of the most desired demographic group as possible. Further, while demographics are the lifeblood of advertiser-supported networks, they mean little to a subscription network beyond strategic planning to develop programming to serve a greater variety of audience members.[12] A subscription service also endeavors to create programs that people talk about (and consequently subscribe to see), more than it seeks to air programs that are widely viewed, but about which there is not a lot of "buzz."

Scheduling Capabilities

Eliminating advertising content also offers subscription networks very different scheduling capabilities, which in turn alters their programming options. Series developed for subscription networks are radically different from broadcast and basic cable series because of the lack of commercials that take up time and necessitate a particular narrative structure. Subscription networks free writers from the inflexible twenty-three or forty-six minutes of narrative time with specifically prescribed commercial breaks that shape the development of some sort of narrative climax. The time difference may seem insignificant, but the seven-minute disparity (an additional 25 percent of narrative) marks a clear distinction between an episode of *Friends* and one of *Sex and the City*.[13] In many cases, this additional narrative time allows for more substantial character and plot development and contributes greater story or character depth.

The elimination of commercial breaks also allows subscription series to develop according to a structure dictated by the narrative, rather than by the need to take commercial breaks at specified intervals. This too frees writers to create distinctive programs. Tom Fontana, writer and executive producer of the HBO series *Oz* explains, "When you don't have to bring people back from a commercial, you don't have to manufacture an 'out.'

You can make your episode at a length and with a rhythm that's true to the story you want to tell."[14] Finally, if a story turns out to be thirty-two minutes in length, it can remain the length that is needed to best tell the story because of subscription networks' flexible program schedule in which it is not uncommon for a show to run five minutes past the hour. Broadcast and basic cable series typically must be altered to fit the time slot, regardless of the detriment to the story.

Audience Expectations

Finally, subscribers have very different expectations of the networks they subscribe to than of broadcast networks, which also enables subscription networks to pursue particular strategies. Subscribers do not expect a full day's worth of new programming in order to derive value from subscription services in the same way they have come to expect new programming throughout the day from broadcast networks. This also contributes to repurposing opportunities and the ability of subscription networks to allocate greater funding to a few programs, rather than spreading that same program budget over the diverse and multiple needs of a broadcast network. A network such as NBC must sell viewers each and every day on a schedule full of programs; a few isolated hits on one night only marginally benefit the network's overall program schedule. By contrast, HBO can rerun content multiple nights per week without the same detriment that a broadcast network such as NBC would experience if it used the same technique.

Further, it is important to remember that the subscription networks exist as just one entity within conglomerated media empires (Time Warner in the case of HBO and Cinemax, Viacom in the case of Showtime).[15] Michael Curtin and Tom Streeter have emphasized media industries' construction of economies of scope as a characteristic of competitive strategies emerging since the replacement of the network era with a multi-channel environment.[16] The conglomerates utilize the subscription networks to reach a certain segment of the audience and return profits, but each also owns a broadcast network, a variety of cable networks, film and television studios, and other media entities that buoy the conglomerate.

These varying measures of success, scheduling capabilities, and audience expectations differentiate subscription networks and provide them with opportunities for programming experimentation less available to those networks that must stake their fate on each and every episode of each and every show. In constructing an industrial explanation for the differentiation of subscription networks' content, it is important to acknowledge that the institutional arrangements do not mandate the programming situation so much as they contribute to composing an environment enabling some experimentation.

Broadcast networks arguably constrain their own creative process through self-censorship to an extent that is in excess of what is needed. Indeed, while a base of subscribers may enable a network to explore non-traditional programming forms and themes, such explorations are possible for advertiser-supported networks as well. Self-determined adherence to established norms often blocks the pursuit of forms that defy conventionality. *Broadcasting & Cable* editor P. J. Bednarski smartly replied to broadcast networks' whining about their inability to compete with *The Sopranos* by opining that it is not the violence and language that separate HBO's approach, but that if the series aired on a broadcast network the basic premise of a "dramatic series about a guy's family business, his eccentric partners, his unrelenting mother, his wise children and his long-suffering wife in suburbia" would be altered so that

> Tony would be handsome and in his late 30s. His wife would be gorgeous and working (architect? pediatrician?), the kids would be wisecrackers (and the daughter would be a sexpot), and Tony's mom would be annoying mainly by doting on him, not by psychologically torturing him. . . . Homes would be stylish. Clothes would be current. Weight would be perfect. And what you'd have, more or less is a typical television show.[17]

Moving into a Post-Network Era: The Changing Strategies of Subscription Networks

In June 2005, HBO's U.S. subscription base equaled approximately 28 million subscribers,[18] and Showtime reached 13 million[19] (of a television universe estimated at 109.6 million households), most of whom paid ten to fifteen dollars per month for each service. Subscribers who obtain the networks through the analog cable lines that have dominated transmission until recently commonly receive a single version of each network. Since the mid-1990s, however, opportunities have increased greatly for those purchasing digital cable tiers or who contract with direct broadcast satellite providers such as DIRECTV and EchoStar. These subscribers often receive a "multiplex" package that consists of eight to twelve channels—in the case of HBO, these include HBO, HBO Plus, HBO Signature, HBO Family, HBO Comedy, HBO Zone, and HBO Latino, as well as both East and West coast feeds of each channel.[20] By multiplexing, subscription networks increase their value to subscribers without substantial additional cost to the network because the content airing on the additional channels is often neither new nor distinct, but common to the networks' library. Many customers receiving the multiplex package pay no more than those who subscribe to the basic network, with fees mostly varying based on the deal negotiated between the subscription service and the cable or satellite provider.

In general, these alternative channels provide subscription networks with outlets to repurpose their original programming and counter-program themselves—that is, they offer different types of programming at the same time, such as a film on HBO and a series on HBO2. Such a niche strategy is valuable because of the diverse interests through which such channels appeal to subscribers. HBO's HBO Plus channel, which it advertises as "It's about Choice," primarily serves as a counter program channel by offering original series opposite films on the main channel and vice versa. HBO Signature, "Entertainment for a Woman's Heart, Mind, and Spirit," follows the trend toward niche targeting by sex, while HBO Family features programs for children, specified as programming that provides "Safe, Commercial-free, Entertainment." Similarly, HBO Comedy features comedy ". . . All the Time." HBO distributes the final two channels less widely, including them more among HBO's international range of channel offerings.[21] HBO Zone, ambiguously promoted as "Freedom of Speech, Sight, and Sound," addresses a young, urban, and African-American audience with more emphasis on music and Black performers—sort of an HBO version of MTV. HBO Latino specializes in "Innovative Programming for Today's Latino" by airing a Spanish-language version of whatever the network is showing on its flagship channel.[22]

The addition of multiple channels of subscription networks may appear to greatly increase the value of a subscription (eight channels for the price of one), but these channels primarily expand the range of opportunities to watch the same programs available with the single channel subscription. HBO recognized the possibility of repurposing programming early; its first multiplex experiments began in 1991 and yielded significant benefits for HBO and cable providers. The premium networks began multiplexing in order to reduce "churn," the number of subscribers who cancel the service—a figure cited at 60 percent annually for HBO.[23] Churn rates are to subscription networks what Nielsen ratings are to broadcast networks; the network's profitability depends on reducing churn in order to spend less on promotions to entice new and additional subscribers. Early multiplex experiments in 1991 (of just two additional HBO channels) reduced customer complaints about HBO's cost from 30 percent to 22 percent and, surprisingly and more importantly, reduced complaints about repeated content from 52 percent to 35 percent.[24] This second figure provided a paradox, as multiplexing actually increases the venues for repeats. But multiplexing offers viewers additional programs at any given time in a manner that reduces their experience of finding content already viewed and consequently decreases a primary cause of churn.

The multiplex channels lead subscribers to believe they are receiving additional services in a manner that appeals to viewers' sense of value. Indeed, this primarily benefits the network with the appearance of goodwill, while not requiring substantial additional production costs. In fact, the unconventional nature of subscription networks' programming, in both theme and length, has made their material more difficult to distribute in traditional syndication markets.[25] Consequently, the additional multiplex channels provide an outlet for programming likely to otherwise collect dust in an archive. As the technological capabilities of the industry have expanded, subscription networks have continued efforts toward value enhancement by giving subscribers more choice through on-demand networks. HBO On Demand comes without additional charge for most HBO subscribers who subscribe to digital cable tiers and contract with multiple-systems operators (MSOs) that have developed the infrastructure to provide on-demand services. HBO On Demand makes a range of movies, series, and specials available to viewers to watch whenever and however they desire. Again, the on-demand technology does not enhance the true range of viewing options, as the movies, series, and specials that can be accessed on-demand typically do not exceed the options regularly scheduled during that month of programming. The value enhancement results from viewers not being limited by a static and finite program schedule. Showtime executive vice president Mark Greenberg reports, "The feedback from our research is that Showtime On Demand creates more stickiness for our subs[-cribers] . . . Showtime On Demand clearly is providing incremental value on both the acquisition and retention sides."[26] In industry terms, "stickiness" refers to the likelihood of subscribers maintaining the service, so that the on-demand option helps draw new subscribers to the network and keep those who subscribe.

Multiplexing and on-demand offerings have allowed subscription networks to derive more value from their existing libraries by enabling them to provide subscribers with more choice without substantially increasing programming costs. As subscription networks have become series and movie producers rather than venues for recirculating theatrical content, their business model has also changed. The increased reputation of subscription networks as valued content producers has allowed less reliance on subscription funding because secondary market sales of their films and series in international and domestic syndication and through DVD release have provided new funding streams. By March of 2004, 20 percent of HBO's revenues were drawn from ancillary markets such as DVD and video sale of its series and films.[27] Such shifts in the basic economic model of subscription networks further distinguish them from advertiser-supported networks. Moreover, the alternative revenue streams also affect the subscription networks' programming strategies.

The series that have been part of HBO's ascendance in the original series marketplace (those since and including *Sex and the City*) also have returned profits to HBO by circulating around the globe on the seventeen program networks that reach subscribers in forty nations and territories and through sales of series to other networks.[28] As of March 2004, *Broadcasting & Cable* projected $113 million in syndication revenue and $75 million in international sales, just for *The Sopranos* property.[29] By January 2005, HBO had negotiated a deal in which cable network A&E will pay $2.5 million per episode for a syndicated run of the series which will increase the revenue from that series by another $195 million.[30] The fact that only 25 percent of U.S. television households subscribe to HBO, and only 31 percent of those households watch even a well-known hit such as *The Sopranos*, means that many viewers have not had access to the original HBO release and compose a substantial syndication or DVD market.[31] Syndication sales are not new to subscription networks seeking to maximize profit from their productions; HBO sold *Tales from the Crypt* as early as 1995.[32] More recently, Showtime sold its drama *Soul Food* within its Viacom family to BET in 2001, and HBO has cleared $100 million in selling the domestic syndication rights for *Sex and the City* to the Tribune station group, cable network TBS, and independent station KRON.[33] Such deals indicate how the networks might recoup the costs of production and make a profit on series in secondary markets. Such opportunities decrease the networks' reliance on subscription fees, which in turn affects the content they can and are likely to produce.

Given the distinctive content of subscription networks' series, DVD sell-through and rental may provide an equally profitable after-market for the network. Before its fifth season, *The Sopranos* already had earned $240 million in DVD sales, and an HBO executive reported that half of the buyers were HBO subscribers seeking to re-watch episodes or build a library —a fact indicating that a sizable audience that has never seen the show remains for syndication viewing.[34] HBO deliberately releases its DVDs to maximize the promotion potential of new seasons of continuing series, and the after-market circulation also serves important marketing functions in driving subscribers back to the network. HBO also has expanded its premium reputation by maintaining a high-definition feed of its flagship channel and offering its original content in high-definition, further expanding the value of the HBO subscription.[35]

The additional revenues secondary markets provide for subscription networks allow them to further pursue the strategies of differentiation that are already enabled by their subscription economic model. The opportunity to recoup costs and expand programming budgets reinforces the mandate of the networks to create programming offering such value and distinction that audiences are willing to pay for it—either through monthly subscription fees or in the sell-through market. These new developments reinforce the ways in which subscription networks operate under a different competitive logic than that which governs their advertiser-supported "competitors."

Conclusion

Proclamations of the excellence of subscription networks' original series often overlook the consequences of these new institutional organizations for foundational principles of U.S. broadcasting and theories built upon assumptions of a mass audience. The potential of U.S. broadcasting as a democratizing form—as it was once theorized—was certainly never fully or even nearly accomplished, as commercial imperatives always have taken precedence over "public interest, convenience, and necessity." Even beyond democratizing ideals, ubiquity and accessibility have been hallmarks of television and central to theories about the role the medium plays in culture. The relative inaccessibility of subscription networks—both in terms of choice and cost—provides additional justification for differentiating them from other forms of television.

Television scholars should be concerned about the consequence of "the best" television being created only once a subscription service creates an economic barrier to access. Such a development suggests in some ways a reinstatement of the distinctions among main floor (premium cable), balcony (basic cable), and those who cannot afford to enter (over-the-air reception) that has otherwise played a limited role in television and its consumption. I acknowledge these distinctions more to note the discrepancy of access and availability than affordability, as the relationship among cable subscription and income level is complicated: many who could afford to subscribe choose not to and we require more detailed data before assuming income is the key or primary determinant. The notion that different television forms carry varying cultural capital is not a new one (as the audience researchers whose respondents recount their PBS viewing can attest); but access to those shows so widely-touted as of highest quality has not been limited by an economic barrier in the past. The economic model of subscription networks mandates that they offer distinctive content—which led them to distinguish their programming in a particular way throughout the late 1990s and early 2000s. As ad-supported cable networks such as FX and even broadcast networks imitate their innovations, the subscription networks will need to find yet other ways to mark their difference and establish their value.

The paradox of subscription networks emerges from the fact that valuable and justifiable reasons exist for assessing them as a form distinct from "television"—at least the version of television known during the network era. There is some legitimacy to the complaints of broadcast network executives that the discrepant regulatory régimes and economic bases make subscription networks a derivative of the television industry rather than representative of its core. Yet at the same time, subscription networks are unquestionably a form of television, and this problem of variable institutional practices and conditions is now the norm for U.S. television. This contemporary "tri-furcation" among broadcast, ad-supported cable, and premium cable is just a stop along the way as distribution models shift radically in response to expanding on-demand and video-sharing capabilities. The growing range of ways to finance and share video content requires assessment of these structures alongside discussion of changes in programming and aesthetic value. Inattention to the radically different context of production and distribution between just ad-supported and subscription networks render meaningless comparisons of programming unless their varied institutional circumstances are addressed (and it is not the freedom in language, sexual explicitness, and violence that is so significant). Accounting for institutional features may not be central to textual analyses, but even examinations of programs must acknowledge variant institutional contexts in a way that makes comparisons across situations problematic. The institutional certainly does not resolutely determine the textual, but it provides a significant factor that evaluations too often under-emphasize.

NOTES

1. The essay expands on Charlotte Brunsdon, "Television Studies," in Horace Newcomb, ed., *Encyclopedia of Television* (Chicago: Fitzroy Dearborn, 1997).

2. See John McMurria, "A Taste of Class: Pay-TV and the Commodification of Television in Postwar America," in this volume.

3. See David Gunzerath, "'Darn That Pay TV!': STV's Challenge to American Television's Dominant Economic Model," *Journal of Broadcasting & Electronic Media* 44, no. 4 (2000): 655–673, 657.

4. Patrick R. Parsons and Robert M. Frieden, *The Cable and Satellite Television Industries* (Boston: Allyn & Bacon, 1998), 54; citing FCC Second Report and Order 1970.

5. Parsons and Frieden, *The Cable and Satellite Television Industries*, 54; see also Megan Mullen, *The Rise of Cable Programming in the United States: Revolution or Evolution* (Austin: University of Texas Press, 2003).

6. Such cases do emerge, as in the cancellation of advertising commitments for the coming-out episode of *Ellen*. A more recent and relevant example can be found in the FX series *The Shield*, which arguably comes closest to replicating an HBO-style series on an advertiser-supported network. Despite the presentation of a cutting-edge depiction of a somewhat rogue police force with high production values, many advertisers pulled out of sponsorship agreements out of fear of being connected with controversial content—although it is the controversial nature of the content that draws viewers comparable to the HBO audience.

7. In my case, I find enough value in my HBO subscription in the original series it produces to continue my subscription. The theatrical films or sports that others may primarily subscribe for offer no value to me.

8. Allison Romano, "Chris Albrecht's Goombas," *Broadcasting and Cable* 132, no. 37 (September 9, 2002): 12–13.

9. Michael Curtin and Thomas Streeter, "Media," in Richard Maxwell, ed., *Cul-*

ture Works: The Political Economy of Culture (Minneapolis: University of Minnesota Press, 2001), 225–250; Michael Curtin, "On Edge: Culture Industries in the Neo-Network Era," in Richard Ohmann, Gage Averill, Michael Curtin, David Shumway, and Elizabeth Traube, eds., *Making and Selling Culture* (Hanover, NH: Wesleyan University Press, 1996), 181–202.

10. Ohmann et al., "Interview with Mark Edmiston," *Making and Selling Culture,* 137.

11. Bruce Owen and Steven S. Wildman, *Video Economics* (Cambridge: Harvard University Press, 1992), 151.

12. When I contacted the network to inquire about the gender percentage of *Sex and the City's* audience for another research project, the executive I spoke with told me that HBO does not even gather such demographic information because it does not sell its audience to advertisers. I really doubt that this is the truth, but it underscores the relative unimportance of demographics for subscription networks.

13. In fact, a 2004 trade press report noted that broadcast networks are requesting that sitcoms last only twenty to twenty-two minutes to accommodate up to ten minutes of advertising per half-hour. See Christopher Lisotta, "Comedies Face New Realities," *Television Week* 23, no. 35 (August 30, 2004): 9.

14. Andy Meisler, "Not Even Trying to Appeal to the Masses," *New York Times,* October 4, 1998, 45.

15. See Deborah L. Jaramillo, "The Family Racket: AOL Time Warner, HBO, *The Sopranos,* and the Construction of a Quality Brand," *Journal of Communication Inquiry* 26, no.1 (2002): 59–75.

16. Curtin and Streeter, "Media," 232.

17. P. J. Bednarski, "Carmela, I'm Home!," *Broadcasting & Cable* 131, no. 20 (May 7, 2001): 24.

18. David Bauder, "HBO Still Has the Gold, But Is It Losing the Golden Touch?" Associated Press, June 19, 2005, retrieved from Lexis-Nexis.com.

19. John Dempsey, "Splashy Bows but a Slow-go," *Variety,* April 18, 2005, 13.

20. The other subscription networks have developed similar multiplex packages. The Showtime Unlimited package features Showtime, Showtime Too, Showcase, Showtime Extreme, Showtime Beyond, Showtime NEXT, Showtime Family Zone, ShoWomen, The Movie Channel, TMC Xtra, Sundance Channel, and FLIX.

21. For example, HBO Latino, HBO Comedy, and HBO Zone are not available through the DirecTV service I receive. They are available in Puerto Rico and the U.S. Virgin Islands, but the DISH Network features Comedy and Latino. The local cable MSO offers the same range of channels I receive on satellite, but offers only either East or West coast feeds, instead of both as available on the satellite.

22. See Katynka Z. Martínez, "Monolingualism, Biculturalism, and Cable TV: HBO Latino and the Promise of the Multiplex," in this volume.

23. Half of this is estimated as "move churn," subscribers who cancel because

they are moving and then re-subscribe at their new residence. John M. Higgins, "Angels, Emmys and DVD," *Broadcasting & Cable* 134, no. 39 (September 27, 2004): 12.

24. Sharon Moshavi, "HBO Multiplex Test Off to Good Start," *Broadcasting* 121, no. 22 (November 25, 1991): 36.

25. In August 2002, HBO created a new executive position of Vice President of Program Distribution, which suggests a new emphasis for the network on maximizing syndication opportunities. See Melissa Grego, "HBO Plots Sex and the Syndie," *Variety*, August 29, 2002, http://www.variety.com/.

26. Shirley Brady, "It's Showtime for Showtime," *CableWORLD*, October 18, 2004, http://www.cableworld.com/.

27. John M. Higgins and Allison Romano, "The Family Business," *Broadcasting & Cable* 134, no. 9 (March 1, 2004): 1, 6.

28. Chuck Crisafulli, "Offshore Assets," *Hollywood Reporter*, November 2002, 38.

29. Higgins and Romano, "The Family Business," 1, 6.

30. Denise Martin, "Tony and His Henchmen Go Artsy," *Variety*, January 31, 2005, http://www.variety.com/.

31. Higgins and Romano, "The Family Business," 1, 6.

32. Cheryl Heuton, "Is There Life After Pay Cable?" *Mediaweek*, May 30, 1994, 14–16.

33. John Dempsey, "Sex Sells Its Sex to KRON for $10.4 Million," *Daily Variety*, October 21, 2003, 5.

34. Higgins and Romano, "The Family Business," 1, 6; Stuart Levine, "DVD Format a Perfect Fit for the Fanatics," *Variety*, November 4–10, 2002, A18.

35. Availability of the HD channel varies by MSO and is often included in the purchase of HD programming tiers from cable and satellite providers.

Chapter 5

Where the Cable Ends
Television beyond Fringe Areas

Lisa Parks

Why is it that in the United States there are 65.4 million homes with cable television subscriptions and only 26.1 million with direct satellite broadcasting services?[1] This numerical disparity involves a complex set of issues ranging from federal policies to product designs, from landscape topographies to population densities, from cultural sensibilities to technological anxieties. Different TV distribution systems have emerged in different parts of the country at different times for different reasons. This essay explores sites beyond cable infrastructures in an effort to develop a more relational understanding of cable and satellite systems that can account for their distinct distribution architectures while considering how and where their histories intersect. Television historians Tom Streeter and Megan Mullen forged an important sub-field of cable television studies focusing on key regulatory and programming issues.[2] Their work has deeply enriched our understanding of cable television, yet too often the medium still gets reduced to terms such as the multi-channel environment, niche marketing, and consumer choice. While these issues have been subject to critique, there are ways of researching cable and satellite television that considers the spatial dynamics of distribution as well.

In their introduction to *MediaSpace* Nick Couldry and Anna McCarthy suggest that "The spatial orders that media systems construct and enforce are highly complicated, unevenly developed and multi-scaled. . . . the development of electronic media is a spatial process intertwined with the development of regimes of accumulation in capitalism."[3] To probe this complexity and unevenness, it is necessary to consider the fringe areas of television and beyond—that is, those places geographically removed from television's operational centers. Historically, television reception in remote

Fig. 5.1. This photo was taken on a road beyond Yucca Valley, California, while looking for the area where the cable ends. Photo by author.

locations was a dubious affair as technicians had to contend not only with vast distances, but unpredictable weather, mountainous topography, and rudimentary technologies. Nevertheless, it was the push for transmission to and reception of signals in outlying areas that ultimately helped to generate nationwide television networks. By the mid-1950s TV signals passed through communities scattered from coast to coast. Since then a variety of distribution systems has emerged, and, as a result, the study of television's content and form cannot be separated from cable, satellite, web, and wireless technologies.

In this essay, I combine approaches from media studies and cultural geography to survey two sites in California that lie beyond the grid of cable television infrastructures.[4] To begin, I examined maps of California cable television coverage and market ownership, but such maps tell us little about conditions within a given locale and require further investigation and explanation. The term "geo-annotation" has recently emerged in relation to the location-based or wireless economy. It refers to the convergence of global positioning system (GPS), wireless telephony, photography, and text technologies, which allows users of mobile devices to identify a location numerically and upload personalized data about it that can be shared with others. One writer describes it as the act of attaching "digital graffiti to real places."[5] I find geo-annotation to be a useful critical practice and model for building knowledge about television's multiple sites and especially those that lie beyond its operational centers. Such a practice might extend site-specific research by scholars such as Lynn Spigel and David Morley, who have explored television "homes" and "homelands"; Anna McCarthy, who has documented television in the tavern, the laundromat, and the airport; or Nick Couldry, who studied TV-fan pilgrimages to various sites in rural England.[6] Expanding our sense of *where* television research might take place, such scholars revealed that the medium's study cannot be reduced to frame-based aesthetics, genre, or narrative analysis alone. Television demands analytical models that can account for its complex spatiality. Adopting geo-annotation as a critical practice might be a way of stretching the spatial imaginary of television studies, of venturing beyond familiar tropes of frame, narrative, channel, and network to consider how television has become part of the built and natural environment.

Using cable coverage maps as a starting point, then, I set out to annotate specific locations with written descriptions, photographs, and critical discussions.[7] The sites that I explore include a state campground in Carpinteria, California, that abuts the Pacific Ocean and an area north of Yucca Valley in the Mojave Desert. My selection of these particular places is related in part to my familiarity with them. I have visited each of these sites several times, and each visit aroused my curiosity about the way that television signals arrive in such places. Having said this, I want to add that this is not a study of television and rural or remote life; rather, it is a set of descriptive sketches of areas that might never even be considered much less written into television's history unless a television scholar had actually visited there. One of the underlying intentions of this essay, then, is to reflect upon the ways we select the legitimate or proper places from which we historicize and critique television in the process of developing a field of study. Because of television's almost ubiquitous presence, critics might find themselves wondering about it in a variety of places, in the most peculiar moments. Thus, there is something at once quite personal and yet, as I hope we shall see, geophysical and atmospheric about the issues I want to explore here. By examining television at the edges of the desert and the ocean, where electronic culture meets a natural divide, I had to confront the medium as yet another natural resource—an electromagnetic spectrum that has historically been parceled and sold off in particular ways.

Looking where the cable ends thus led to a broader set of questions about television's different forms, places, and limits. It also enabled me to reflect upon the transparency of television distribution architectures, to identify nascent industrial strategies in formation, and to consider television in relation to natural resources and environmental conditions.

"Fringe Areas"

In U.S. television history, the concept of the "fringe area" emerged in the late 1940s and early 1950s as people across the country began purchasing television sets, installing antennas on rooftops, and tuning in signals from local transmitters. TV viewers located 25 miles or more from transmitters typically experienced reception problems, such as weak or snowy pictures. Consequently, these regions became known as "fringe areas."[8] Eager to bring those on the outskirts into the fold of clear television reception, the electronics industry manufactured an entire line of products to ensure viewers in fringe areas could receive a strong signal. A company called Trio crafted a series of Zig-Zag antennas with models for "suburban," "near fringe," "fringe," and "ultra fringe" areas.[9] Another company sold an antenna called Fringebeam, promising it provided "clear picture in the 100–200 mile range."[10] The Yagi, invented by Japanese physicist Hidetsugu Yagi during World War II, was known as "one of the best TV fringe area antennas,"[11] and models with names like Skyline and Tel-a-Ray also appeared on the market.[12] Edlie Electronics, a New York–based firm, designed a special chassis called the DX 630 enhanced "for excellent long distance [200 mile] television reception beyond the 'Fringe' Area" and claimed it had ten times the sensitivity of the standard TV receiver and did not require boosters or special antenna arrays.[13]

While those in fringe areas could purchase or improvise systems to receive a signal, many citizens lived far beyond these zones, hundreds of miles away from TV stations and transmitters. Electronics magazines during the period regularly published articles about the challenges of television reception in remote areas, advising TV servicemen and tinkerers how to negotiate noise, snow, ghosts, and other forms of interference caused by terrain and weather; they also provided a litany of heroic tales about technicians who devised makeshift systems to capture signals from hundreds of miles away. A 1951 article in *Radio and Television News* reported on Lawrence Pickell (known as "Pick") who created a system in the mountain community of Longmont, Colorado, to capture a signal from Omaha, Nebraska, more than 500 miles away.[14] His reception log indicated he had captured signals from 24 stations around the country including Greensboro, Detroit, and Seattle. Other issues of the magazine provided lists of men in remote areas and stations they managed to capture from 400 to 1,200 miles away. As technicians in the early 1950s experimented with remote reception, Western Electric Company installed a coast-to-coast microwave relay system that included 107 towers dispersed across every type of terrain. A feature article in *Radio and Television News* described the treacherous construction and displayed photos of mountaintop towers in Nevada and Utah that would help move signals to remote areas.[15] In 1953, as part of an effort to provide television services in remote regions, Pennsylvania Congressman Alvin Bush urged the establishment of "unattended satellite TV stations in communities too small for regular TV broadcast stations" and indicated Sylvania Electric Products was working on just such a plan to make automated TV stations available to small, isolated communities, particularly those in hilly or mountainous areas.[16]

At the same time that new gadgets, heroic technicians, and transcontinental infrastructures all worked to push television signals beyond fringe areas during the early expansion of broadcast television networks, some remote communities were implementing community antenna systems. The history of community antenna TV is fairly well-documented in television studies. What we still know relatively little about, however, are the overlapping histories of television distribution—of over-the-air, cable, satellite and web-based platforms. I offer this brief discussion of the fringe area to emphasize the parallel if unequal tracks on which television distribution technologies have emerged. In one sense, the deep history of cable television surfaces only when we consider early television's development beyond the fringe area. If this is true, then perhaps we must look beyond the cable to understand the history and spatial dynamics of satellite distribution. Only by moving away from television's operational centers and to its outer limits can we begin to imagine a relational broadcast historiography that accounts for the medium's complex and uneven developments in different locations. Such an historiography also must consider more carefully the industries—whether antennae manufacturers or direct satellite broadcasting companies—that form around and target consumers on the outskirts of television's centers, the development of distribution systems that negotiate terrain and weather, the installation of transcontinental infrastructures, and the way television relates to territorial control.

Since distribution infrastructures are dispersed across territories, it is helpful to examine coverage maps that represent them. Doing so allows us to formulate a sense of their scale and reach, and brings fringe areas out in bold relief. The 1998 map of commercial television stations in the state of California, for instance, reveals the overlapping densities of signals in the air in certain regions and emptiness in others. The Cable Television Franchise area map of California shows color-coded ownership of all the franchise-areas in the state and also clarifies which areas do not have cable service. Maps of DIRECTV and DISH Network footprints show that their signals can be received throughout the North American continent. While such maps are useful in identifying sites and patterns of distribution, they reveal little about the material conditions within those sites, and they thus demand annotation in the form of written descriptions, photographs, video, or other representations that may situate television within a given locale and, at the same time, expose this locale's irreducibility. The following sections feature experiments with geo-annotation in two sites of Southern California that lie beyond cable services—a residential area north of Yucca Valley in the Mojave desert and a state campground on the coast in Carpinteria. The goal here, of course, is not to provide an entire history of each place, but rather to generate sketches of the televisual landscape and to use those sketches to raise questions for the field of television studies.

Television in the Desert

Yucca Valley, California, is a community of just under 17,000 people. Its cable franchise is owned by Adelphia Communications, a company that has cable contracts in small and large communities throughout California and the United States. Yucca Valley is 112 miles from Los Angeles and sits in the 5,200-square-mile Morongo Basin of the southeastern part of the Mojave Desert. The area is referred to as "the gateway to California's outback."[17] It is the type of place where repossessed vehicles sit for sale on the side of the road, where the Kmart has been closed down (but Wal-Mart and the 99 cent store remain open), where billboards peel off in the hot sun or simply lie empty, and where pawn shops and bail bond brokers seem as pervasive as fast food restaurants. Those who live in Yucca Valley do so for a variety of reasons. It is much less expensive than other places in Southern California. Their employment requires them to be here. Or for one reason or another they are drawn to life in the desert. Beyond Yucca

Fig. 5.2. An abandoned homestead shack in the Mojave desert. Photo by author.

Valley lies a vast territory, part of which is preserved as Joshua Tree National Park and another part of which hosts the largest U.S. Marine Corp base in the world—Twenty Nine Palms Air Ground Combat Center. In 2000, the per capita income was $16,053.[18] Yucca Valley is traversed by non-stop military and tourist traffic, and the local economy is supported by a handful of major supplier stores, fast food restaurants, and gas companies, including Arco AM PM Mini Mart, Barr Lumber, Big Lots, Circle K, Del Taco, Food 4 Less, G&M Oil, Hutchins Motor Sports, JC Penney, Little Caesars, McDonald's, Phelps Chevrolet Nissan, R&R Chevron, Sears, Sizzler, Stater Brothers, Super One Food Store, Vons, Wal Mart, Walgreens, Yucca Valley Chrysler Center, Yucca Valley Ford Center.

The Morongo Basin, where Yucca Valley is located, had been home to Chemuevi and Serrano Indians until the 1880s, when their populations suffered a small pox epidemic. During the late nineteenth and early twentieth centuries, ranchers and homesteaders moved into the area. In the first half of the twentieth century, if a citizen wanted a piece of land out in the desert, he or she simply had to build a tiny homesteader shack on the land. Many people moved to the area after World War II and construction companies began to specialize in building 12-by-16-foot homestead shacks. In 1954 one couple bought 5 acres for $125,[19] which was inexpensive for California property at the time. Many of these makeshift structures are still standing, boarded up and abandoned, and there is now a local movement to destroy them, as these empty shacks are interspersed with plots filled by larger homes and compounds, some of which have driveways or yards filled with old vehicles and equipment that has landed out in the desert to bake.

When I drove north of Yucca Valley, beyond Adelphia's cable lines, I found a scattering of homes equipped with satellite dishes. The inhabitants of such areas are the early adopters of satellite television and have used satellite reception for nearly thirty years. During the late 1970s and early 1980s residents here and elsewhere began using five-foot diameter C-band satellite dishes to downlink (or more accurately intercept) the raw feeds of new cable networks such as HBO, TBS, and CBN.[20] These dish-owners proved the viability of a direct satellite broadcast industry, and trade organizations implicitly relied on people in remote areas to be able to make the case before the FCC that there were citizen-consumers living beyond the cable infrastructure who needed television service. In 1986, HBO began encrypting its signal so that dish-owners could no longer downlink it for "free," and other cable networks soon followed suit. Some viewers figured out how to use descramblers, and others became subscribers of Primestar in 1991 or DIRECTV in 1994 or DISH Network in 1997.[21]

The ruins of the early satellite TV age now settle on the desert floor as big dishes have been replaced and upgraded with the smaller ones used by DIRECTV and DISH Network. It is quite common in this area to see several generations of satellite dishes placed on the same property. In the illustrations a large C-band dish sits next to at least one eighteen-inch dish and also near an over-the-air antenna. This placement of multiple dishes on some homes is striking, given the controversies that sprouted up around satellite dish installation in urban and suburban residential districts during the 1990s. In some cases property owners and renters were forbidden from placing satellite dishes because of aesthetic concerns.[22] But the aesthetic issue of the satellite dish is not such a pressing concern in the desert where there is so much space that some properties are filled with mounds of old cars, scraps of metal, or old appliances, including old TV sets; eventually, these objects settle in the sand. While there remains a handful of C-band subscribers, most big dishes are no longer in use. Some have even been repurposed as birdbaths or flowerpots. In the nearby sculpture garden of artist Noah Purifoy, a big dish is the centerpiece of one of his sculptures made out of repurposed materials.

Communication scholar James Carey has suggested that with "satellite

Figs. 5.3 and 5.4. These residences have multiple generations of television receiving equipment, including C-band and 18-inch satellite dishes and over-the-air antennae. Photos by the author.

Fig. 5.5. An old satellite dish is the centerpiece in one of Noah Purifoy's sculptures in an outdoor gallery north- east of Yucca Valley. Photo by author.

communication there occurs a thrusting out of cultures into new regions of space."[23] Yet people settled in this region long before the age of satellite television. Perhaps the question to explore, then, is how different distribu- tion technologies intersect with the cultural, economic, and environmen- tal histories of a place. We need to recognize the ways community an- tenna, cable, and satellite distribution infrastructures co-exist, compete, and/or begin to supersede one another.

Even though I was in an area where the cable presumably ended, I encountered warning signs posted by the General Telephone Company

Fig. 5.6. A General Telephone Company sign indicating a
cable route in the desert. Photo by author.

throughout the area indicating the presence of "underground telephone
cables," which reminded me that there are often ways of keeping distribution architectures invisible and beyond public concern, whether these involve cables buried underground, satellites out in orbit, or large antenna
towers disguised as trees.[24] This topic is treated in *Splintering Urbanism* by
Stephen Graham and Simon Marvin, who critically examine networked
infrastructures of transportation, telecommunications, and energy and
analyze "how the wires, ducts, tunnels, conduits, streets, highways and
technical networks that interlace and infuse cities are constructed and

used."[25] Making visualization a crucial aspect of their study, they use photographs, maps, drawings, and other illustrations to expose massive infrastructures that often go unnoticed or become naturalized as part of urban space itself.

Such work is relevant to the study of television because distribution architectures typically fall within the rubric of electrical engineering or urban planning as opposed to critical studies. But if television technology is a historically shifting form and set of practices, then it is necessary to consider more carefully how the medium's content and form change with different distribution systems. Claiming to know what "cable television" is involves not only being able to identify it as a panoply of programming streams or a multi-channel environment, but also being able to understand and differentiate its distribution mode from others, whether these be broadcast, satellite, web, or wireless—systems that have themselves emerged in places in different ways.

It is on the peripheries that television's capitalistic structures can be seen stretching their potentials and assuming new forms. What we find beyond the cable is the formation of an increasingly automated and ubiquitous system of television distribution. Unlike cable providers, satellite carriers such as DIRECTV and DISH Network do not have to maintain local physical infrastructures, send technicians to residences, or sustain public access facilities. Most transactions between the satellite carrier and the consumer transpire either by telephone or onboard the satellite. The main costs for direct satellite broadcasters are the satellites themselves. In 2005, there were over 26 million satellite subscribers in the United States; because of the high cost of launching and maintaining satellites there are only three providers, two of which, DIRECTV and the DISH Network, serve the vast majority of subscribers. As Satellite Broadcasting and Communications Association explains to potential consumers, "Unlike cable and DSL, we are not limited in our ability to serve any geographical location. Although rural America served as the core area of early satellite adopters, recent survey information shows that 70 percent of new satellite television subscribers come from areas where cable is available, which indicates that satellite is increasing its suburban and urban penetration."[26] Urban, suburban, and rural designations are made redundant by satellite distribution since anyone with properly installed reception equipment can receive a television signal anywhere within the North American footprint. If we examine the coverage maps, we find that DIRECTV can always pick up where Adelphia (or any other cable provider) left off. DIRECTV's footprint turns the North American continent into one giant distribution field, making the continent coterminous with commercial television since all lands in the continental United States can be permeated by its signals.

This statement may sound extreme, but by formulating the issue in such a way, I want to return to a consideration of the electromagnetic spectrum as a natural resource that, as we know, has historically been allocated and sold off to private interests, not unlike the way in which the federal government authorized the parceling and sale of desert lands to ranchers and homesteaders in the early twentieth century. Legal scholar Patrick Ryan suggests that the "electromagnetic spectrum is among our largest natural resources. However, while the past few decades have seen a rich body of environmental law develop for other natural resources, this movement has largely passed over the electromagnetic spectrum."[27] Arguing that the spectrum should be put back in a public trust, he explicitly connects his discussion to the homestead movement: "In 1862, the government passed the Homestead Act and gave away thousands of acres to citizens, who would acquire property rights by living on, using, and improving the land for a certain amount of time. . . . The giveaways," Ryan continues, "that have taken place in the broadcast spectrum over the past 75 years have reached their limits, and action must be taken to reverse them."[28] The satellite distribution of commercial television makes these issues all the more pressing because the footprint has a coverage area so vast that it shifts us beyond Raymond Williams' famous formulation of "mobile privatization" and toward a system of ubiquitous privatization. The technology of the satellite transponder, the ground station, and the dish can be arranged to make virtually all North American lands penetrated by DIRECTV's signal. Thus, the distinction between territory and commercial television becomes altogether obscure.

Satellite distribution also makes cable television infrastructures and services redundant. This is why during the mid-1980s the cable industry led a negative public relations campaign against emerging direct satellite broadcasting companies. The cable providers claim that the structure of their industry generates more potential for economic competition and for more players in the market, but because of the deregulation of the media industries really only a handful of cable companies own and control most of the cable markets in the country. In addition, cable operators claim that their services reflect the demands of local constituencies since franchises have to be renegotiated every five years, but in many markets citizens are simply unaware of these requirements, and, as a result, public, educational, and governmental (PEG) channels have been eliminated or scaled back. More recently, the cable industry has scrambled to maintain viability by becoming a digital television and Internet service provider.[29]

California, because of its large population and size, has 2,048,170 satellite subscribers—more than any state in the country. At the same time, it has some of the most densely populated cable markets. In this sense, it can be understood as a microcosm of the nation with its combination of densely and sparsely populated areas, its inclusion of many players/competitors in the cable market, and its diverse topography. The coverage maps show overlapping circles in the western and central part of the state that disappear in the east. The state serves as an interesting site for thinking about histories of television distribution more generally not only because it is crisscrossed by so many different systems, but also because the state is so vast that some regions did not receive signals until the age of satellite broadcasting.

Indeed, by looking beyond the cable we are able to identify phases of technical experimentation, competing distribution systems, markets that have not yet been tapped or targeted in the same way as broadcast or cable-rigged areas. Studying television in such places also involves rethinking television as a natural resource—not unlike lands, forests, or oceans—that have been expropriated by corporate entities. It also involves developing critical approaches that account for the issue of distribution, which is not just a passing fancy between production and consumption, but introduces a complex set of technological, cultural, and historical issues in its own right. Only by investigating the multiple sites of television can we begin to specify the arrangements of its distribution.

Television Near the Ocean

On most summer weekends, Palm Avenue, the street leading to the Carpinteria State Beach campground, is backed up with a thick line of recreational vehicles waiting in the residential streets of this small coastal town for a prime spot on the Pacific. Carpinteria has not always been a haven for RV types. It is named after the Chumash Indian tribe that operated a large canoe-building and carpentry shop on this stretch of coast. Most of the remaining Chumash Indians have since moved north and east. Carpinteria is now a small community of Hispanic migrant workers as well as a tourist destination for beach lovers, recreational vehicles and camping enthusiasts. The campground sits just beyond the line of Cox

Fig. 5.7. The beach at the Carpinteria State Campground. Photo by author.

Communication's Carpinteria cable grid. Cox has had the cable franchise in Carpinteria since 1985, and the city renewed its contract again in 2002. There are houses to the north along this stretch of coast that receive cable television, but since this land is owned by the state of California, it, like all California beaches, is defined as a public property and sits beyond the cable. This does not mean, however, that people do not watch television on this beach. On the contrary: most RV owners hook up to the city's electrical system and end up receiving signals through a satellite dish or over the air antenna.

In the electrified section of the campground, RVs line up and are sandwiched neatly into their stalls. Sometimes American flags wave from their rooftops. Handcrafted wooden signs dangle from their awnings. Vibrant plastic flowers sit on tabletops. Stuffed animals appear in RV windows. And strings of electric lights dimly illuminate campsites at night. Almost always a satellite dish or over-the-air antenna is propped on the top of those RVs equipped with electricity. The vehicles have names like Zanzibar, Trailblazer, Prowler, South Winds, and of course Winnebago, and visitors come from far and near. Those who visit the campground spend their days sitting on the beach, swimming, fishing, surfing, shopping in

Fig. 5.8. (*top*) RVs lined up tightly at the Carpinteria State Campground. Photo by author.

Fig. 5.9. (*bottom*) Stuffed animals appear in the window of an RV and a satellite dish lurks at the lower right. Photo by author.

Fig. 5.10. Over-air antenna on RV (background). Satellite dish in front of RV (foreground). Photo by author.

nearby tourist joints, and in the evening many of them socialize or watch television.

As a site beyond the cable infrastructure, this place is distinct in a few ways. First, the act of watching television occurs inside a recreational vehicle as opposed to within a stationary structure such as a home, a tavern, or an office building. In this sense, RV television might be best understood as an inversion of the drive-in movie. Rather than looking out the window to watch a film next to other cars, the viewer turns inward to watch a screen inside the RV by him- or herself. This melding of entertainment and transportation is, of course, not new. As Lynn Spigel points out in her work on portable TV, there has long been a preoccupation with fantasies of mobility in discourses surrounding television technology.[30] Currently, we are witnessing a host of similar mobile forms of leisure ranging from in-flight entertainment to mini-van gaming systems, from cell-phone-

delivered satellite television to podcasting.[31] And this trend is both manifest in and exemplified by the mobile home. Describing its cultural history
in *Home on the Road*, Roger White note that "When couples and families
began to explore the rural and wilderness areas as a pleasure activity, they
used the automobile as a buffer, transferring the sophisticated furnishings,
technological systems and daily routines of home to the healthful attractions, scenic splendors and deprivations of the outdoors."[32]

The second distinction of campground television is that it is integrated within a transient place where visitors are constantly coming and
going, equipment is being assembled and taken apart, people are unpack-
ing and packing. Despite the solidity of the RV (especially when compared
to the tent), life in the campground is fleeting. Visitors can stay for a maximum of only seven days. The weather changes, sands shift, fog drifts in,
tides go high and low, and sometimes even an earthquake rattles the
ground. The transitory microphysics that characterize this place suggests
that television's history could be considered more carefully not only in relation to its domestic and social contexts but also to its natural surroundings as well. It is a little uncanny to find television's airwaves landing at
the very place where ocean waves can be seen and heard smacking the
shore. But maybe this situation recovers a history of nautical television yet
to be written. After all, part of broadcasting's deep past did unfold on the
high seas.[33]

Rather than assume campers would not watch television on the beach
or in any other natural space for that matter, RV manufacturers have
transformed the vehicles into entertainment systems. There are now television sets, stereos, antennas, and satellite dishes with names like "Tracvi-
sion," "MovinView," and "Winegard" made specifically for RV use and
costing from $90 to $2,700. RV trade associations report that there are 7
million homeowners who also own RVs and as many as 30 million RV enthusiasts who regularly rent or borrow motor homes. As RV-Net reported
in 2004, "Despite record-high gasoline prices nationwide this summer,
June's RV shipments were the highest for any June since 1977. For the first
six months of 2004, shipments totaled 196,200 units compared to 163,500
units during the first half of 2003."[34] Between May and August 2004, RV
campers spent $3.8 billion in the local communities they visited.[35] Televi-
sion producers and networks have also recognized RV travelers as a niche
market with highly specialized interests. In 2001 the Affinity Group began
producing a program entitled *RVtoday* that has since aired seven days a

Fig. 5.11. This is the web interface of television series *RV Today*, which airs on the Outdoor Life Network.

week on the Outdoor Life Network. A segment in one episode even instructs viewers how to use television in the RV.

I can't help but imagine that just as community antenna television was the prototype for the cable television industry, so, too, could motor home television become the model for a shifting television distribution paradigm—one based upon ubiquitous privatization. In this case, by going where the cable ends, not only do we find a new niche demographic—RV enthusiasts—already being targeted by cable channels (Outdoor Life Network), but also different structures of television reception and consumption becoming apparent as well. DIRECTV now encourages residential subscribers to register their RVs and to purchase mobile subscriptions so that they can access programming either at home or on the road. Satellite distribution turns open space into trade routes and viewers into moving targets. As Anna McCarthy astutely reminds us, the "quite different notions of what television is that emerge from site to site offer strong evidence of the ideological flexibility of television as an apparatus capable of linking everyday locations and their subjects to wider abstracted realms of commerce, culture and control in any number of ways."[36] Near the ocean, television not only interlinks campers to broader systems of commerce and culture (whether through *RVtoday* or The Weather or Travel Channels), but also enables us to recognize how satellite distribution works to rearticulate the medium as a recreational vehicle itself, transcontinental, disentangled from cables, modular and flexible, as free as an open road, as accessible as public lands and oceans.

And yet, if cable doesn't end in the desert, where I found underground telephone cables, nor does it really end on the California coast either. The California Coastal Commission, which regulates the use of coastal zones, handles requests from telecommunications and cable companies to lay transoceanic cable for global communication systems. By thinking about television near the ocean, then, I not only encountered different distribution systems, but also began to detect other networked global communication infrastructures that are largely invisible in the public sphere. There are now 15 cable "landings" along the California coast owned by companies such as Alcatel, AT&T, MCI-Worldcom, and Tyco that link the continent to areas throughout the Pacific. And while transoceanic cable lines create underwater communication and trade routes, there are economies visible above the water as well. When users of the Carpinteria State Beach look toward the western horizon, they glimpse eight oil platforms in the distance. The same resource once used by Chumash Indians to seal their wooden boats (tar) is now extracted from the depths of the sea and can be used to move RVs from place to place. Moreover, at the campground headquarters there is a stack of free magazines called *RV Journal* sitting next to an educational placard that reminds campground visitors "Chumash and other California Indians utilized trade routes not only to transfer material goods but also as a communication system. Thus, concepts, ideas, and information were shared over great distances . . ." How are such places, the resources that define them, and the infrastructures built though them organized to create communication systems? In this essay, I have tried to create a media geography that binds television to desert homesteads, coastal campgrounds, transoceanic cables, oil platforms, recreational vehicles, and Native-American history (if only in passing) as a way of emphasizing the irreducibility of television and of place. By looking at television where the cable ends, we can situate the technology in different spatial and historical contexts and thereby imagine new conceptual frameworks. I want to close by briefly highlighting some of them.

Conclusion

First, we need more critical and historical research on television distribution technologies. Too often histories of television unfold in the form of a linear progression from over-the-air to cable broadcasting, from cable to satellite delivery and more recently to the web and wireless.[37] While these histories play out in unique ways in different places, they are also overlapping, and the timing of their emergence is not rational or progressive. Further, television distribution practices both structure and are structured by different spatial dynamics. Satellite television in the desert is a different technological and cultural formation than over-the-air television in the city or cable television in the suburbs. Studying television's differences across these locales can challenge deterministic understandings of the medium in its past and its future. It may also require the kind of close analysis of networked infrastructures that Graham and Marvin develop in their study of urban regions. Adapting their approach to visually document and consider the different technologies and sites of television distribution may be a useful place to start.

Second, television's historical emergence intersects with the history of the environment (and I want to distinguish this history from what is known as "media ecology," which focuses on issues of morality and taste). As part of the electromagnetic spectrum, television technology is a natural resource, and there is much to contemplate about the history of its definition and treatment under the law as such. While research by Harold Innis, Armand Mattelart, James Carey, Herbert Schiller, and others has historicized and critiqued the ways communication technologies, capitalism, and imperialism have impacted territories, it may be possible to map media geographies and environments in ways that stress what Dick Hebdige calls the "incandescence of the particular."[38] Perhaps by returning to a more materialist definition of television, we can begin to map its spatial and environmental history alongside its programming, textual, and ideological structures and thus to understand the medium not only as a sphere of representation but also as a set of dispersed distribution infrastructures that have been built upon the earth and that have, in turn, inscribed it with particular cultural topographies. Pursuing further studies of television distribution would demand a more interdisciplinary approach that considers television's emergence in relation to fields such as geography, meteorology, and environmental studies, to name a few. It would also involve foregrounding television's complex place not just in landscapes or territories, but in criticism and history as well.

Finally, although we need to look at maps to understand the history of television's distribution infrastructures, they are ultimately insufficient. Annotating numerically structured geographies with knowledge practices such as on-site investigation and personal reflection may work to generate new spaces of critical and historical inquiry. Only by visiting several specific sites was I able to identify a new niche market, find out that the cable doesn't really end, that distribution infrastructures are designed to be transparent, that television signals constantly pass through Native Americans' lands. The location of television in everyday life is irreducible, and in this sense, given that the technology is so dispersed, involves so many different settings, programs, technical constellations, and financial matters, television studies is faced with a fundamental problem of description. One way to begin to address it may be through annotation, which comes from the Latin *annotatus* and means "to add notes to." In order to acknowledge and assess television's complex place in the world, we need to continue to fill its multiple sites with notes.

NOTES

Author's note: I would like to thank James Hay, Dick Hebdige, Vicky Johnson, Cynthia Chris, and Anthony Freitas for their helpful suggestions and feedback as I worked on this essay.

1. Federal Communications Commission, Twelfth Annual Report, Annual Assessment of the Status of Competition in the Market for the Delivery of Video Programming (released March 3, 2006, MB Docket No. 05-255), 5.

2. Thomas Streeter, *Selling the Air: Commercial Broadcasting Policy in the United States* (Chicago: University of Chicago Press, 1996); and Megan Mullen, *The Rise of Cable Programming in the United States* (Austin: University of Texas Press, 2003).

3. Nick Couldry and Anna McCarthy, "Introduction," *MediaSpace: Place, Scale and Culture in a Media Age* (London: Routledge, 2003), 1.

4. Like Anna McCarthy, I am interested in developing a multi-site approach that, as she puts it, can broaden our "sense of the spatial dynamics of the medium and its technological forms." See McCarthy, *Ambient Television* (Durham: Duke University Press, 2001), 226.

5. Jim McClellan, "Get Caught Mapping," *Guardian,* March 27, 2003, http://www.guardian.co.uk/ (accessed on October 22, 2004).

6. Lynn Spigel, *Make Room for TV* (Chicago: University of Chicago Press, 1992); David Morley, *Home Territories: Media, Mobility and Identity* (London: Routledge, 2000); McCarthy, *Ambient Television*; and Nick Couldry, *Media Rituals* (London: Routledge, 2003).

7. This essay builds upon a practice discussed in my essay "Plotting the Personal: Global Positioning Satellites and Interactive Media," *Ecumene: A Journal of*

Cultural Geographies 9, no. 2 (2001): 209–22. Expanded version published in Ursula Biemann, ed., *Geography and the Politics of Mobility* (Vienna: Generali Foundation, 2003).

8. Milton S. Kiver, "The Television Booster," *Radio and Television News*, October 1951, 35.

9. Trio ad, *Radio-Electronics*, August 1953, 81.

10. Ad for Fringebeam, *Radio-Electronics*, November 1953, 136.

11. Harold Harris, "The Yagi Angenna," *Radio and Television News*, October 1951, 66.

12. Tel-a-Ray System ad, *Radio-Electronics*, January 1951, 75.

13. Edlie Electronics ad, *Radio and Television News*, October 1951, 120.

14. J. A. Stanley, "No Television in Your City?" *Radio and Television News*, December 1951, 40.

15. William Alberts, "Bell System Opens Trans-Continental Radio-Relay," *Radio and Television News*, October 1951, 40–41.

16. "Satellite Systems," *Radio-Electronics*, November 1953, 6.

17. "Yucca Valley," California Travel and Tourism Commission website, http://www.visitcalifornia.com/state/tourism/tour_htmldisplay.jsp?sFilePath=/tourism/htdocs/welcomecenters/yucca.html (accessed November 2, 2004).

18. "Yucca Valley Business Information," Yucca Valley Chamber of Commerce website, January 23, 2004, http://www.yuccavalley.org/directory/bus_info.htm (accessed November 2, 2004).

19. Hugo Martin, "Tug of War Over Desert Homestead Shanties," *Los Angeles Times* November 1, 2004, A1.

20. The 1984 Cable Communications Act legalized private reception of unscrambled satellite television programming. Between 1981 and 1985 the "big dish" C-band market began to take off. By 1986 the cable industry began a negative press campaign against the nascent DBS industry, and signal encryption in the cable industry became more widespread, which eliminated "free" satellite TV reception by the 1990s.

21. See Satellite Broadcasting and Communications Association Facts & Figures, reprint of article "C-Band Subs Steamrolled Again," *Skyretailer*, December 2, 2004, http://members.rogers.com/4dtv/facts.html (accessed December 18, 2004).

22. For a discussion of this issue in the urban context of London, England, see Charlotte Brunsdon, "Satellite Dishes and the Landscapes of Taste," *New Formations* 15 (Winter 1991): 23–37.

23. James W. Carey, *Communication as Culture: Essays on Media and Society* (New York and London: Routledge, 1992), 170–71.

24. For a discussion of wireless infrastructure and transparency see my essay "Underneath the Antenna Tree" in the online project "Objects of Media Studies," curated by Amelie Hastie for *Vectors* 3 (Spring 2006), http://www.vectorsjournal.org/issues/03_issue/objectOfMediaStudies/.

25. Stephen Graham and Simon Marvin, *Splintering Urbanism: Networked Infrastructures, Technological Mobilities and the Urban Condition* (London and New York: Routledge, 2001), 8. I am grateful to James Hay for encouraging me to make this connection.

26. "Frequently Asked Questions," Satellite Broadcasting and Communications Association website, http://www.sbca.com/mediaguide/faq.htm (accessed November 2, 2004).

27. Patrick S. Ryan, "Application of the Public-Trust Doctrine and Principles of Natural Resource Management to Electromagnetic Spectrum," *Michigan Telecommunications and Technology Law Review* 10, no. 2 (2004): 285–372.

28. Ibid., 305.

29. According to the trade journal *Multichannel Online,* the only reason satellite subscribers would switch to cable is weather-related reception problems. "Weather or Not to Switch to Cable," *Multichannel News* (October 25, 2004), http://www.multichannel.com (accessed November 6, 2004).

30. Lynn Spigel, "Portable TV," *Welcome to the Dreamhouse* (Durham: Duke University Press, 2001), 60–103.

31. For an interesting analysis of in-flight entertainment, see Nitin Govil, "Something Spatial in the Air: In-flight Entertainment and the Topographies of Modern Air Travel," in Couldry and McCarthy, eds., *MediaSpace,* 233–52.

32. Roger B. White, *Home on the Road: The Motor Home in America* (Washington and London: Smithsonian Institution Press, 2000), 1.

33. See chapters four and seven in Susan Douglas, *Inventing American Broadcasting, 1899–1922* (Baltimore: Johns Hopkins University Press, 1987).

34. "National Survey Indicates RVers and Campers Spend Billions of Dollars," Affinity Group, Inc., press release, RV.Net website, August 30, 2004, http://www.rv.net/ (accessed November 2, 2004).

35. Ibid.

36. McCarthy, *Ambient Television,* 227.

37. For a critique of these linear histories see Lynn Spigel and Jan Olsson, eds., *Television after TV: Essays on a Medium in Transition* (Durham: Duke University Press, 2005).

38. Dick Hebdige, *Hiding in the Light* (London: Routledge, 1989), 230.

Part II

Channels

Introduction

In its brief history, television has experienced an explosion of networks. Especially since the implementation of fiber-optic and satellite digital delivery, channel availability in the United States has mushroomed from a few broadcast channels to hundreds of channels and on-demand programming. The expansion of channel capability has led, and continues to lead, many to claim that cable television offers extraordinary possibilities for communications technologies.[1] In this sense, television is framed in "the rhetoric of the electrical sublime"—a rhetoric that represents new technologies as salves to society's ills.[2] As discussed earlier in this volume, people from a range of political and economic philosophies have pointed to cable's multichannel capability as a means to break free from the broadcast dominance of the major networks; to open new and more diverse channels, programs, and ownership possibilities; and to create a media that is open, representative, and responsive to the public's entertainment, political, and social needs and wants.[3] Many, particularly those on the left, hoped the expansion of channel capability would usher in media that addressed a wide range of social issues from diverse viewpoints. Others saw the opening of more channels as opportunities for ownership and for profit outside or alongside of the broadcast environment.

Diversity and diffusion in ownership, channels, and programming continue to be touted as the never-yet achieved promise of cable. This section takes up the issues of development and transformation of networks and channels on cable television. Even before the All Channel Receiver Act of 1962 that required UHF tuners to be included in all sets and thus opened channels 14–83, the opening of more of the electro-magnetic spectrum for the proliferation of channels held the promise of more diversity in ownership, content, and audiences beyond that of the major broadcast networks of ABC, CBS, and NBC. This promise of viewing bounty was repeated with the introduction of early cable channels and again with each generation of improved wiring or delivery format, including today's 500-plus channel capability of digital cable and on-demand programming. Indeed, the opening of local UHF stations, the advent of the Fox network in 1986, of Univision in 1986, of Telemundo in 1987, and of both The WB (named for its parent company, Warner Bros.) and UPN (the Universal Paramount Network, owned by Viacom) in 1995 (which subsequently combined to form the CW Television Network in 2006), and the multi-channel environment of cable did result in more and more diverse channels, owners, and programming as well as the targeting of narrow and, arguably, underserved audiences.[4] In fact, the number of cable channels more than quintupled in the decade between 1994 and 2004, from 106 networks to 531.[5]

Does this quantitative explosion of channels represent a similar qualitative transformation in the channels available? Are there more channels with more programming and more diverse ownership, with different audiences and different business strategies? The answers to these questions are yes and no. While many channels have launched with innovative programming, business plans, and audience targeting, the same few big corporations have controlled much of the cable industry. Vertical integration is the ownership of the means of both production *and* distribution, which, in the cable industry, usually indicates investment by a cable system (or multiple-system operator, or MSO) in cable channels. Vertical integration reached as high as 53 percent in 1994, but dropped to about 22 percent by 2005.[6] This drop did not result as much from the entry of new owners into the market as from shifting patterns of ownership that put more channel equity into the hands of huge media conglomerates that are heavily invested in other parts of the media industry, including broadcasting (such as General Electric's NBC, Viacom's CBS, Disney's ABC, and News Corp's Fox), but that do not also operate MSOs.

In other words, as channel capacity increased over time, with systems that had long relied heavily on analog systems being rebuilt to facilitate new digital tiers, the number of new players in the market did not grow proportionately to the number of new outlets. The majority of new channels have been taken over by corporations already established in cable. For example, the Hearst Corporation started Lifetime and A&E in 1984 and later launched the History Channel in 1995, the Biography Channel in 1998, and, eventually, a handful of digital spin-offs; it also owns a 20 percent stake in ESPN. Likewise, well-leveraged networks have used on-demand services to offer multiple viewing opportunities for their programming. In such instances, new channel capacity has not always been assigned to new industry players, but rather has provided new territory into which established players could expand with marginally differentiated —and more of the same—product. In this way the variety of channels available to us is the result of a process that is akin to product differentiation wherein products that differ in some aspects mostly perform a similar role. Take the variety of toothpastes offered by two or three large consumer product companies, for instance: while there are slight variations in function, delivery, and taste, as toothpastes per se, they are all very similar; what the differences in the products represent are attempts to gain valuable market shares from one or two competitors; and in this way similarities tend to override differences.

A comparable trend is apparent from the way in which the broadcast networks and other media megaconglomerates gradually gained control over dozens of cable channels. A key player in this trend is Time Warner. The media giant already controlled a large stable of print, sports, music, movie, broadcast, and television companies, including HBO, when it acquired Ted Turner's media and other entertainment properties such as CNN and TNT in 1996; now, Time Warner also operates the MSO Time Warner Cable. But while Time Warner is among the largest media corporations in the cable business, four other media companies—Fox, ABC, NBC, and CBS—also control large numbers of cable channels. Fox, a subsidiary of Rupert Murdoch's News Corporation, currently, enjoys affiliation with FX, the National Geographic Channel, and the Fox News Channel, sports channels, and more. Disney, which owns ABC, also owns full or majority stakes in ESPN, ESPN2, Lifetime, ABC Family, and the Disney channels. NBCU, whose parent company is GE, owns or has a financial interest in MSNBC, CNBC, the National Geographic Channel, Bravo, USA, and Sci-Fi. And, since its acquisition by Viacom in 2000, CBS now has links to MTV Networks (MTV, MTV2, VH1, Nickelodeon, Spike, Logo, CMT, and Comedy Central), BET, Showtime, and The Movie Channel. Clearly the ownership of cable channels is quite complex, involving sprawling empires. Just as clearly, a pattern of ownership is manifest: simply put, it involves more and more media assets, including these channels, being owned by fewer and fewer corporations.

Thus far, this discussion only accounts for the U.S. portion of these businesses. Fox shares its parent company, Australian-founded News Corporation, with foreign versions of its channels and major DBS providers in the United Kingdom, Europe, Asia, and the Australia-Pacific region. Viacom (through MTV Networks) has extensive overseas channels as well.

Time Warner—largely through HBO and CNN—also has extensive, and relatively long held, international channels. This account of ownership and cross ownership, both domestically and internationally, begins to outline the very significant barriers that new networks face in trying to enter into a crowded cable terrain that is dominated by colossal international media corporations. Many of the channels entering the market, whether they are backed by an established network, are independent of these networks, or operate in some combination, face incredible challenges in placement on cable systems and in attracting and maintaining audiences, subscribers, or advertisers. Often channels have had to reformat to try to bring in and hold audiences. Increasingly, channels are finding that in order to maintain audience share, let alone grow, they must include costly original programming, rather than simply rerunning or repurposing available content.

The essays that follow address the strategies of these channels, as well as the difficulties they have had, in attempting to carve out new and different business, programming, and audience models within the for-profit media environment. Several of the authors in this section outline how innovative and alternative programming structures have given way to more traditional models of launching and maintaining an economically feasible channel. These chapters explore both the successes and the limitations in fulfilling the diverse and often conflicting dreams for the multichannel environment.

The section begins with Cynthia Chris's examination of the creative financing and programming strategies of The Discovery Channel in its efforts to build an audience and attract advertisers. As Chris points out, Discovery was among the first channels to offer MSOs a financial stake in the company in exchange for carriage in their markets. In addition, the channel drew on inexpensive libraries of stock footage and leased programming to revive the commercially moribund genre of wildlife, as well as to experiment in early forms of reality programming. Within just a few years, Chris shows, new channels designed around overlapping concepts, Animal Planet and National Geographic, would enter global markets under vastly changed conditions.

Cheri Ketchum, in her essay on Food Network, details the channel's innovative strategy of product placement, celebrity chefs, inexpensive programming, and integral commercial web content to capture an audience and become one of the fastest self-sustaining channels in cable history. Like Discovery, Food Network took a segment of programming that was historically consigned to public television or non-prime-time commercial television and built an entire channel around it. Ketchum's essay illustrates how the channel's strategies for commercial success have resulted in a depoliticization of food—its production, distribution, and consumption—which contributes to maintaining the channel's upbeat atmosphere, its advertisers, and its revenues. Ketchum argues that while eschewing the controversial for far less politicized programming is not unique to the Food Network, the implications for better understanding something as integral to our survival as food are immense.

As both Chris's and Ketchum's essays demonstrate, as the market for cable expansion approaches saturation, new and existing channels cannot rely on new cable customers to fill in the audience lost to other channels in the growing cable universe, but must work to draw viewers from these other channels. Additionally, television, like cinema, finds itself increasingly competing with computers, the Internet, and videogames for audiences' attention. In this environment, channels pursue a number of strategies to build and maintain audiences. One of these is to develop channels around particular programming themes, as The Discovery Channel and Food Network have. Another is to develop channels around audiences, identities, and demographic segments. Channels targeting demographic groups—African Americans, Latinos, Asians, children, and lesbians and gay men, to name but a few—have emerged. This increasingly narrow targeting is in keeping with larger marketing trends in which marketers seek to divide consumers into finer and finer segments in hopes of maximizing the effects of their advertising dollars by appealing to consumers or audiences who are most likely to buy or watch their product. Since these channels are meant to generate profit through subscriptions, advertisements, or both, the niches to which they appeal are those presumed to be large, have money, and be heavy media consumers. However, these channels do not simply appeal to already existing audiences; rather, they help to form and define those audiences.

Among the earliest of cable channels targeted to a particular demographic was Black Entertainment Television (BET). As Beretta Smith-Shomade outlines in her essay, BET was crucial in proving the viability of niche marketing cable channels on the basis of identity. She argues that BET "provide[d] the means for specific forms of black identity politics and representation to receive validation." Founded by Robert Johnson in 1980, BET was financed in the then-novel approach of teaming with cable distributors. Since its launch, BET has consistently captured the largest share of the African-American cable audience,[7] and, until 2003 and its sale to Viacom, was among the nation's largest African-American-owned businesses. However, as Smith-Shomade argues, BET represents both the promise and the disappointment of for-profit television in representing Black people and their calls for justice and equality within the nation. While showcasing some of the most dynamic and influential names in African-American entertainment, the channel also routinely draws on pervasive and persistent cultural tropes of blackness in ways that reify dominant stereotypes and rely on economically exploitative relationships between channel owners and performers to seek wider and whiter audiences. Further, Smith-Shomade argues—pursuing a theme picked up in the essays by Anthony Freitas and Sarah Banet-Weiser—the audience and the community that are targeted by the identity-based channel are also exploited by it. BET sells the Black audience to advertisers as a commodity and in doing so, not only depoliticizes the audience by transforming it into a market but also thereby privileges business over social, political, and cultural concerns.

While a channel for Black audiences did not launch until the 1980s, Spanish-language programming has long been a staple of U.S. television. While local stations began broadcasting original and dubbed Spanish-language programs in the mid-1950s, and a national presence for Spanish-language television began building in the 1960s, it took off in the 1980s with the launch of Univision and Telemundo. It was also in the 1980s that HBO began its first, limited foray into Spanish-language services, but as Katynka Z. Martínez points out in her essay in this section, the Latino or Hispanic market turned out to be more complex than HBO originally envisioned. When HBO first entered the Latino market in late 1988, its executives were reluctant to devote many resources to Spanish-language programming within the United States and simply provided a second audio program (SAP) in Spanish. Throughout the 1990s, as HBO attempted to draw in Latino viewers, the niche market category "Hispanic" proved problematic for them. As Martínez argues, HBO (like other marketers) failed to comprehend that this group was made up of people who shared a language, but often shared little else in terms of history, culture, geography, economics, and social location. Acquiring and producing programming for such a diverse group and positioning it as "Latino" thus became a challenge. Although HBO still relies on dubbing of either theatrical or HBO original productions (such as *The Sopranos*), Martínez shows how HBO has attempted both to construct a cohesive Latino audience out of diverse cultures and to sponsor original programming on HBO and in other venues as a means of attracting and maintaining a foothold with the Latino audience.

Anthony Freitas also examines the formation of channels targeted to a particular demographic in his study of gay and lesbian themed networks. Investigating the financial and social issues encountered by companies attempting to develop gay and lesbian channels in several nations, Freitas illustrates how these channels, in seeking a purportedly wealthy and un(der)tapped audience, have consistently stumbled in their attempts to build profitable business models and attract significant audiences. Their somewhat novel but problematic funding schemes of subscription-plus-advertising, the lack of original content, and the resistance from cable operators, advertisers, and viewers to gay-themed networks hobbled early attempts to create self-sustaining channels. Moreover, as Freitas points out, while these channels counted on the interests of an already existing audience, they, like HBO, also had to create their audience from a varied and diverse group that did not necessarily easily fit into one marketing or audience category. For some of these gay and lesbian networks, the audience has been slow to materialize, forcing them to rethink many of their assumptions about their target audience, their marketing, their funding, their programming, and their distribution.

The issues of identity, community, and television as well as nation are explored further in Sarah Banet-Weiser's essay on Nickelodeon, the final chapter in this section. Banet-Weiser argues that the channel addresses children as informed and active participants—"empowered citizens." Indeed, Nickelodeon, claims that its programming for children empowers kids by "respecting" them. But, as Banet-Weiser argues, this empowerment is as much a product focused on the network's brand identity as it is a form of civic engagement or personal virtue. "Nick kids" are members not just of the larger national citizenry, but also of the "Nickelodeon Nation." In this way, Banet-Weiser contends, Nickelodeon very successfully adopts the language of political and personal empowerment to promote its own brand.

NOTES

1. See for example, François Bar and Jonathan Taplin's essay in this volume.

2. James W. Carey and John J. Quirk, "The Mythos of the Electronic Revolution," in James W. Carey, ed., *Communication as Culture: Essays on Media and Society* (Boston: Unwin Hyman, 1989), 23.

3. Thomas Streeter, "Blue Skies and Strange Bedfellows: The Discourse of Cable Television," in Lynn Spigel and Michael Curtin, eds., *The Revolution Wasn't Televised: Sixties Television and Social Conflict* (New York: Routledge, 1997), 221–36.

4. Raoúl Cortez founded KCOR in San Antonio in 1955. In 1961, it became KWEX, a component of the Spanish International Network (SIN), owned by the Mexico-based Telesistema Mexicano. In 1986, to comply with FCC rules barring foreign ownership of stations, SIN was sold to Hallmark and rebranded Univision. In 1992, Univision was sold again to a group of Miami-based investors who took the network public, rapidly expanded its programming and reach, and in 2006 resold it to other private investors for more than $11 billion. Andrew Sorkin, "In Late Twist, Univision Accepts Bid," *New York Times*, June 27, 2006, http://www .nytimes.com/; Handbook of Texas Online: Spanish Language Television: http:// www.tsha.utexas.edu/handbook/online/articles/SS/ecs1.html (accessed January 22, 2006).

5. FCC, Twelfth Annual Report: Annual Assessment of Competition in the Market for the Delivery of Video Programming (released March 3, 2006, MB Docket No. 05-225), 7.

6. Ibid.

7. R. Thomas Umstead, "Black Cable Viewers Cast Eyes Beyond BET," *Multichannel News*, February 13, 2006, http://www.multichannel.com/.

Chapter 6

Discovery's Wild Discovery
The Growth and Globalization of TV's Animal Genres

Cynthia Chris

In the 1970s and into the 1980s, nonfiction wildlife filmmaking reached American television audiences largely in the form of low-cost, syndicated half-hours, such as *Mutual of Omaha's Wild Kingdom,* and highbrow series and specials, such as *Nature* and *National Geographic,* featured by the Public Broadcasting System (PBS).[1] Wildlife could be found by viewers only on the fringes of an industry dominated by three commercial networks. By the end of the 1980s, however, the wildlife genre served not only as a flagship of The Discovery Channel's innovative and profitable programming strategy, but also as part and parcel of a widespread proliferation of animal TV in various forms. How can we account for Discovery's savvy use of documentaries, especially the wildlife genre, and the general new prominence of animals on TV?

These shifts in the status of a genre resulted substantially from a matrix of changing conditions in the television industry that influenced how both broadcast and cable networks financed, scheduled, and distributed TV programming, and even how they conceived of TV form and content. As shown elsewhere in this volume, in the 1980s and 1990s, cable- and satellite-TV subscriptions increased at a rapid pace the United States, and the number of channels offered by cable systems grew. Many new channels (such as Discovery) started as relatively small operations that depended on economical means of acquiring or producing programming to fill their schedules, frequently turning to nonfiction genres that typically cost a fraction of dramatic and comedy series. Most would relinquish independent ownership to established media giants in order to survive.

The Discovery Channel entered the market by means of an innovative set of alliances with already established corporations. It also played a role in repositioning nonfiction throughout the industry, as well as in recom-modifying wildlife programs that commercial broadcasters had all but phased out as irreversibly formulaic and unprofitable. The Discovery Channel's success with wildlife was so convincing that in 1996, Discovery teamed with the British Broadcasting Corporation (BBC) to spin off Animal Planet, a channel devoted entirely to companion animals and wildlife. A year later, in 1997, National Geographic countered with its own channel featuring science, nature, and adventure programming. In each case, the competitive media environment nudged these entities in two directions: first, as they moved away from proclaimed early commitments to a documentary form that would educate as it entertained, they moved toward less didactic nonfiction genres such as game shows, talent shows, and "reality-based" programs; second, these channels would go emphatically global, seeking to expand into newly privatizing markets outside the United States. This chapter examines the launch and growth of these three channels, which are paradigmatic of trends in the industry over these decades.

The Discovery Channel

On June 17, 1985, The Discovery Channel's premiere reached only 156,000 homes in an improbable, undercapitalized entry into the increasingly competitive cable market. Within a decade, it became one of the most widely distributed cable programming services and the cornerstone of a mid-sized media conglomerate. Discovery Communications, Inc. (DCI) now controls four analogue and nine digital cable channels in the United States. Variations on these services are distributed to a combined total estimated at one billion subscribers in 160 countries. In addition, Discovery controls a chain of retail stores; home video, book, music, and multimedia publishing interests; and more.[2] Seeking to reach movie as well as television audiences, DCI created Discovery Channel Pictures to produce the feature *The Leopard Son* (1996) as well as IMAX films such as *Africa's Elephant Kingdom* (1998); it also created Discovery Docs, which co-produced Peter Gilbert's *With All Deliberate Speed* (2004), on the Supreme Court's 1954 *Brown v. Board of Education* decision, and Werner Herzog's *Grizzly Man* (2005). Discovery's brand extensions were not in themselves unusual, except, perhaps, in the extent to which they emanated from a strategically prescient upstart in a media environment that is mostly ruled by old

mega-conglomerates. In this way, Discovery has been both typical and exceptional, conducting business-as-usual in terms of the standard practices
of the cable television industry and contributing to innovations in industry structure and practices, including the new centrality of wildlife in the
brand-identity of a commercial network.

Entering the cable marketplace during a period of rapid growth and
change in the industry, Discovery was unique at the time in its reliance on
documentaries and on natural history, science, exploration, and related
content believed by most programmers and potential investors to be irre-
trievably unprofitable. No other network, not even PBS, devoted so much
of its schedule to documentaries. In its early years, some industry professionals touted Discovery as an example of the best that television had to
offer: Ajit Dalvi, a Discovery board member and vice president of Cox Cable, commended Discovery for offering "educational programming . . . in
the true sense," and TCI Senior Vice President John Sie boasted that the
network would "repatriate [sic] science and technology" to a general public lacking in knowledge of these subjects.[3] But Discovery has not always
continued to represent itself as an educational programmer. Instead, Discovery has cultivated the entertainment value of nonfiction, in nature and
in its other content categories. As DCI executive Chris Moseley declared
after the network had been on the air ten years, "[education] isn't why
people are drawn to our brand."[4]

Nonetheless, education was a key ingredient of Discovery's original
concept and its sales pitch to early investors. In 1982, John Hendricks began to orchestrate an increasingly complex partnership that would finance
the network's launch. Attracting investors was difficult, since the scant
presence of documentary on television at the time seemed to prove that
the form lacked commercial value. According to industry lore, Hendricks'
breakthrough came when he secured a letter of support from the legendary newscaster Walter Cronkite, who had hosted the science-oriented
documentary series *Universe* (1980–1982, CBS).[5] Bolstered by such a prestigious recommendation, the company began to attract the interest of
other cable industry players. In Discovery's first (but short-lived) vertical affiliation with a carriage provider, Westinghouse's Group W Satellite
Communications acquired a 6 percent share in exchange for $3 million
worth of marketing and satellite transmission services.[6] The Discovery
Channel went on the air in 1985, offering twelve hours of programming
daily, much of which consisted of older product previously aired by National Geographic or PBS.[7] Charging no affiliate fees to cable systems that

agreed to put Discovery on the air, the channel depended solely on advertising revenue.

In 1986, a year after launch, Hendricks brokered an innovative deal with Tele-Communications, Inc. (TCI), United Television Cable Corp., Cox Cable Communications, and Newhouse Broadcasting Corp. (later Advance/Newhouse). Each of these multiple-system operators (MSOs) acquired a 10 percent stake in Discovery and agreed to carry the network, brightening Discovery's prospects on three counts.[8] First, the injection of cash allowed the network to begin to acquire programming not yet aired on other channels. This new programming, in turn, allowed Discovery to expand its day from twelve to eighteen hours and, therefore, to sell more advertising.[9] Second, the MSOs' commitments to carry the channel led to non-ad revenue in the form of fees when, in 1987, Discovery began to charge cable systems two to five cents per month per subscriber. Larger affiliates paid the least, privileging the systems vertically affiliated with Discovery. Third, the deal pushed Discovery into 13.5 million homes, surpassing the threshold at which the A. C. Nielsen Company would, at the time, begin to monitor a channel's viewership and making Discovery far more intelligible and appealing to advertisers.[10]

Why did the MSOs risk investing in a fledgling network? The short answer is that they had both the power and the motivation to assure that Discovery would attain carriage sufficient to attract significant numbers of viewers and advertisers and could thereby raise the value of their investment. Some observers gloated that the MSO bailout of Discovery was "one of the brighter moments in cable history" and (erroneously) that it constituted "the industry's first equity alliance between programmer and operators."[11] But Discovery's exploitation of these vertical affiliations demonstrates the potential anticompetitive hazards of a market in which such affiliations are rampant: the MSOs are likely to add channels in which they invest to their systems preferentially, as well as to favor them in placement, pricing, and promotion over unaffiliated networks or those owned by competitors. After Discovery became the fastest-growing cable network in 1987 and 1988, the MSOs increased their investments until they controlled some 97 percent of its equity.[12] The Discovery Channel quickly became one of the most widely distributed cable channels and by 2004 reached 88.6 million subscribers, more than any other cable channel in the United States.[13]

While Discovery developed a niche audience and MSO executives crowed over its virtues, some critics thought of Discovery in its early years

Fig. 6.1. The Discovery Channel promoted its annual Shark Week in August 2006 using academic insignia and a shelf of library books with invented titles, as tongue-in-cheek markers of the potentially educational value of the programming.

as "a repository for off-PBS reruns, a showcase of stale fare that nevertheless managed to attract an audience."[14] But as the network's finances stabilized, it became less dependent on acquired material.[15] Some of its first original specials were nature-oriented, such as *Ivory Wars* (1989) and *In the Company of Whales* (1992), and they garnered excellent ratings.[16] These successes, and the popularity of Shark Week—a special weeklong event, in which the Discovery Channel replaces prime-time programming with documentaries on sharks, aired annually since 1988—positioned wildlife as a centerpiece of the Discovery brand identity.[17] For the 1995–1996 season, Discovery consolidated nature programming under the title *Wild Discovery*, an hour-long series that was aired five nights per week in prime time and repeated ("looped" in industry jargon) in late-night hours. During its first month, Discovery's Nielsen ratings in the timeslot increased by 50 percent.[18] In many ways, with its eponymous title and desirable timeslot, *Wild Discovery* served as the channel's flagship during this season. But

Fig. 6.2. (*top*) Detail of the Discovery Channel's Shark Week ads, August 2006.
Fig. 6.3. (*bottom*) An ad placed on the sides of buses in New York City touted The
Discovery Channel's Shark Week in 2006; common occupations, likely held by bus
riders, are contrasted with the "shark chummer."

Fig. 6.4. Another Shark Week 2006 ad also used themes of employment.

that was not the only wildlife programming to be found on the channel. Alongside *Wild Discovery* and Shark Week, Discovery featured predators in *Fangs!* (1995–1998). The weekly series, which the *Wall Street Journal* called a "bloodbath," was Discovery's highest rated in 1995.[19] Discovery's animal programming—which formed a key part of a schedule that also included gender-targeted lifestyle programming in daytime, and mechanics, engineering, and crime themes in prime time—proved so exploitable that the network gave the genre an outlet all its own, to be called Animal Planet, in 1996.

What advantages did wildlife present over other documentary themes? Observers within the television industry have pointed convincingly to two clusters of largely economic factors: costs (relative to potential revenue) and shifting audience profiles. Documentaries, especially those with nature or wildlife subject matter, are inexpensive, durable, and flexible products, compared to many other TV genres. Their budgets vary tremendously but are typically only a fraction of the cost of fictional TV genres:

"no expensive actors, directors, writers, or sets. The animals and plants are nonunion, not entitled to residual payments, and work for nothing."[20] Unlike newsmagazines or documentaries on topics in which knowledge and practices change frequently, such as medicine and health or current public affairs, nature documentaries have an unusually long "shelf-life." Programs are rerun season after season without losing audiences, and their footage is frequently reused in different programs.[21]

The genre's audiences, while modest in size, have been reliable. They even grew in the mid-1990s, and not only for *Wild Discovery*. Audience share for TBS's *National Geographic Explorer* grew almost 50 percent in 1995, and for PBS's *Nature*, 30 percent in 1996.[22] Further, the genre's small audiences often sport demographic characteristics attractive to advertisers. In 1996, *Forbes* described the audience for *National Geographic Specials* as "male viewers aged 25 to 54, a typically tough crowd to reach with TV."[23] Other observers have claimed that increasing "interest in the environment" also paved the way for cable programmers to embrace the wildlife genre in the late 1980s and early 1990s.[24]

Settling into a newly comfortable financial situation, Discovery sought further growth through diversification of its domestic television activities. In 1989, Discovery joined other networks and some MSOs in developing Cable in the Classroom, an educational media transmission service, and in 1991, it purchased The Learning Channel. At the same time, DCI set its sights overseas, fulfilling the logic of commercial media that, as Nicholas Garnham argues, seeks to expand not by providing viewers with new choices but by finding new markets to saturate with its product.[25] In television, global expansion became possible in the early 1980s, according to Dan Schiller, as a result of "unremitting U.S. pressure, supranational initiatives within the European Union and the World Trade Organization, shifting affinities among national elites after the fall of Soviet socialism, and hardly least, the explosion of Internet systems."[26] Thus, in formerly public and state-controlled markets, deregulation, privatization, and commercialization eased the entry of entities like Discovery into global markets, frequently through partnerships with established local networks. This strategy facilitates cost-sharing for new productions, accesses a staff already experienced in local culture, and opens the door to markets where foreign investment is limited by law.

In 1989, DCI launched Discovery Channel–Europe, a network that lost money for six years, but that allowed Discovery to test programming and modify marketing strategies before seeking to expand throughout Asia,

the Americas, and elsewhere.[27] Discovery Networks Asia, based in Singapore, launched in 1994 and met with quick success. By 1997, The Discovery
Channel ranked as "Asia's No. 1 cable channel . . . [most] watched by adults
ages 25 to 64," according to surveys conducted in Tokyo, Jakarta, Manila,
Taipei, and other urban centers where viewers in these markets—especially the elite top 10 to 15 percent sought by networks for their advertisers
—are concentrated.[28] In India, The Discovery Channel launched in 1995,
initially in English, and then added programming feeds in Hindi and
Tamil.[29] Elsewhere, Discovery encountered more formidable barriers to
entry, including limitations on foreign ownership of media in some nations. For example, in 1995, hoping to establish early ties with the potentially huge but still relatively closed markets of China, Discovery initiated
a season-long programming exchange with the state-run China Central
Television (CCTV).[30]

In 1993, when the United States, Mexico, and Canada enacted the North
American Free Trade Agreement (NAFTA), the Mexican broadcaster Tele-
visa entered into a joint venture with DCI to produce a version of The
Discovery Channel for Latin America.[31] Canada, having negotiated a "cultural exemption" to NAFTA, kept in place media regulations designed to
protect Canadian economic and cultural sovereignty, especially from dependence on the United States.[32] Discovery partnered with the Labatt
Brewing Company to launch a network, its equity share restricted by law
to only 20 percent.[33] Hendricks' boast, "We hope to blanket the world by
late 1995 or early 1996," was hardly an exaggeration.[34] By 1997, Discovery
reached cable and satellite subscribers in 145 countries.

In several sub-Saharan African nations that have provided the setting
for countless wildlife (and ethnographic) films, Discovery established the
Discovery Channel Global Education Fund in 1997. Collaborating with
other corporate donors, nongovernmental organizations (NGOs), local
governments, and the World Bank, the fund supplied dozens of learning centers at schools and community centers in South Africa, Tanzania,
Uganda, and Zimbabwe with TV sets, VCRs, cable hook-ups or satellite
dishes, as well as solar power where electricity is otherwise unavailable,
and access to programming. TV reaches relatively small portions of these
countries' populations, especially in rural areas—for example, at the time,
only a quarter of Zimbabwean households' owned sets.[35] But Discovery's
donations train rural teachers in the use of media in the classroom, and
acclimatize rural populations to media use. The donations are also made
in anticipation of eventual economic development in these markets.

Animal Planet

In 1996, seeking to expand by means of product differentiation, Discovery undertook another growth spurt in both domestic and global markets, launching a raft of digital channels and the basic-cable network, Animal Planet, which was promoted in ad campaigns as the "all animals, all the time" channel.[36] Its expansion depended on DCI's ability to divert funds from the profitable Discovery Channel and The Learning Channel to buy Animal Planet's way into carriage, offering MSOs $5 to $7 per subscriber to add the channel in 1996 and 1997.[37] Within two years, Animal Planet reached 37 million subscribers, averaging 200,000 viewers in prime time and occasionally, a Nielsen rating of 0.5, surpassing channels such as VH1 and MSNBC.[38] Globally, growth depended on an alliance between Discovery and the BBC, which gave the British public entity's commercial division, BBC Worldwide, a 20 percent interest in the U.S. version of Animal Planet and 50 percent of the network in other markets.[39] In 2004, seven years after launch, Animal Planet reached 85 million U.S. homes, and 126 million in other countries.[40]

Animal Planet programming extends the success that Discovery had with wildlife. Most of its traditional wildlife programming features adventurous naturalist-hosts who lead the viewer through vicarious encounters with wild (and sometimes captive) animals. One of Animal Planet's most successful series has been *The Crocodile Hunter*, launched in 1996 and featuring Australian naturalist Steve Irwin, who was stung fatally by a stingray in 2006, while shooting an underwater scene for a new series. The show attracted a record one million viewers in the United States alone to the first night of a week-long "Croc Week" marathon in 2000.[41] Irwin also appears in spin-offs with varied concepts, including *The Crocodile Hunter Diaries* (2002–), *Croc Files* (2003, for Discovery Kids), *New Breed Vets* (2005–), and the feature film *The Crocodile Hunter: Collision Course* (2002). Similarly, naturalist Jeff Corwin has appeared in a number of Animal Planet series, beginning in 2000 with *The Jeff Corwin Experience* and followed by *Jeff Corwin Unleashed* (2003) and *Corwin's Quest* (2005–).

But much of Animal Planet's programming incorporates elements of other TV genres, developing "reality" concepts with animal themes and blurring the boundaries between fiction and nonfiction. These shows exemplify the pervasive industry logic that a network can never have enough of the latest successful thing. Thus, for example, when *Animal Court* premiered in fall 1998, it featured Judge Joseph Wapner, formerly of the syndicated *People's Court* (1981–1993), who heard cases involving breeding rights, boarding bills, and other animal-related disputes. In 1998, the network set *Emergency Vets* in a Denver animal hospital; the reality show was based on a BBC series. Launched in 1999, *The Planet's Funniest Animals* was based, like *America's Funniest Home Videos* (1989–, ABC), on home videos sent in by viewers. In 2002, *Animal Precinct,* in the style of Fox's long-running reality series *Cops* (1989–), sent a camera crew along as officers answered calls about abused animals in New York City. As *Animal Precinct* became one of Animal Planet's most popular shows, *Animal Cops: Detroit* (2002), *Animal Cops: Houston* (2003), *Miami Animal Police* (2003), and *Animal Cops: San Francisco* (2005), followed with more of the same in different settings.[42] Other shows mimic widespread TV trends, such as the repopularization of the talent show in *Pet Star* (2002–).

The National Geographic Channel

In the 1990s, televisual representations of animals proliferated not only on The Discovery Channel and Animal Planet, but also from a new competitor, the National Geographic Channel. The same economic advantages that spurred the wildlife genre's surge in the cable boom of the late 1980s played roles in the globalization of the genre, as did, importantly, the assumption that its content is culturally nonspecific—that is, that nature (and representations of nature) transcend ideology. According to Animal Planet General Manager Clark Bunting, "There is interest in the Siberian tiger whether you're in America or Cuba or Czechoslovakia [sic] or anywhere in-between . . . It [wildlife programming] crosses borders and media platforms better, and more economically, than any other form of TV."[43] Here, Leo Braudy's argument concerning generic "portability"—or the production of stories for an American population so diverse that these stories also appeal to audiences in other nations—and its role in facilitating the globalization of Hollywood film is pertinent to this context, in which TV executives point to wildlife as universally appealing, when packaged with only the minor modifications to narration.[44]

The assumed universal appeal of nature and wildlife themes (and related science and adventure themes) shaped National Geographic's decision to develop a 24-hour cable channel featuring nature, science, exploration, and adventure programming, which was announced not long after DCI and the BBC made public their plans for the global joint venture

Animal Planet.[45] National Geographic enjoyed advantages that distinguished its channel from most upstarts: a partnership with NBC that eased access to global TV markets, as well as brand equity revered by media consumers. This rested primarily on the circulation of the *National Geographic Magazine* (published continuously since 1888) and on a long-time television presence;[46] the *National Geographic Specials* aired on various commercial broadcast networks from 1964 to 1975, and then on PBS until 1994. In 1995, the nonprofit National Geographic Society reorganized its television unit as a for-profit, taxable enterprise to separate commercial activities from—and therefore protect—the tax-exempt status of its core operations.[47] National Geographic Television (NGT) promptly returned the *Specials,* after two decades on PBS, to NBC.[48] Some of its television work was already long ensconced in commercial territory: *National Geographic Explorer* first aired on Nickelodeon in 1985, then moved to TBS in 1986, and on to CNBC and MSNBC in 1999; by 2005, *Explorer* became part of the National Geographic Channel's own line-up.

The first joint launch by NBC and NGT, whose international division became known as National Geographic Channels Worldwide (NGCW), was the National Geographic Channel UK in September 1997, followed within a year by channels in Australia, Ireland, Finland, Poland, and throughout Scandinavia.[49] To secure widespread carriage, the NBC-NGT venture turned over a 50 percent stake in its U.K. and Scandinavian launches to News Corporation's British Sky Broadcasting (BSkyB).[50] In July 1998, NBC, facing disappointing ratings abroad for its situation comedies and talk shows, turned over some of its channels in Europe and all of NBC Asia to National Geographic. The revamped channels immediately reached 11 million subscribers in Europe and 7 million in Asia.[51] Some new cable launches were coordinated with the introduction of local-language editions of *National Geographic Magazine,* which previously had published only in English and Japanese.[52] These new editions quickly ranked among the most popular subscription magazines in Spain, Greece, Israel, and elsewhere.[53] In four years, the National Geographic Channel claimed to reach 160 million households in 160 countries worldwide.[54]

In light of the capacity of multinational media corporations to engage in a kind of cultural imperialism by flooding new markets with product that suits economic agendas more fully than it satisfies consumer demand, it is difficult to assess with certainty the degree to which their networks have been embraced by diverse populations worldwide. Thus, the question, not fully answerable here, remains: to what extent are television viewers in global markets exercising consumer sovereignty, and to what extent are they choosing among a constrained set of formats and contents imposed with little regard to local relevance? National Geographic's perspective on entry into global markets resembles Animal Planet executives' assumptions that Anglo-American aesthetics and values are universal and that the points of view of off-screen narration, on-screen hosts, and behind-the-scenes filmmakers are uninflected by their own cultural experiences. Thus, NGCW President Sandy McGovern expressed confidence in the venture's globalization project: "This kind of programming transcends borders. It's not considered American or British or whatever. Unlike general entertainment or comedy, it goes easily from country to county."[55] But for both Discovery and National Geographic, global TV launches required local tweaking. Following Discovery's lead in India, where National Geographic first launched in 1998, the channel airs in English, Hindi, and Tamil. But unlike its competitor, and also unlike other versions of the National Geographic Channel, this one takes India itself as the subject of one evening per week's programming. In 2001, National Geographic began to cultivate local production as a source of new content, calling for digital video submissions to the adventure series *Up For It* and entering into a joint venture with Singapore's Economic Development Board to assist young Asian filmmakers.[56]

In the United States, the National Geographic Channel faced a scarcity of open channels on which to launch. Eventually, after NGCW and NBC assigned a one-third stake in the U.S. venture to News Corporation's well-leveraged (via its widely distributed FX and Fox News Channel) Fox Entertainment Group, the National Geographic Channel found carriage on U.S. cable systems early in January 2001. While the channel foregrounded the wildlife filmmaking with which it has long been associated, its schedule also includes other kinds of programming. Like Animal Planet, the National Geographic Channel has also added more playful variations on the genre, as in the Kratt Brothers' *Be the Creature* (2003–2005); it has also featured companion animals as subjects, as in *Dogs with Jobs* (2000–2004) and *Dog Whisperer* (2004–). In addition, historical, scientific, and cultural themes round out the National Geographic Channel schedule. Some series have veered into sensationalistic territory: *Taboo* (2003–), for example, explores controversial or marginalized social practices in episodes focusing on death, sex, tattooing, piercing, pain, and more. Other series use scientific approaches to cultural phenomena, as in *Science of the Bible* (2005–), which applies forensic archaeology to the Bible; and still other series seek to expand the boundaries of science, as in *Is It Real?* (2005–), which explores paranormal phenomenon from "Bigfoot" legends to UFO sightings.

Not content to contain its media activities to a single cable channel, National Geographic Television's investments sprawled. In 1996, National Geographic Television became a partner in Destination Cinemas, which co-produces and distributes large-format films and operates a chain of IMAX theaters. NGT itself began to co-produce large-format films, beginning with the dramatized *Mysteries of Egypt* (1998), a record-setting hit for the IMAX format, grossing nearly $100 million; this was followed by *Lewis and Clark: Great Journey West* (2002), *Roar: Lions of the Kalahari* (2003), and *Forces of Nature* (2004), among others. National Geographic's feature film division invested in *K-19: The Widowmaker* (2002, directed by Kathryn Bigelow), and partnered with Warner Independent Pictures to buy North American distribution rights to the French documentary *March of the Penguins*; the film became the surprise hit of 2005, grossing $70 million within three months of release.[57]

Conclusion

Shifting paradigms in the wildlife genre demonstrated the heated competition among television networks in the 1980s and 1990s. While commercial networks had little use for wildlife programming after the syndication boom of the early 1970s, as channel capacity expanded, audiences fragmented, and advertising revenue dispersed, the genre—slow to go out of date, relatively inexpensive, recyclable in a variety of contexts—fulfilled network needs for a relatively inexpensive product with which to fill their schedules, amidst a proliferating number of channels competing for not only viewers but also advertisers' dollars. Nature migrated with ease from its prestigious but small presence on PBS to its strategic and regular deployment by The Discovery Channel. Eventually, in efforts to develop new networks with coherent brand identities, niched audiences, and global reach, Discovery teamed with the BBC to launch Animal Planet, a round-the-clock source of representations of animals; National Geographic, a competitor in natural history and adventure programming, partnered with NBC and News Corporation subsidiaries to found its own cable channel. Following industry trends, these leading producers and outlets for wildlife programming did not constrain their activities to domestic markets, but sought new revenue streams in recently privatized markets worldwide. Intrigued by the genre's successes and seeking to cut programming costs, even broadcast networks and cable channels not previously amenable to nonfiction animal genres experimented with cheaper reality-based series, miniseries, and specials and turned at times to animals, as in *The Trials of Life* (1991, TBS), *The World's Most Dangerous Animals* (1996–1997, CBS), *When Animals Attack* (1996–1997, Fox), *Man vs. Beast* (2003, Fox), *Miracle Pets* (*Animal Miracles,* 2001–2004, PAX TV), and *Wildboyz* (2003–, MTV).

During this period of recommercialization and globalization, The Discovery Channel defied industry wisdom that an all-documentary format could not succeed. Discovery programming choices were applauded by some and decried as a crass commercialization of its educational mission by others. In its early years, industry observers noted that the network allowed critical analyses (especially on environmental issues, as in Michael Tobias's 1989 *Black Tide,* on the Exxon Valdez oil spill) that were usually absent from programming dependent on advertising, corporate sponsorship, or government funding.[58] Later, some would be led by programming trends to condemn some of Discovery's programming as tarnish on its reputation. According to *Broadcasting & Cable* Editor-in-Chief Harry A. Jessel,

> TV is awash in UFOs, alien abductions and just about every other para-normal phenomenon you can think off [sic]. But one of TV's greatest unexplained mysteries . . . is why such programming dreck has found homes on networks like Discovery and The History Channel. . . . [Recently] Discovery aired an hour on the afterlife (with zombies!) that careened so violently between fact and fiction that it made you want to question every claim on every show on the network.[59]

Jessel's concern—that The Discovery Channel had veered away from documentary veracity and toward cheap reality-based formats, foregoing the "edu-" in the old "edutainment" model and the natural world for the supernatural and the speculative—would be echoed in criticism of its high-end collaborations with the BBC, such as *Walking With Dinosaurs* (1999), that filled in gaps in the evolutionary record, or the irretrievable absence of knowledge about the behaviors of extinct animals, with fictions masked as fact.[60] Likewise, Animal Planet hybridizes traditional wildlife and pet-care formats with all manner of other TV genres, trading instruction for action (and, whenever possible, a contest with winners and losers): game shows, talent shows, police-chase shows, court TV, talk shows, and psychic readings, each looped around the clock and aired across the globe.

Similarly, by the third season of the documentary competition that the National Geographic Channel had launched in Singapore, the call for submissions suggested a shift in focus away from National Geographic's conventional approaches and toward "disaster . . . riddles of the dead . . . killer instincts . . . What we're not looking for: factual educational storytelling."[61] But for every gore- and gagfest (such as Shark Week and *Dirty Jobs,* which first appeared in 2003 as a few specials exploring professions whose workers must encounter all manner of excrement, vermin, and hazardous materials, from sewer inspectors to coal miners, returning as a hit special in 2005–2006), there is also a thoughtful consideration of historical or current events (*With All Deliberate Speed; Global Warming: What You Need to Know,* a 2006 Discovery Channel Special). And some of the most provocatively promoted titles contain factual glimpses of natural and social phenomena and even provide useful information (consider, for example, Discovery's sibling channel TLC's nutrition-focused *Honey, We're Killing the Kids!,* which premiered in 2006), albeit at times cloaked in melodrama, blood, and guts. Once a darling of industry executives eager to claim cable's worth in the public interest, The Discovery Channel, with National Geographic on its heels, has competed for fractions of ratings points with spectacle and sensationalism. Is any other means possible? Thus far, amid competition for viewers in advertising-supported media, other strategies are not on the television horizon.

NOTES

1. For an extended discussion of the wildlife film and television, see *Watching Wildlife* by Cynthia Chris (Minneapolis: University of Minnesota Press, 2006); this chapter is derived from and extends material in chapter 3.

2. "Twenty Years of Discovery Quality" (interactive timeline), http://corporate .discovery.com/brands/discoverychannel.html (accessed August 30, 2004); Kelly Barron, "Theme Players," *Forbes* 163, no. 6 (March 22, 1999): 53.

3. Richard Tedesco, "Discovery Looks to Build a Programming Empire," *Cablevision,* June 18, 1990, 22.

4. Michael Wilkie, "'100 Cabooses, Two or Three Engines,'" *Advertising Age* 66, no. 13 (March 27, 1995): S4.

5. Mark Lewyn, "John Hendricks: The Conscience of Cable TV," *Business Week,* August 31, 1992, 67.

6. "Discovery Channel Gets Four Backers," *Broadcasting* 110, no. 26 (June 30, 1986): 32–33.

7. Simon Applebaum, "Business Strategy Refinements Spur Prospects at The Discovery Channel," *Cablevision* 12, no. 1 (September 29, 1986): 29; Lewyn, "John Hendricks," 68.

8. Laura Landro, "Four Companies to Buy Cable Learning, Gaining Control of Discovery Channel," *Wall Street Journal,* February 27, 1989, B4.

9. "Discovery Buy," *Broadcasting* 111, no. 6 (August 11, 1986): 14.

10. "Discovery Channel Gets Four Backers," 33; Applebaum, "Business Strategy Refinements Spur Prospects at The Discovery Channel," 29.

11. Tedesco, "Discovery Looks to Build a Programming Empire," 22. TCI and Time Warner acquired about 17 and 15 percent, respectively, of Black Entertainment Network (BET) before its 1980 launch; see "A Walk on the Acquisition Side with TCI," *Broadcasting* 111, no. 5 (August 4, 1986): 70–71.

12. In February 1989, United and most original investors sold their shares to TCI, Cox, and Newhouse; Simon Applebaum, "MSOs Consolidate Grip on Discovery Channel," *Cablevision* 13, no. 13 (March 13, 1989): 19–20. Discovery's ownership consortium remained stable for ten years, with TCI/Liberty Media holding a 49 percent share; Cox and Advance/Newhouse each holding not quite 25 percent each; and Hendricks in control of an undisclosed figure just under 3 percent. In March 1999, three years after the Telecommunications Act of 1996 lifted the 1970 prohibition on cross-ownership of local telephony and cable systems, AT&T acquired TCI in a merger valued at nearly $60 billion; see Price Colman, "AT&T-TCI Merge Starts New Era," *Broadcasting & Cable* 129, no. 11 (March 15, 1999): 39.

13. See Federal Communications Commission, First Report: Annual Assessment of the Status of Competition in the Market for the Delivery of Video Programming, in *FCC Record 9*, no. 25 (Washington, DC: Federal Communications Commission, 1994), 7599; see also FCC, Eleventh Annual Report: Annual Assessment of the Status of Competition in the Market for the Delivery of Video Programming (released February 4, 2005, MB Docket No. 04–227), 146.

14. Michael Bürgi, "Cable's Promised Land," *Mediaweek* 5, no. 13 (March 27, 1995): 28.

15. By the 1995–96 season, Discovery produced or co-produced 400 hours of material, a figure more than doubled to 952 hours for 1996–97. See Jim McConville, "Discovery Boosts Originals," *Broadcasting & Cable* 126, no. 18 (April 22, 1996): 48.

16. Tedesco, "Discovery Looks to Build a Programming Empire," 22–26; Penny Pagano, "Catch of the Day," *Cablevision* 16, no. 22 (May 4, 1992): 75.

17. On Shark Week, see Stephen Papson, "'Cross the Fin Line of Terror': Shark Week on the Discovery Channel," *Journal of American Culture* 15, no. 4 (Winter 1992): 68.

18. Kate Fitzgerald, "At Discovery, It's Survival of the Fittest," *Advertising Age* 66, no. 48 (November 27, 1995): 20, 22.

19. Mark Robichaux, "Hunger for Mayhem? TV's Nature Shows Offer You a Big Bite," *Wall Street Journal*, August 7, 1995, A1.

20. Robert La Franco, "Actors Without Agents," *Forbes* 157, no. 7 (April 8, 1996): 83.

21. Louise McElvogue, "Running Wild," *Los Angeles Times*, February 13, 1996, D1.

22. Ibid.

23. La Franco, "Actors Without Agents," 82–83.

24. Ann Japenga, "It's a Jungle Out There," *Los Angeles Times*, April 11, 1991, F5; Patricia O'Connell, "Outlets Bloom for Features," *Variety* 86, no. 7 (September 27, 1993): 76.

25. Matt Stump, "Cable Ready to Go Back to School," *Broadcasting* 119, no. 10 (September 3, 1990): 38; and Nicholas Garnham, *Capitalism and Communication: Global Culture and the Economics of Information* (London: Sage, 1990), 43–44.

26. Dan Schiller, *Digital Capitalism: Networking the Global Market System* (Cambridge, MA: The MIT Press, 1999), 203.

27. Discovery executives reported to the press that entering the global marketplace required trimming content that might violate religious or dietary customs; finding translators for dubbing and subtitling in local dialects; and adapting market research techniques to social and discursive practices; see Wayne Walley, "Programming Globally—With Care," *Advertising Age* 66, no. 37 (September 18, 1995): 1–14.

28. Discovery Networks Asia added Animal Planet in 2000, which quickly became one of the five most-watched networks in some Asian markets; see Gerald Chuah, "Discovery Remains Asia's No. 1," *New Straits Times*, December 17, 2002, 4. A spin-off, Discovery Travel and Adventure (DTA), also rapidly gained viewers' attention; see Samuel Lee, "73% of S'pore Elite Watch Cable TV," *The Straits Times*, January 25, 2003, http://www.lexisnexis.com (accessed January 21, 2003). See also Samuel Lee, "Wright Way to Travel," *The Straits Times*, October 25, 2002, http://www.lexisnexis.com (accessed January 21, 2003).

29. Harsha Subramanian, "Discovery's Not Just Nature and Wildlife," *Hindu Business Line*, August 29, 2004, http://www.blonnet.com/catalyst/2002/07/18/stories/2002071800010100.htm (accessed August 29, 2004).

30. Michael Bürgi, "Through the Wall: Discovery Channel Places a Season's Emphasis on China," *Mediaweek* 5, no. 24 (June 12, 1995): 32.

31. John Sinclair, *Latin American Television: A Global View* (Oxford: Oxford University Press, 1999), 43, 165, 167.

32. Ibid., 20; Stephen D. McDowell, "The Unsovereign Century: Canada's Media Industries and Cultural Policies," in Nancy Morris and Silvio Waisbord, eds.,

Media and Globalization: Why the State Matters (Lanham, MD: Rowman and Littlefield, 2001), 119.

33. Meredith Amdur, "Canada Names 10 New Cable Service Licensees," *Broadcasting & Cable* 124, no. 24 (June 13, 1994): 28.

34. Rich Brown, "New Frontiers for Discovery," *Broadcasting & Cable* 123, no. 40 (October 4, 1993): 38.

35. Helge Rønning and Tawana Kupe, "The Dual Legacy of Democracy and Authoritarianism," in James Curran and Myung-Jin Park, eds., *De-Westernizing Media Studies* (London: Routledge, 2000), 163. Learning Centers have also been established in Mexico, Peru, and Romania; see Global Education Partnership, Projects in Action: Action Map, http://www.discoveryglobaled.org (accessed December 8, 2004).

36. The digital channels launched in 1996 were Discovery Science, Discovery Kids, Discovery Home & Leisure, and Discovery Civilization. Discovery Health, Discovery Wings, Discovery en Español, and BBC America, a joint venture with the BBC, followed in 1998. Meanwhile, DCI began to acquire equity in Paxson Communication's Travel Channel in 1997, a takeover completed in 1999, and in the fall of 2002, NBC filled its two-and-a-half hour Saturday morning day part with Discovery Kids; see Lisa de Moraes, "NBC Hands Discovery Keys to the Kid-dom," *Washington Post*, December 7, 2001, C7; "NBC Rents Saturday Morning to Discovery," *Pittsburgh Post-Gazette*, May 9, 2002, C6.

37. John M. Higgins, "Start-ups Burn DCI's Cash," *Broadcasting & Cable* 128, no. 13 (March 30, 1998): 53. The practice of launch support originated in 1995, when Rupert Murdoch's News Corporation offered cable systems $13.88 per subscriber to add the Fox News Channel to their lineups; see John M. Higgins, "Discovery's Big Stretch," *Broadcasting & Cable* 129, no. 4 (January 25, 1999): 106. Now commonplace, the practice hiked the cost to launch—and therefore, the barriers to entry—for new cable programming services.

38. Warren Berger, "A Family Channel That Even the Pets May Enjoy," *New York Times*, March 1, 1998, 36.

39. Carlos Grande, "Three's Company for Animal Planet," *New Media Markets* 16, no. 46 (December 17, 1998): 1. BBC Worldwide operates as a separate for-profit division from which profits are returned to the parent company, compensating for its decreased access to public funds, cut during the Thatcher era, and increased exposure to competition; see "Come In, the Water's Lovely," *Economist* 346, no. 8049 (January 3, 1998): 63; and Edward S. Herman and Robert W. McChesney, *The Global Media: The New Missionaries of Corporate Capitalism* (London: Cassell, 1997), 167–68.

40. See Discovery Communications, Inc., Businesses & Brands: U.S. Networks: Animal Planet, http://corporate.discovery.com/brands/animalplanet.html (accessed August 29, 2004).

41. Jim Rutenberg, "Carnival of the Animals: On TV, It's Raining Cats and Dogs, Not to Mention Pot-Bellied Pigs," *New York Times,* June 29, 2000, E4.

42. Neal Koch, "As Police Shows Thrive, Cable Grabs Its Share," *New York Times,* December 25, 2002, E4.

43. Ray Richmond, "Critter TV Not Just for Cable Anymore," *Variety,* June 30– July 13, 1997, 27.

44. Leo Braudy, "The Genre of Nature: Ceremonies of Innocence," in Nick Browne, ed., *Refiguring American Film Genres: History and Theory* (Berkeley: University of California Press, 1998), 280. See also Walley, "Programming Globally," 1–14.

45. Louise McElvogue, "National Geographic, NBC Form Cable TV Partnership," *Los Angeles Times,* December 5, 1996, D4.

46. Ibid.; Paul Farhi, "Big-Game Hunting," *Washington Post,* December 14, 1998, 13; David Ignatius, "Geographic: Exploring Renewal," *Washington Post,* May 17, 1999, A19.

47. "National Geographic to Shed TV Unit," *Los Angeles Times,* August 1, 1995, D2.

48. Alan Bunce, "National Geographic Specials Migrate Back to Network TV," *Christian Science Monitor* 87, no. 38 (January 20, 1995): 13.

49. Hugo Davenport, "Does Late Launch Matter If Your Brand Is Strong?" *New Media Markets* 15, no. 42 (November 20, 1997): 5–6; Michael Katz, "Geo, Carlton Sign Production Deal," *Broadcasting & Cable* 128, no. 17 (April 20, 1998): 66.

50. Lawrie Mifflin, "NBC Europe and Asia Channels to Carry National Geographic," *New York Times,* August 21, 1998, D8.

51. Ibid.

52. Steve Donahue, "It's a Jungle Out There: A Big Year for National Geographic," *Electronic Media* 17, no. 42 (October 12, 1998): 1, 14.

53. Paul Farhi, "Mapping Out a Greater Society," *Washington Post,* January 27, 1997, business sec., 12–14; Ignatius, "Geographic," A19.

54. National Geographic Channels International, http://www.nationalgeographic.com/channel/intl/index.html (accessed September 3, 2004).

55. Ibid.

56. "Are You Up for It?" *Hindu,* June 24, 2001, http://web.lexis-nexis.com (accessed January 23, 2002); David Boey, "National Geographic, EDB Set Up Fund for Film-making," *Business Times Singapore,* December 19, 2001, 6.

57. Martha McNeil Hamilton, "Mapping New Territory; National Geographic Society Adds to Revenue with Wider Licensing of Its Logo," *Washington Post,* June 25, 2001, E1; see also Farhi, "Mapping Out a Greater Society," 12–14; Constance L. Hays, "Seeing Green in a Yellow Border," *New York Times,* August 3, 1997, F12; Dorren Carvajal, "Compared with Their Filmmakers, Penguins Have It Easy," *New York Times,* September 28, 2005, E2.

58. Larry Jaffee, "Plugged In Producers: A Guide to Working with Cable Net-

works," *The Independent* 14, no. 5 (June, 1991): 26–28. Alexander Wilson observed such elisions in National Geographic's treatment of grizzly habitat loss, which failed to name its causes or imagine its reversal, apparently for fear of offending sponsors; see his *The Culture of Nature: North American Landscape from Disney to the Exxon Valdez* (Cambridge, MA: Blackwell, 1992), 142–43.

59. Harry A. Jessel, "Fusing Fact and Fiction," *Broadcasting & Cable* 131, no. 10 (March 5, 2001): 15.

60. See Karen D. Scott, "Popularizing Science and Nature Programming: The Role of 'Spectacle' in Contemporary Wildlife Documentary," *Journal of Popular Film and Television* 31, no. 1 (Spring 2003): 29–35.

61. National Geographic Channel Asia website, NGCI-EDB Documentary Fund 2004 Submission Guidelines, http://www.ngcasia.com/asiafilmmaker (accessed August 29, 2004). National Geographic also continued to obtain independent documentary fare through its All Roads Film Project, a festival launched in the fall of 2004 to feature the work of "indigenous and underrepresented minority-culture filmmakers"; see "National Geographic All Roads Film Festival Premieres in Los Angeles and Washington, D.C., in October," press release, September 8, 2004, http://press.nationalgeographic.com/pressroom/ (accessed October 2, 2004).

Chapter 7

Tunnel Vision and Food
A Political-Economic Analysis of Food Network

Cheri Ketchum

In the 1980s, television executives would have probably scoffed at the idea of a channel devoted entirely to food. Cooking shows were relegated mostly to public television channels or the occasional weekend afternoon slot of a commercial station's offerings. Though programs like Julia Child's had been successful and were relatively cheap to produce, they weren't considered prime-time material. However, in 1993, Food Network was launched with the hope that there was room for this type of niche programming in the expanded channel offerings of cable. In the beginning the network struggled. But through its selection and creation of sexier, more adventurous, and more polished programming, it was a success by the end of the decade and appealed to some of the most attractive television audience segments—the young and the wealthy.

One way the Food Network accomplished this was through honing in on "lifestyle" programming, which involved carefully highlighting middle-class and upscale consumer fantasies 24 hours a day. The money to do this came from Scripps, which took majority control in 1996. Investing millions of dollars in new programs in the late 1990s, the firm saw Food Network as a crucial component in its plan to create a lifestyle television empire. Since then, by carefully negotiating the boundary between infomercials and conventional television, Food Network has become an identifiable brand that is used to attract both viewers and customers. However, not surprisingly, the network fails to address any social issues around food or concerns over nutrition, safety, and ethics.

In this essay, I examine some of the political and economic underpinnings of Food Network in an effort to provide a better understanding of how media discourses pertaining to food in cable television are constrained by the requirements of capital. Though there is a great deal of political-economic research that addresses how capital hinders expression, very little academic work has addressed the commercial underpinnings of lifestyle-oriented channels like Food Network. This network is also interesting because it was fairly exceptional in the cable business in its initial plan to rely almost exclusively on advertising revenue (instead of subscription dollars from cable companies), to provide most of its own content, and to abandon its original narrowcasting plan. Through presenting data on Food Network's business maneuverings and programming choices, I show how economic factors have shaped, and limited, this network's offerings.

Food Frights

In order to make informed choices about food, the public must hear debates regarding the social and environmental implications of such choices. Accordingly, food issues have been prominently featured in the media. In 1963, Rachel Carson first published *Silent Spring,* a book that documented the harmful effects that the pesticide DDT had on humans and wildlife. Public outcry led to a ban on DDT in the United States in 1972. Seventeen years later, in 1989, a *Sixty Minutes* segment on the dangers of Alar was followed by a 20 percent decline in apple sales.[1] The apple industry blamed the public's concern for a $50 million loss of revenue.[2] According to a 1988 survey conducted by the Food Marketing Institute (a Washington-based trade association), 95 percent of those surveyed claimed to be concerned about chemical residue in their food.[3] In the 1990s, reports on mad cow disease scared some people away from beef, at least temporarily.[4] More recently Eric Schlosser sold millions of copies of *Fast Food Nation,* his exposé on the fast food restaurant industry,[5] and the documentary *Supersize Me,* which focuses on the health implications of eating at McDonalds, was one of the most successful documentaries of all times.[6] The popularity of these two recent works demonstrates that food is an important topic for a media system to address in terms of both satisfying public interests and serving public needs for information. And public concern about the U.S. food supply is persistent. A 2004 Roper poll found that 70 percent of those surveyed believe that exposure to pesticides is risky.[7] Confirming these fears, the EPA ranks pesticides as the third leading cancer risk, among the 24 hazards that it regulates.[8]

In addition to prominent fears over possible food contamination, there have been ongoing concerns about the health risks of a high-fat, high-calorie diet. By the early 2000s, many nutritionists and media outlets were claiming obesity was an American epidemic.[9] Marion Nestle, among other scholars, compares new political struggles around food to the debates around tobacco in the 1990s.[10] This analogy is bolstered by government statistics that show weight-related illnesses are just behind deaths linked with tobacco. According to one study, in 2000, obesity-related illnesses caused as many as 400,000 deaths, more than 16 percent of all deaths.[11] In comparison, tobacco use killed 435,000 people, 18 percent of all deaths.[12]

Less on the public radar, but equally important, are concerns related to global environmental destruction and workers' exposure to harm in agricultural production. Environmental concerns include worries over the destruction of basic resources like air, soil, and water, and chemical risks to both humans and animals.[13] Even less publicized are the often deplorable working conditions for those in the food and agricultural industry, especially for migrant laborers. Each year, between 10,000 and 20,000 farm workers are poisoned by pesticides,[14] and in 1997 farm workers suffered from 140,000 disabling injuries.[15] Outside of the fields, jobs in the meat-packing industry are among the most dangerous in the nation, with over 40,000 injuries per year.[16] In dairy production, workers suffer chronic strains, repetitive stress syndrome, lower back pain, and asthma, which is caused by exposure to noxious fumes.[17]

Within this context of both public concern about food safety issues and a problematic food production system, Food Network was launched. The network's programming systematically ignores any critiques of the U.S food system in favor of celebrating food consumption. Food Network is a case of a wider trend of the mass media's avoiding serious conversations about the environmental risks or ethics of contemporary capitalism.[18]

Political Economy and Democratic Media

This analysis is grounded in the theoretical work of critical political economy. Vincent Mosco explains that political economists are generally interested in studying control and survival in social life and maintains that control processes "are broadly political in that they involve the social organization of relationships within a community."[19] Peter Golding and Graham Murdoch contend that any analysis should begin by looking at the organization of property and production as the foundation for all cultural production.[20] Mosco further notes that political economists who study communication analyze historical forces, seek to understand a social totality and embrace moral philosophy (or making claims about the social good).[21] Though sometimes denigrated by other scholars for being economic determinists, Mosco insists that political economists see economic factors as just one of many forces that shape social life and cultural production.[22]

There is a long history of communication research that has studied the impact that corporations and sponsors have on the production of content. In relation to news, NBC's parent company GE has been accused of pressuring the news divisions' coverage of the first Gulf War.[23] And an ABC journalist has accused Disney of rejecting stories critical of Disney's labor practices.[24] In another case, Fox news journalists in Florida were asked to change their critical story about the potential dangers of the genetically engineered recombinant bovine growth hormone (rBGH) given to cows to increase milk production after the journalists received flak from corporate heads and Monsanto, the company that produces the hormone.[25]

These cases are examples of how the pursuit of profits and the avoidance of controversy can affect news. However, the type of instructional and lifestyle-oriented programming offered by Food Network and other cable channels (including The Learning Channel, The Travel Channel, or Discovery) has not been comprehensively analyzed for its political-economic foundations or the programming that results from its commercial commitments. Critical analysis is needed not only in hard-hitting news, but also in mundane discourses about cooking dinner. Ideologies also play out here, even though such daily practices are often seen as apolitical.

Like all commercial television networks, Food Network has considered the desires of advertisers from the outset. Part of the new network's appeal to investors was its plan to draw in sponsors with restaurant reviews, nutritional information, and fitness news—all themes with identifiable consumers.[26] Though this is part of a long pattern of profit-seeking in U.S. television, today's multichannel cable world has created even more pressure to appeal to an attractive, well-defined, consumer market and, consequently, more attention to demographics. Such concerns influence a network's ability to form alliances with more powerful industry players, its programming choices, and its brand identities. They also define the audience only as consumers. As a result, in the case of Food Network, it has been difficult to raise issues around food that might engage the audience as citizens—citizens who could be concerned about the risks and ethics of food (or other commodity) production.

As early as the 1980s, cable networks began to promote their audiences to sponsors based on the quality of viewers as much as or even more so than the quantity of viewers. Such early attempts at targeting a narrow set of consumers in a multichannel environment meant there would now be important similarities between the newer forms of cable television programming and the older media of magazines in terms of conceptualizations of audiences and content selection. In their social history of advertising, William Leiss, Stephen Kline and Sut Jhally contend that in the early 1970s, when magazines began to develop more highly segmented formats, content was produced in an effort to create a lucrative environment for the reception of advertisements and only well-defined consumer groups were deemed worthy of targeting.[27] Consequently, content became even more bland and uncontroversial in an effort to create pleasant feelings in the audience and to make ads seem even more attractive and exciting. For their part, advertisers likewise became less interested in specific ideas than in formats and stylistic techniques because they, too, wanted to establish a particular mood rather than heightening interest in engaging content.[28] Television also saw a strong push to better mesh content and commercials, which is a key part of Food Network's market plan.

In 1999, the then-president of the Food Network, Eric Ober, said he wanted to do "*all* things food"[29] (emphasis added). However, programming choices indicate that the network has a narrow definition of "all things." In fact, reflecting the general tone, Sara Moulton, host of the Food Network show *Sara's Secrets* and executive chef of *Gourmet* magazine, said of food, "Let's face it. It's not controversial. It's about the only thing in our lives right now that isn't."[30] But food is not necessarily inherently uncontroversial. Rather, it is network executives who have made programming choices that ensure that food is rarely controversial on Food Network. And this strategy, moreover, is related to the marketing of goods. Although the *network* does not sell goods directly to the public in its programs, as this would bring the taint of transforming show into infomercials, it subtly refers viewers to its website, and more recently its shopping channel, where goods are sold directly.

Food Network illustrates both common and uncommon trends in the cable industry. Like other cable ventures, the network was developed by a well-established television insider, Reese Schonfeld—founding president of the Cable News Network (CNN)—who understood the importance of identifying a clear consumer market. Media industry connections meant the network was able to attract the capital necessary to grow. But Food Network was also unique in its initial plan to get by solely on advertising funding, to produce the majority of its programming, and to mix previously disparate television genres to attract both "foodies" and others. This includes mixing cooking instructions with live talk show, game show, or travel genres. Finding that a lucrative market niche could be achieved by targeting a variety of audiences, including those from the middle and working classes and men as well as women, Food Network has also been able to attract sponsors that range from José Olé frozen entrees to the Olive Garden and Mercedes Benz. Through these programming choices, the network cultivated new audiences for food-related programming and created new ancillary economic opportunities for itself. This was a sharp contrast to its early years of reliance on conventional instructional cooking shows.

The History of Food Network

There is a long history of cooking shows in both radio and television, but most have been produced for non-commercial stations because the major networks did not see them as moneymakers. On the other hand, public stations found these shows attractive because they were inexpensive to produce and food-related sponsors were readily available. The biggest breakthrough for television cooking shows occurred in 1963 with the unexpected success of an unlikely celebrity chef: a middle-aged woman named Julia Child who cooked French food. The success of Child's program, *The French Chef*, on PBS provided two important lessons. The first was that cooking shows could attract a male audience. Jeremy Iggers claims that before Child, gourmet cooking was thought to involve the "odd fellow, effete, a bit of a snob, and almost always a bachelor."[31] After, it could be for the "masculine" man too. Indeed, Child's friendliness, as well as her willingness to reveal her own mistakes and improvise, made her endearing to a wide audience. The second important realization, arising from the fact that her featured items would frequently sell out in stores the day after her shows aired, was that the show could sell cookbooks, food products, and cooking supplies. The success of Child's show led PBS affiliates to develop and/or air other cooking shows, including *The Frugal Gourmet, The Galloping Gourmet*, and *Yan Can Cook*.

While PBS was airing several cooking shows, the commercial networks primarily used them as filler between other shows up until the late 1970s. Then, in the early 1980s, many local affiliates began running syndicated cooking shows during the daytime or weekend hours. Networks were probably attracted to this programming because of its low cost and usefulness in fulfilling FCC requirements about airing local and/or educational programming. But it was not until the increased demand for content and target audience segmenting brought about by the rapid expansion of cable in the 1980s that cooking shows became more of a mainstay of commercial television.

Seeing a potential gap in domesticity-themed programming, the early Food Network first attempted to reproduce the conventional cooking show genre. With relatively low budgets for programming and the potential to sell advertising time to food-related businesses, executives felt they could make a profit from developing content for what they considered to be an underserved market niche—upscale women. But early on the network had difficulties attracting viewers. Food Network's history demonstrates how the search for capital encouraged the creation of a programming environment that celebrated the pleasures of consumer culture more broadly.

In the early 1990s, a representative from the Providence Journal Company approached Schonfeld about developing a cable network devoted entirely to food.[32] When asked if it made sense to him, Schonfeld responded yes and that it worked because "food and packaged goods is [sic] the largest advertising category. Of the country's top 100 advertisers, 45 are food-related."[33] Eric Schlosser claims that by 2001, the fast food industry alone spent about $3 billion annually on television advertising.[34]

The Providence Journal Company, a somewhat diversified media firm (with holdings in newspapers, cable television systems, broadcast television stations), shopped the idea around and was able to attract a range of other media investors, including the Chicago Tribune Company, Scripps-Howard, Continental Cable, and Landmark Communication.[35] These companies provided capital in exchange for equity in the company. Providence would air the network on its cable system (Colony Communication) and make deals with other multiple-system operators (MSOs) to get Food Network into households. A central strategy was to offer to provide the network for free if the cable system agreed to air the network to 80 percent of its subscribers.[36]

Food Network was launched in November 1993. The owners anticipated spending between $50 and 60 million before the network turned a profit, which it hoped to do only in its fourth year—1997.[37] Such a delay, as well as substantial multi-year losses, are expected in starting up a new network, and they underscore both the necessity of big backers and the ability to take enormous financial risks to become competitive in the cable industry. Thus, as other authors in this volume point out, economic and other barriers to entry make it difficult for new independent cable channels to succeed.

After a year, things did not look good for the network. Though it reached between 10 and 11 million cable subscribers in 1995 and landed a national advertising contract with a grocery chain, overall ratings were dismal, and advertisers were turned off. That year cable analysts were questioning its viability, saying that cable was already filled with niche programming and that not all these channels could succeed.[38] But its prospects greatly changed with the phenomenal success of a cook who would eventually be known simply as "Emeril." His success prompted the network to change its narrowcasting strategy and to attract younger people and males.

In 1996, just prior to Emeril's rise to superstardom, Food Network hired a new programming director, Erica Gruen, to transform the network's offerings.[39] Soon after signing on, she launched its website, foodtv.com, which is frequently promoted on the channel. In terms of programming, she sought to acquire and create innovative food media. Her attitude, as she put it, was to "dump the has-been chefs. Out with the talking-head-behind-a-stove. Stop lecturing viewers and entertain them!"[40] For Gruen, programming about food needed to be light and funny and to engage with viewers in ways that would connect food to consumer "lifestyles."

Instead of just running cooking shows, Food Network would henceforth concentrate on two areas—programming that was instructional and programming that would frame food as part of living a comfortable "good life."[41] Despite Emeril's success, the future network president Judy Girard contended that only a small percentage of the audience was interested in cooking shows.[42] Therefore, Food Network would take people into fantasy worlds of vicarious consumption, where everyone had access to nice food and beautiful surroundings, and it accomplished this goal by improving its production values and focusing more on what it called "category television," a term I elaborate on later in this essay.[43]

Multichannel News described the changes in a 2004 article likening the transformation of the network to the remodeling of a kitchen, where linoleum floors are replaced with marble. Eileen Opatut, senior vice president of programming, told *Multichannel News* that in the late 1990s, the network began emphasizing "stories" about food and finding hosts that audiences could "latch onto." As early as 1997 it began investing in programming to get hosts out of the kitchen and hiring producers who had done work on channels like Discovery, A&E, and the History Channel. Opatut said that the network now shoots all of its programs like sports events. Even if something is in a frying pan, she wants "to get [it] into that picture."[44]

Such changes required funding, which came from a takeover by EW Scripps (formerly Scripps-Howard). Though ratings were still poor in 1997, Scripps saw Food Network as an important element in a new strategy the company was adopting—namely, cornering the market for lifestyle and home-improvement programming. An early investor in Food Network, Scripps bought Home and Garden Television (HGTV) and sought majority control of the food channel in 1996. Then, in 1997, Scripps made a deal with A. H. Belo Corp., a company that had taken over the Providence Journal Company, to acquire 56 percent of Food Network. In exchange, A. H. Belo would be given two television stations and $75 million.[45] By 2003, Scripps owned 69 percent of Food Network, with the Tribune Company owning the remainder.[46]

Following this deal, the network took advantage of rules passed in the 1992 cable act, which required that cable systems make arrangements with local TV stations to carry their signals.[47] Scripps owned ten broadcast television stations, covering 10 percent of the national market.[48] In exchange for cable systems offering HGTV, these stations would deliver Scripps' broadcast network's feed to the cable system for free. Scripps recruited 54 other television stations to barter the retransmission for carrying HGTV. In exchange these stations would receive a percentage of HGTV's advertising revenue.[49] Scripps soon asked MSOs to offer both HGTV and Food Network as a package. Its size and holdings gave it an advantage here.

By the early 2000s, Scripps was trying to create an easily identifiable and highly specific viewer and an advertiser-friendly media brand devoted to food, calling its narrowcasting strategy "category television."[50] According to Harry Jessell, this meant that Scripps began attempting to "inhabit" the category of home and lifestyle as a media genre and capitalize on it as much as possible in a process that has further involved building companion websites and developing merchandising. Seeing HGTV as a way of tapping into a huge group of sponsors in the do-it-yourself home-improvement industry, which had previously advertised almost exclusively in magazines and newspapers, Scripps pursued its efforts to build a media empire devoted to selling domestic lifestyles—efforts to which the acquisition of Food Network contributed. At that point, the target audience for the network was both the wealthy and those interested in acquiring the trappings of wealth through buying various middle-class and upscale commodities or learning about food.[51]

In 1999, executives at Food Network had made a calculated decision to target a wider audience, which would range from lower-middle-class to upper-class females *and* males.[52] Their programming directive also changed, becoming *personality*-driven, rather than *cuisine*-driven, as this would establish a more intimate relation to the audience.[53] Although the avoidance of controversial issues was not listed as a rule, given the content offered, it does seem as though food controversy were frowned upon in favor of happy talk about food, cooking, and dining. Before 2004, there were no programs that addressed health issues, and to this day few address any risks in food production or consumption. Even the network's documentaries, which as a genre are more derivative of news programs, usually simply celebrate famous food and restaurants.

In 2001, Food Network announced plans for several new shows that were part of their efforts to make chefs, especially those who are male, seem "cool." This strategy was designed to target males under 35. A new tagline, "taste the adventure," was used to classify a set of programs that showcased various food "voyages," including, for example, those of chef Anthony Bourdain, author of *Kitchen Confidential,* who would go out in search of "extreme cuisine" from around the world. In addition to making the network safe for men, this new programming was also edgier and more entertainment-oriented.[54]

After years of struggle, Food Network is now a financial success. In 2001, distribution grew by 31 percent to 71.5 million and ratings increased 26 percent.[55] Its ratings were up another full 5 percent between 2003 and 2004, with growth of 18 and 15 percent among adults 18–49 and 25–54, respectively.[56] In 2004, the network attracted an average of 550,000 households specifically during prime time each day, which was a 16 percent increase from the previous year.[57] Today, its estimated reach is 85.6 million.[58]

Even with this somewhat small viewership, Scripps' combined cable stations are ranked fifth (behind Discovery Communications, Turner, ESPN and Fox) by cable systems in terms of their "contribution to revenues and profitability."[59] Along with HGTV, Food Network brought $204 million in profits to the parent company in 2003, which was up 64 percent from the previous year.[60]

Food Network is also now an easily recognizable cable brand. David Everitt explains that as more channels are available, overall network satisfaction becomes more important than an individual program's ratings.[61] The network has ranked number one for having the most liked hosts and on-air personalities.[62] In terms of quality, it was ranked in people's top-five favorite cable programming services.[63] Robert J. Thompson, Director of the Center for the Study of Popular Television at Syracuse University, claims that "Food Network has been more successful than anyone else at creating a varied lineup centered around a single topic."[64]

The Food as Consumer Lifestyle Network

The channel has also become closer to the early goal conceptualized by Reese Schonfeld, wherein "the editorial looks like the ads and the ads look like the editorial."[65] This might be good for advertising revenues, but it does not allow for any critical commentary in the programs. Even an Emeril fan, interviewed in the trade press, admitted that the show often seems like an infomercial.[66] In fact, the network appears to be attempting to occupy a space between infomercials and "reality" programming. Fearing people would turn off the channel if Emeril were to start selling knives outright, it instead directs viewers to its website, where products can be bought directly. In the future, Food Network might direct people to its shopping channel.

According to Louis Chunovic, Food Network grew up with the idea that food equals cooking.[67] Now, however, food equals lifestyle, and the network has changed its orientation accordingly. Insofar as lifestyle involves "a discourse of the self related to certain kinds of consumption," according to Carolyn Voight, it creates a narrow sense of identity,[68] and this is the effect of Food Network's attention to lifestyle. As Adam Rockmore, the network's vice president of marketing in 2002, put it, the network sought to attract viewers "who might not know a wok from a waffle but who just might tune in for cuisine-themed travel shows like *Follow That Food* or *A Cook's Tour*."[69] There was no sense that viewers might be interested in hearing social or political information about food.

While being fairly innovative in their programming choices and advertising campaigns, Food Network and Scripps have also attempted to branch out into other areas and diversify their revenue streams. The network's companion, ad-supported website is constantly promoted on the show and has become very successful. Between the shows themselves and the website, where viewers log on afterwards to download recipes or chat with chefs, viewers engage in what the Cabletelevision Advertising Bureau calls a "converged experience."[70] Moreover, when viewers log on seeking further instruction, they see a range of banner advertisements and promotions for a variety of celebrity chefs and products. According to Scripps, the website, foodtv.com, now averages 20 million page views and 1 million unique viewers each month, making it one of the more popular online sites.[71] While people go to foodtv.com mostly for recipes, they also encounter a "marketplace," where the network sells kitchenware, apparel, and food, as well as cookbooks written by its hosts. Recently, the network has also begun providing instructional videos on demand[72] and has developed plans to launch magazines.[73]

Through the network, Scripps has been able to better realize synergy. The company has asked its broadcast television networks to produce content related to food, travel, and home decorating that could be repurposed for the cable stations, and vice versa. Scripps also recently launched two other cable networks—Fine Living and Do It Yourself (DIY)—both of which were heavily promoted on both Food Network and HGTV and became part of Scripps' proclaimed attempt to be "the leader in lifestyle programming." By 2003, after just two years of being launched, DIY had reached 26 million homes and Fine Living 20 million homes.[74] That same year Scripps announced that it would begin airing DIY programming on its broadcast networks in an attempt to attract new viewers to the network.[75] The company claimed to have great success with its promotions for HGTV on these same stations.

In 2002, Scripps also acquired a 70 percent controlling interest in a home shopping network called Shop At Home.[76] Then, in 2003, Scripps bought the channel outright for $285 million.[77] The company planned to make 90 percent of the goods "lifestyle-related" and to rely heavily on Food Network chefs to promote products that would be available on the channel.[78] Scripps hopes that Shop At Home will allow the company to build a television commerce business, which will provide "innovative solutions for marketers."[79] It will also cross promote all of its ventures and sell goods on all of its channels.

Because of this focus on creating programming that will generate sales of featured goods, there are very few opportunities to speak about contemporary food issues. The network claims to be creating food programming from "every conceivable angle, including nostalgia, travel, trivia, competition, contests, and more."[80] However, "more" does not include any critical programming, which would allow for a diversity of voices to address issues related to contemporary food production and consumption. Though surveys mentioned earlier show that much of the public is concerned about a myriad of food issues, airing such concerns, which could alienate the food industry and other sponsors as well as potential online customers, would become a liability to the network. Without any public service commitments, it can instead focus on realizing profits in new ways.

In addition to producing various revenue streams for Scripps, the network has also helped transform the popularity, and the financial opportunities, for up-and-coming-chefs. Celebrity chefs on Food Network cheerfully present their idealized lives and make money from selling their goods.[81] For example, Jamie Oliver, host of *The Naked Chef*, sold more than 80,000 copies each of his first and second books and endorses T-Fal cookware.[82] Emeril sells cookbooks, cookware, and a line of spices and prepared sauces and endorses over 150 products.[83]

Rachel Ray, host of *30 Minute Meals*, *$40 A Day*, and *Inside Dish with Rachel Ray*, fully embodies the new philosophy of celebrity chefs, wherein cooking is just one component of her larger marketing campaign. For example, as Allison Romano notes, *30 Minute Meal* viewers bombarded the network with inquiries about what knives Ray used;[84] subsequently, the network made a deal with Ray and Wüstof, the knife manufacturer, to sell the products on foodtv.com. Current Food Network President Brooke Johnson said that this is a perfect example of how the network wants to operate,[85] and now it is making plans to develop more merchandise with its talent. Lisa Ekus, a public relations executive, claims that television chefs are "hot" media items because "consumers can see them, taste their food, feel like they are really getting something from them."[86] This situation is just what Food Network had wished for: an environment where programming itself becomes a type of commercial—for food, for lifestyles, and for the celebrity chef who represents and promotes both.

Conclusion

Overall, Food Network demonstrates what strategies and programming choices were made to assure that the channel became viable. First, it had to come up with large amounts of capital from big media firms and build a set of identifiable consumers who were attractive to advertisers. Second, it became part of a larger conglomerate, Scripps, which leveraged its ownership of both broadcast stations and HGTV to extend its reach on cable systems. Third, the network altered the conventions of older food media to reach a "young, modern, urban" audience. This audience was assumed to be apolitical and concerned only with experiencing a pleasant environment or being entertained. Encouraging viewers to obtain and use a wide range of consumer goods and promoting restaurant experiences to confer symbolic status on individuals, programs served the interests of the network, its sponsors, and consumer society more generally.

Food Network is also an excellent example of how a network can be part of a conglomerated corporation (Scripps) and help it achieve synergy. Food Network produces over 90 percent of its programming, with minimal costs, and Scripps mines both it and its broadcast networks for new shows to use across its empire. Through the Food Network, Scripps also controls a network that has one of the most useful websites related to a television station because it provides desired information: full recipes, which you cannot really get while watching the programs themselves. With the additional purchase of a shopping channel, Scripps will be in a good position to sell branded goods directly to the consumer over both cable and telephone wires.

While Food Network's business and programming strategies have been effective at building a viewing base and the network has become profitable, its success has come in part from ignoring the negative elements of our food production and consumption, as well as critical perspectives on both. Consumers and citizens need to be aware in order to make informed choices in terms of what is safe to consume and understand the practices their choices support. Instead, the network builds a vision of a utopian consumer culture where we all eat together, no one really works or cleans, no one gets sick, and no one suffers the consequences of our food industry, which is increasingly destructive to human health and the environment. To be sure, it is hard to imagine a cable channel that would be critical of consumption when such a channel must become part of already existing networks that must promote products and consumption to survive.

Nonetheless, the success of critiques of the food industry such as the film *Supersize Me* and the non-fiction book *Fast Food Nation* demonstrate that there is a viable market and demand for this type of content. The question is, does fear of alienating advertisers keep commercial networks from providing this kind of programming?

NOTES

1. Phillip Shabecoff, "Apple Industry Says It Will End Use of Chemical," *New York Times*, June 2, 1989, A1.

2. Ibid.

3. Nancy Harmon Jenkins, "Nutrition and the Young Chefs," *The New York Times*, April 16, 1989, Section 6, 51.

4. In a much-publicized episode of *Oprah!* host Oprah Winfrey said that mad cow disease had "'stopped me cold from eating another burger.'" See David Usborne, "Oprah's biggest beef; BSE was the subject," *The Independent* (London), June 10, 1997, 2.

5. Eric Schlosser, *Fast Food Nation: The Dark Side of the All-American Meal* (Boston: Houghton Mifflin, 2001). The book was on the *New York Times'* paperback best sellers for over 100 weeks straight. See "Paperback Best Sellers," Book Review Desk, *New York Times*, September 26, 2004, 28.

6. According to the website Box Office Mojo, *Supersize Me* is the fourth highest-grossing documentary of all time. See http://www.boxofficemojo.com/genres/chart/?id:documentary.htm.

7. Organic Consumers Association, "Poll Finds Americans Prefer Family Farms," May 3, 2004, http://www.organicconsumers.org/organic/roperpoll050704.cfm (accessed November 11, 2004).

8. Kerry Hannon, "Pure and Unadultered," *U.S. and News and World Reports* 129 (May 15, 1995): 86.

9. See Amanda Spake, "Hey, Maybe It's Not a Weakness. Just Maybe . . . It's a Disease," *U.S. News & World Report* 138 (February 9, 2004): 50; and Peg Tyre, "Getting Rid of Extra Pounds," *Newsweek* (December 8, 2003): 62.

10. See Marion Nestle, *Food Politics: How the Food Industry Influences Nutrition and Health* (Berkeley: University of California Press, 2003).

11. Associated Press, "Diet and Fitness: Obesity Nearly as Deadly as Tobacco in United States," MSNBC.com (March 9, 2004), http://msnbc.msn.com/id/4486906/ (accessed November 9, 2004).

12. Ibid.

13. On resource issues, see Manuel Altieri, "Ecological Impacts of Industrial Agriculture and the Possibilities for Truly Sustainable Farming," *Monthly Review*

50 (July/August 1998): 60–71. On risk to humans and animals, see Theo Colburn, Diane Dumanoski, and Jonathon Peters Myers, *Our Stolen Future: Are We Threatening Our Fertility, Intelligence, and Survival? A Scientific Detective Story* (New York: Penguin, 1997). Colburn et al. point to research that has made links between exposure to the agricultural chemicals and intersexed offspring, low fertility rates (often due to low sperm counts), deformities, and death in both animals and humans.

14. National Institute for Occupational Safety and Health, "Pesticide Illness & Injury Surveillance," http://www.cdc.gov/niosh/topics/pesticides/ (accessed September 11, 2004).

15. Iowa Farm Safety Council, "From the President: A Rearview Mirror Look at Safety," The Council's Chronicle: http://www.abe.iastate.edu/Safety/PDF/Ccv3_3 .pdf (accessed September 11, 2004).

16. Schlosser, *Fast Food Nation*, 172.

17. Rebecca Clarren, "Got Guilt?" *Salon* (August 27, 2004), http://archive.salon .com/news/feature/2004/08/27/dairy_farms/ (accessed September 1, 2004).

18. See David Edwards, "Can We Learn the Truth about the Environment from the Media?" *The Ecologist*, January–February 1998, 18–22; Charles Anderson, "Missing the 'Big Story' in Environment Coverage," *Nieman Reports* 56, no. 1 (Winter 2000): 45–47; and Morton Mintz, "The Sound You Hear Is Silence," *Nieman Reports* 54, no. 2 (Summer 2000): 60.

19. Vincent Mosco, *The Political Economy of Communication: Rethinking and Renewal* (Thousand Oaks, CA: Sage, 1996), 26.

20. Peter Golding and Graham Murdock, "Culture, Communications, and Political Economy," in James Curran and Michael Gurevitch, eds., *Mass Media and Society* (New York: Arnold, 1991), 11–30.

21. Ibid., 17.

22. Mosco, *The Political Economy*, 70.

23. Robert McChesney, *Rich Media, Poor Democracy* (Urbana: University of Illinois Press, 1999), 52.

24. Ibid.

25. David Croteau and William Hoynes, *The Business of Media: Corporate Media and the Public Interest* (Thousand Oaks, CA: Pine Forge Press, 2001), 74.

26. Anonymous, "Leach to Host Celebrity Cooking Show," *Houston Chronicle*, July 28, 1993, Food, 1.

27. William Leiss, Stephen Kline, and Sut Jhally, *Social Communication in Advertising: Persons, Products, and Images of Well-being* (London: Routledge, 1997), 114.

28. Ibid., 122.

29. Kim McAvoy, "Food Network," *Broadcasting & Cable* 129, no. 42 (October 11, 1999): 42.

30. Anonymous, "Rethinking Food TV: Bright personalities, world travel and high production values make for a reinvented food network," *Multichannel News* 24, no. 46 (November 24, 2003): 8A.

31. Jeremy Iggers, *The Garden of Eating: Food, Sex, and the Hunger for Meaning* (New York: Basic Books, 1996), 29.

32. Rod Granger, "New TVFN Net Food For Thought," *Multichannel News* 14, no. 16 (April 19, 1993): 20.

33. Anonymous, "Leach to Host Celebrity Cooking Show," 1.

34. Schlosser, *Fast Food Nation*, 47.

35. Richard Katz, "TV Food Moves to Improve Programming and Lures Ads," *Multichannel News* 15, no. 47 (November 21, 1994): 51.

36. Christopher Stern, "Television Food Network Develops Strategy for Wider Carriage," *Broadcasting & Cable* 123, no. 23 (June 7, 1993): 50.

37. Rod Granger, "New TVFN Net Food For Thought," 20.

38. Ibid.

39. Scripps, "Scripps Cooks With Second TV Network" (updated July 27, 2002; cited August 15, 2002), http://www.scripps.com/shnews/fall97/covfeat/fall97cov.html.

40. Katrina Booker, "Selling Cookbooks to Non-cooks," *Fortune*, July 6, 1998, 35.

41. Mark Meister uses this phrase in his analysis of Food Network. See "Cultural Feeding, Good Life Science, and the TV Food Network," *Mass Communication and Society* 4, no. 2 (2000): 165–83.

42. John Higgins, "Food Preps New Menu," *Broadcasting & Cable* 130, no. 11 (March 13, 2000): 50.

43. Harry Jessell, "E. W. Scripps: Building, Growing with HGTV," *Broadcasting & Cable* 128, no. 9 (March 2, 1998): 22.

44. Anonymous, "Rethinking Food TV," 8A.

45. Donna Petrozello, "New Recipe at Food: Ober Replaces Gruen," *Broadcasting & Cable* 128, no. 46 (November 16, 1998): 81.

46. Yahoo Finance, "Tribune Company," http://biz.yahoo.com/ic/11/11508.html (accessed January 22, 2005).

47. Rich Brown, "Time Warner to Try 10-Channel Tier; Networks Hope Low Rates Will Buy Them a Place on Expanded Lineups," *Broadcasting & Cable* 124, no. 50 (December 5, 1994): 6.

48. Karissa Wang, "Scripps Helps Do It Yourself With Synergy," *Electronic Media* 21, no. 14 (April 8, 2002): 35.

49. Jessell, "E. W. Scripps," 22.

50. Ibid.

51. Donna Petrozello, "Food Net Varies Its Fare," *Daily News* (New York) (July 5, 1999): 75.

52. Jim Cooper, "Food Net Sets the Table," *Mediaweek* 9, no. 14 (April 5, 1999): 12.

53. Lewis Beale, "A Network Thrives on TV Dinners," *Daily News,* June 8, 1997, 6.

54. Stephanie Thompson, "The Biz: Extreme Cuisine at Food Network," *Advertising Age* 73 (February 4, 2002): 29.

55. Ibid.

56. Mark Berman, "Buzz Columnists: The Programming Insider," *Mediaweek* 14, no. 52 (December 28, 2004), http://www.mediaweek.com/ (accessed January 21, 2005).

57. Corie Brown, "Food Shows Are Making Chefs into Stars," *Valley News: The News Source of the Upper Valley,* July 28, 2004, http://www.vnews.com (accessed January 22, 2005).

58. One TV World, "2005 Cable Network Profiles: Food Network," http://www.onetvworld.org/cgi-bin/cab_profiles2.cgi?record=29&image.x=16&image.y=7 (accessed January 22, 2005).

59. Jack Myers, "Discovery Tops Cable Industry for Contribution to Revenues and Profitability," *Jack Myers Report* (July 17, 2003), 2, http://jackmyers.com/pdf/07-17-03.pdf (accessed January 22, 2005).

60. Scripps, "2003 Annual Report: To Our Shareholders," http://www.scripps.com/2003annualreport/shareholder_letter/ (accessed January 22, 2005).

61. David Everitt, "Ratings Don't Tell the Full Cable Story," *Media Life Magazine,* January 30, 2002, http://www.medialifemagazine.com/ (accessed November 9, 2004).

62. Cabletelevision Advertising Bureau, "Cable Network Profiles: The Food Network," http://www.cabletvadbureau.com/02Profiles/foodprof.htm (accessed April 17, 2003).

63. James Hibberd, "Ratings not key to satisfaction," *TelevisionWeek* 23, no. 50 (December 13, 2004): 7.

64. Honolulu Advertiser, "Cable Food Network a Scrappy, Profitable Success" (January 18, 2001), http://the.honoluluadvertiser.com (accessed August 1, 2005).

65. Richard Katz, "TV Food Moves to Improve Programming and Lures Ads," *Multichannel News* 15, no. 47 (November 21, 1994): 51.

66. Janice Okun, "Perpetual Motion Man: Nobody Kicks It Up a Notch Like Emeril," *Buffalo News,* September 22, 1999, Food Section, 1C.

67. Louis Chunovic, "Food Network Is Tweaking Its Recipe," *Electronic Media* 21, no. 5 (February 4, 2002): 7.

68. Carolyn Voight, C. *You Are What You Eat: Contemplations on Civilizing the Palate with "Gourmet" (Magazine).* Master's Thesis, McGill University (1997), 53.

69. Chunovic, "Food Network Is Tweaking Its Recipe," 7.

70. Cabletelevision Advertising Bureau, "Cable Network Profiles."

71. Scripps, "Doing it right where you . . . shop," http://www.scripps.com/annrpt/99/nofrills/themes/mainpages/shopmain.htm.

72. Daisy Whitney, "How-To Series Find On-Demand Home," *Electronic Media* 21, no. 4 (January 28, 2002): 10.

73. Jessell, "E. W. Scripps," 22.

74. Scripps, "2003 Annual Report."

75. Wang, "Scripps Helps Do It Yourself With Synergy," 35.

76. Andrea Lillo, "Merger Gives Scripps 100 Percent of TV Shopping Network," *Home Textiles Today*, January 5, 2004, 17.

77. Scripps, "2003 Annual Report."

78. Allison Romano, "Wants a Bigger Slice," *Broadcasting & Cable* 134, no. 36 (September 6, 2004): 10. Shop-At-Home is also available in some markets through a broadcast signal.

79. Scripps, "2003 Annual Report."

80. Scripps, "2003 Annual Report: Exploring New Angles on Three Squares," http://www.scripps.com/2003annualreport/ideas_in_motion/food_network.html (accessed January 22, 2005).

81. Natalie Danford, "The Way to a Nation's Heart . . . ," Book News, *Publishers Weekly* 248, no. 49 (December 3, 2001): 20.

82. Ibid.

83. Romano, "Wants a Bigger Slice," 10.

84. Ibid.

85. Ibid.

86. Andy Cohen, "Look Who's Cooking Now," *Sales and Marketing Management* 153, no. 12 (2001): 30–35.

Chapter 8

Target Market Black
BET and the Branding of African America

Beretta E. Smith-Shomade

> In the early years there were people who thought that
> Black Entertainment Television, with the emphasis on
> Black, was too narrow and was not a way for anyone to
> promote a business.
> —Robert L. Johnson, BET founder, *Celebrating*
> *20 Years: BET Black Star Power*

In November 2000, BET founder, President, and CEO Robert L. Johnson sold the then twenty-year-old Black Entertainment Television to Viacom, Inc. This sale hoisted Johnson into the billionaire's club while simultaneously eliminating one of the few black-owned media entities. Before America and its scholars had time to fully appreciate the impact of the sale and ponder its legacy, Johnson won his 2003 bid for the new Charlotte (North Carolina) professional basketball franchise. His establishment as the first black NBA owner swept virtually all critical attention away from BET to celebrate Johnson's inclusion (and by extension, all of black America) into the world's exclusive club.[1] Yet, the triumph, tragedy, and triviality of BET continues to exist as the unexplored space of black America's "American Dreams," which are rooted within the promises of capitalism, black entrepreneurship, and cultural relevance assumed with representation.

U.S. communications regulatory bodies (along with other areas of the federal government) have consistently paid lip service to the notion of public interest and public trust as an impetus for and by-product of black media development. From the 1970s, their overtures translated into several different FCC policies established to create equality and opportunity, particularly for women and people of color. Yet in 1995, Congress repealed the minority tax certificate program that gave tax benefits to media companies that sold to minority vendors. The 1996 Telecommunications Act further crippled minorities' ability to participate in ownership because it raised the maximum ownership allowable by a single owner. It also lifted several restrictions to open competition among the telephone, cable, satellite, broadcast, and utility companies—industries where African-American ownership exists only in limited cases.

As cited by the National Association of Black-Owned Broadcasters, 2005 found African Americans controlling only 16 of the 1,243 U.S. broadcast television stations, or just over 1 percent. And with the enormous costs in acquiring and maintaining technology, the trend of consolidation, and continued federal deregulation, this number is likely to decrease. Given this economic and technological scenario, BET has operated with virtually no competition. While several cable networks addressing the same demographic entered the market, including the World African Network in the mid-1990s, the Major Broadcasting Cable Network (changed to Black Family Channel) in 1999, and TV One in 2005, none has yet gained enough capital, cable space, and/or subscribers to effectively challenge BET.

BET emerged within a representational landscape that barely acknowledged African-American existence. The closest thing to national television programming targeted at blacks had been local Sunday morning community program slots (i.e., *Tony Brown's Journal*), a few black sitcoms, and *Soul Train*. Yet, Johnson's idea for BET followed in a fairly long tradition of African Americans valuing black information dissemination along several media fronts. While the name "Black Entertainment Television" expressed the network's intention, the company's marketing relied on several circulating discourses for support, including the legacy of the black press to expose white injustices upon blacks, the call for black business ownership, the diversity promise of the cable industry, and the view of representation as a sign of equality. Furthermore, Johnson's entrepreneurship and vision developed with knowledge of African-Americans' craving for representation and their assumption of capitalism's value for black communities. His 1980 promotional tape alerted cable operators and advertisers to the dearth in black visualization and promoted BET as both a much-needed amelioration and an ignored business opportunity.

Thus, for over twenty-seven years, Black Entertainment Television has positioned itself as the quintessential spot to reach African-American consumers. It sits as the hub of African-American cultural identity—presented as a destination of (only) choice for black folks wanting a televisual reflection. BET takes credit for enhancing the careers of many African-American singers, musicians, and actors. The growth and spread of hip hop has certainly helped propel this notion. Yet it is BET's particular practice of making black folks synonymous with consumer products that merits consideration and study. Examining the nexus of African Americans as both BET consumers and the consumed of cable television, this chapter addresses what it means to make and visualize blackness as a brand.

The Business of BET

On January 25, 1980, the beginning of the ninth decade of the twentieth century, Black Entertainment Television launched. In that same year, Ronald Reagan became the fortieth U.S. president; "ghetto-blaster" found its way into the English dictionary; the U.S. boycotted the Moscow summer Olympics; and Richard Pryor sustained third-degree burns from a cocaine mishap. By this time, the United States had entered a reactionary and conservative period where the radical blackness of the 1960s had become passé, and assimilation and advancement had turned into mantras for corporately mobile buppies and yuppies.

At the beginning, Black Entertainment Television served as a much-needed platform for forwarding black imagery and talent. But with this success, imitators began to address its market. By 1990, BET outlined its three major business imperatives. First, it aimed to become the dominant medium used by advertisers to target the black consumer marketplace. Second, it would become the dominant medium engaged in the production and distribution of quality black-oriented entertainment and information to cable television households. And third, it wanted to use the powers of the medium to contribute to the cultural and social enrichment of the network's viewing audience.[2] In Johnson's quest to position his company as the preeminent supplier of black Americans to advertisers, he targeted three areas for expansion: entertainment, information, and leisure products.

To that end, he created the networks BET on Jazz: The Cable Jazz Channel (changed to BET J) and partnered to create Action Pay-Per-View and BET Movies/Starz3. In terms of the printed word, BET Holdings acquired *Emerge* newsmagazine (later renamed *Savoy* and refocused more on entertainment and less on political or socio-cultural insight). The company introduced *YSB* (a magazine targeted at teens), *BET Weekend* (a monthly newspaper insert), and BET.com (in partnership with Microsoft). All three of the print publications no longer exist. In terms of leisure products, Johnson created the BET Sound Stage Restaurant in Largo, Maryland, to capitalize on the BET brand and expanded that venture to downtown Washington, D.C., as well as Disney World and Las Vegas. All of these ventures were constructed as co-productions with other corporations, thereby lessening BET's financial outlay, while allowing it to retain content control.[3] In addition, BET created a clothing line, a credit card, and a book-publishing arm—all in the name of expanding the brand.

Despite the many seemingly progressive expansions that can be seen as a sign of black business development, most criticism and complaints about BET center on the network's program offerings. Since their introduction to the network in 1981, music videos stand as its predominant programming staple.[4] Music videos come free to music networks from record companies—ready-made promotion for and exploitation of the labels' artists. In addition to airing music shows (programs built around music videos), BET also presents a comedy program, public affairs shows focused on news and somewhat on religion, and off-network comedies and dramas that feature black characters and black themes, as well as a few of its own original film projects added since 1999. Nonetheless, in 2006, nearly 70 percent of BET's programming consists of music videos. But rather than capitalizing on black artists' placement in American society as innovators, BET chooses instead to play follow the leader to its corporate mate, MTV.

In fact, some suggest that BET's programs now directly emulate those of MTV. So for example, *Where I Live* on BET resembles *Cribs* on MTV, and BET's *106 & Park* is similar to *TRL–Total Request Live* on MTV. BET's news programming, once more centered on African-American and African concerns, now resembles the news of the day (with a focus on celebrity news) as defined by majority media outlets such as CNN, the Associated Press wire service, and *The New York Times*. The main distinction is that BET's news is conveyed by a young, black anchor. Moreover, while BET canceled its long-standing news programs *BET Tonight with Ed Gordon* and *Lead Story,* along with its program *Teen Summit* in 2003, two years later, during its very first upfront presentation, BET reduced its news commitment to intermittent briefs—MTV style.[5] This is not progress. In

2006, BET stands as a company strengthened by the backing of corporate leviathan Viacom but programming no more innovative than it had prior to becoming a "member of the family."

In a statement about the sale to Viacom made in 2000, Johnson glowed: "the time is now where strong African-American brands with tremendous value should put themselves in league with strong general market brands that can add even greater value . . . I see this as a positive development . . . And I hope one day there will be other African-American companies that create the kind of value created at BET."[6] Organizations, including the National Association of Black-Owned Broadcasters, applauded the sale as a very positive development, while others called it an end of an era. Yet with its focus on branding it is the phrasing, not only with consumers but with black folk as consumer products, that requires critical re-thinking about the relationship between African Americans, capital, and culture.

So What Exactly Is Branding?

Branding exists as part of a corporation's marketing mix—an aspect of its business plan. For the past two decades, the mantra to brand has swept businesses across the country and world. The brand (the trade name, distinguishing image, symbol, or slogan) gives value and meaning to what is offered or sold. Marketer Jeremy Sampson says that a brand is ultimately "a relationship that secures future earnings by securing customer loyalty." Long-time marketing consultants Al Ries and Jack Trout contend that "[p]ositioning [or branding] starts with a product. A piece of merchandise, a service, a company, an institution, or even a person . . . But positioning is not what you do to a product. Positioning is what you do to the mind of the prospect. That is, you position the product in the mind of the prospect."[7] While many of Ries' arguments have been taken to task, this vision of marketing is still widely regarded and heralded in businesses and across marketing firms.[8] It also seems to resonate with the way BET has branded itself for advertisers and audiences.

Before BET, no Lifetime for women, no Nickelodeon for kids, no PrideVision for gays and lesbians existed anywhere in TV land. UHF stations boasted Univision beginning in 1961, but nothing existed on VHF or cable that dealt specifically and exclusively with the targeting of identity. To be sure, both CNN and MTV (channels that began around the same time as

BET) have proven that narrowcasting based on people's interests can be a profitable venture. BET, however, goes beyond narrowcasting to provide the means for specific forms of black identity politics and representation to receive validation. That is, BET finds and offers a method for reinscribing attributes of the slave trade within contemporary codes of image-making and profit. Let me explain.

BET targets demographic black. On the surface of this targeting thrust, it applies a seventeenth-century definition of blackness—a theory that one drop of blood can constitute one's identity. This colloquial logic considers anyone possessing an African ancestor, African American. In U.S. law, the idea was legislated to officially forbid race mixing. In fact,

a mulatto was considered to be of lower status than her White parent and was excluded from the White race and absorbed into the Black race . . . A statute passed by the Virginia legislature in 1662 . . . forty-three years after the first Africans arrived, shows the early importance of drawing broad boundaries around the Negro race. Undoubtedly in recognition of the fact that most interracial fornication occurred between White men and Black women, the law provided: [C]hildren got by an Englishman upon a negro woman . . . shall be held bond or free only according to the condition of the mother . . . Significantly, this law broke with the traditional English common law rule that the children follow the status of the father. Instead it provided that children born of a Black mother and a White father would follow the common law applicable to farm animals—the child would follow the status of the mother . . . [And as] early as 1705, the Virginia legislature, in a statute prohibiting interracial marriage, provided an ancestrally based, biological, mathematical definition of who was Black, to include the child, grand child or great grand child of a negro meaning anyone who was one-eighth Black.[9]

Such logic seems to also resonate in BET. Thus, visually, the extremely light-skinned, bi-racial Alicia Keys, as well as the never-going-to-pass Missy Elliott both find themselves illustrated as part of the racially "Black" part of the BET moniker. Many African Americans accept this characterization and identify, understand, and embrace all persons of African ancestry. As Julie Dash's character Esther (Rosanne Katon) suggests in the 1983 *Illusions,* "we can always tell." However, while this embracing of racialized visual signifiers of African America could be construed as progressive and inclusive, it actually validates the racist assumptions about identity inscribed in the one-drop rule and applies it across generations. Claiming anyone or anything associated with blackness as its own, BET goes further in taking up a racist visual (presumed bloodline) categorization of black folks and uses this as a point for selling to advertisers, as well as to consumers, whether such consumers identify as black or not.[10] Thus, in a way that is consistent with the construction of whiteness, BET uncritically reinscribes blackness as a state of physicality rather than as a specific culture forged by various contributors.

On a scale never seen prior, BET promotes and presents African Americans as a product. It sells black folks like any other merchandise—pop, detergent, or shoes. Cross-media promotions, variant product disseminations, and its insistence on claiming everything black as its purview make this move abundantly clear. African Americans become the customers consuming BET's music videos and *Bobby Jones Gospel* and the product to be consumed; they are both purchased by advertisers and displayed on cable. This type of target recognition finds support in various African-American communities as a sign of progress. But in her treatment of the construction of a far-flung U.S. Latino market in *Latinos, Inc.,* cultural anthropologist Arlene Dávila presents observations that bear equally on the making of one big essentialized black market: "We can never lose sight of the fact that consumers are not all conceived as equally rational and hence 'empowered' within a politics that prioritizes consumption; just as their needs are not considered worthy of attention by the powers that be. . . . Latinas are undoubtedly gaining visibility. . . but only as a market, never as a people, and 'markets' are vulnerable; they must be docile; they cannot afford to scare capital away."[11] The same can be said about African Americans and the programming of BET.

BET tapped into the contempt for and counteracted the failure to distinguish black folks by mass marketers. While BET now claims to target specifically 18–34-year-old African Americans—a very specific and narrow demographic, it has also aired targeted programs for both younger and older audiences. As Chairwoman and CEO Debra L. Lee states, "People think because we target African-Americans that we're narrowcasting, but the truth is, we can't narrowcast. We can't be MTV and only appeal to a particular age group, because we have all age groups. We have to be full service."[12]

Over the years, BET has crafted various marketing slogans to characterize the network and identify with its consumers. At the beginning of the twenty-first century, it was "Now, That's Black." In 2005, the brand phrase was "It's My Thing"—a phrase suggesting that BET was carefully avoiding making the logo "it's a black thing" so that it could overtly attract audiences beyond black ones. But a slogan that was developed in the late 1990s by vice president of Creative Services Veronica Hutchinson, and the one that has had the most resonance, is "BET, Black Star Power." This tag line implies that BET helps create successful artists and that its audience plays a role by supporting the network and the artists it features. But an observation by sociologist E. Franklin Frazier provides a different way of thinking about this relationship. It is that black artists' "prestige is [owed] partly to the glamour of their personalities, but more especially to their financial success, which is due to their support by the white world."[13] Thus, although BET's black target audience may contribute to the success of BET artists, questions concerning the role of the network's white financial base also arise. So, too, does the meaning of the phrase "Black Star Power." Whose definition of blackness is being offered up? What constitutes success and power in the racialized context by which BET and the larger world operate?

The "Black" of BET

African Americans on television represented a very small percentage of all people featured in television narrative when BET started. Before *The Cosby Show* (which began in 1984) and *The Oprah Winfrey Show* (1986), African Americans appeared almost exclusively in comedies (as early as the 1950s, in the *Amos 'n' Andy* and *Beulah* programs) and then only in limited amounts of these. Critical reports from various groups such as the United States Civil Rights Commission, the Annenberg School of Communication of the University of Pennsylvania, and the National Black Media Coalition dogged the television industry for its stereotypical representations of blacks and other minorities, as well as women. It was in this climate that Johnson, a former National Cable Television Association (NCTA) lobbyist and communications director for the Washington, D.C., chapter of the National Urban League, pitched his idea of creating a black cable network to John Malone of Telecommunications, Inc. (TCI), who bought the idea. With TCI as the financial underwriter, Johnson launched the network aimed at African-American audiences. One way to think about "Black Star Power" is that the entertainers important to African Americans receive validation and thus power on BET. Or perhaps, indeed,

BET possesses power due to its ability to showcase recognized and up-and-coming black stars. But since entertainers or black stars constitute a form of capital in Hollywood, as well as a means for profit well beyond, the slogan "Black Star Power" can also produce revenue for BET. Functioning as commodities, black artists can help put BET literally "in the black" financially.

On the other hand, the phrase "Black Star Power" may even pose a mild threat as it invokes the radicalness of the 1960s and 1970s Black Power movement, minus the star. From yet another, albeit related perspective, the branding could imply an "in your face" attitude to others—MTV, the big four award programs, the networks—that consistently render African Americans invisible, non-existent.[14] Whichever of these two readings is privileged, the slogan "Black Star Power" may encourage a certain black solidarity, but this solidarity is once again compromised by BET's financial underpinnings. Furthermore, both of these readings ignore the fact that Black Entertainment Television sells black bodies to advertisers. Transforming these bodies into consumers, BET and its advertisers then encourage these consumers to consume themselves by watching TV. In a paradoxical development, this scenario is being mirrored by other new cable entrants—such as ImaginAsian TV or JTV ("not just news, not just Jews")—wanting to target and market to specific identity demographics.

BET's programming has consistently failed to reflect the diversity of African-American culture in its catering to essentialized notions of blackness —notions that over time UPN and the WB (now combined as CW) have regurgitated well. Similar to the criticism hurled against the earliest white-produced programs featuring black characters, criticism of the programming of BET calls attention to the ways that the network panders to mainstream (and increasingly in-house) expectations of African Americans. The programming output of BET makes painful connections to the legacy of Eurocentric "ideology" pervading the behind-the-scenes control of this formerly black-owned entity. For example, BET offerings such as the talk show *Oh, Drama!,* the reality show *College Hill,* the music video program *Cita's World,* and most of the music videos it airs, especially the ones seen on *Uncut,* invoke a very specific and narrow classist aesthetic (lower socio-economic speech, dress, living conditions, goals), equating success with material acquisitions (and women) and reinforcing inane and male-centered presuppositions—even in programs led by women.[15] As sociologist Herman S. Gray argues, "black cultural expression [has come] to occupy some of the nooks and crannies of the global and domestic cultural marketplace . . . This new global structure still does not so much dictate the content of Black cultural production but rather it establishes the very terms within which such products (including those that are counter-hegemonic) are produced, financed, and exhibited globally."[16] Thus, this aspect of black living becomes the predominant one to illustrate all of black America. The production of such imagery exists within a specific duplicitous framework.

The notion that public trust is a foundation of media businesses is disputed by media scholar Laurie Ouellette. She asserts that in early policy formation, "the term minority was used in broadcast reform discourse primarily as a euphemism for educated white people with uncommonly sophisticated cultural tastes . . . race and class [were] for 'other people.' "[17] What has been profitable for the BET network fails to conform to the middle-class taste that the pubic trust relies on but certainly has exemplified capitalism's hunger for success. Given this interpretation, the goals of BET could only ever be profit—and invoking activist Malcolm X, by any means necessary. Furthermore, as representation continues to be hotly contested, I've argued elsewhere that quantity of images remains an issue despite the seeming plethora available.[18] Thinking about these conundrums leads to the area that BET has received only limited critical examination since its sale to Viacom, who is BET's target market now?

Target Market White? BET and the Rise of Hip Hop

If, as cultural theorist Stuart Hall argues, nothing meaningful exists outside of discourse—that meaning is derived from both what you expect to see and what is actually there, what reading, what interpretations are mainstream (a.k.a. white) audiences making of BET?[19] What is being added, searched for, or identified in the absence of a larger context? In other words, how is the "what is there" made up for "what is not"? As mentioned earlier, at any given time, BET both argues that it tries to serve all of African America and that it only targets the 18–34 demographic of music-buying black folks—with the position depending on which advertiser considers buying the network's programming. This duplicity of engagement with the black audience leads to a conclusion that narrowcasting for an ethnic audience may be only a first step in both defining a market and gaining a wider audience with common interests that advertisers like and want—even if that interest is in a specific ethnic group of people.[20] The maturation of BET has forced many to reconsider what the commodification of blackness means to the viewing pleasure and the satisfaction of whites, as well as to the construction of whiteness itself.

According to the BET website, African Americans constitute 78 percent of its audience while Caucasians account for only 18 percent. Yet increasingly, white viewers seem to be more of a priority for the network. In her work on BET as a supertext, media scholar Ayanna Whitworth-Barner argues that "BET facilitates the racial formation of the wigga by validating the nigga. On a macro-level the network has involved our [economic] structure, capitalism, to defend its creation of the wigga; on a micro-level, BET involves stereotypes that are already set in place [for] America to push the race envelope by exploiting both Blacks and whites."[21] The "wigga" is a white "nigga" or, in the twenty-first century, anyone white invested in supporting hip hop. In a 2005 focus group that consisted of predominately white students at the University of Arizona, some of their impressions of BET gave credence to this designation.

As one 18-year-old student said, "'I love rap. It's my favorite music in the whole world, and I know I'm the whitest white girl in the whole world. But I don't know why. I watch BET all the time . . . I love how it's like black right, black power.'" The students interviewed take BET's music focus and the imagery from music videos as the personification of black culture. For them, hip hop occupies the space of, the actual embodiment of, contemporary blackness—an embodiment forged with ideas of accessibility and co-optation or cooperation, which are integral to its articulation and definition.

Like BET itself, which was founded with financial support coming primarily from out-group sources, hip hop has paradoxically received most of its commercial valuation and validation from white folks, both from its inception and currently. Writer and now BET employee Nelson George notes, "The buppies of . . . black music departments of the early to mid-'80s and programmed radio stations were still putting time into Michael Jackson clones or the latest act from Minneapolis . . . They didn't understand, respect, or support hip hop."[22] BET began and operated with a black middle-class and conservative ideology. Hip hop, with its lower-class aesthetic and voice, existed outside of that parameter. And while this mindset emerged at the initiation of most black musical forms, it is ironic (or perhaps typical) that it held sway with black executives whom many people—black or otherwise—would expect to know better in the late twentieth century.

Nonetheless, as hip hop has grown, so has the BET network—in a definitive two-step dance. The de facto space for national visualization for hip hop—or at least for mainstream, media-friendly hip hop—became BET. Journalist Brett Pulley maintains that "BET became the center stage where black America could witness this profound cultural shift on a daily basis. Viewers began to regard BET as their very own outlet."[23] Beyond the nation, Pulley argues that BET put the insular musical form of rap on a world platform. Many black entertainers—even beyond hip-hop artists— agree with this assessment. In its twenty-year self-published reflection, *Celebrating Twenty Years: BET Black Star Power,* board of directors member Denzel Washington maintains that "BET has presented a complete picture of us to the world . . . whether in news or entertainment, BET has provided opportunities to create, to lead, to soar. It has proven that 'Black Star Power' exists on both sides of the camera."[24] Whether one agrees with this characterization of the network, it is clear that BET has both provided a consistent platform for and benefited from black music performance to an unrivaled degree.

Speaking more directly to the connection of BET to hip hop, pioneer and entrepreneur Russell Simmons maintains that BET was "critical to the expansion of hip-hop culture." He also adds, "You've got to look to them to preserve those art forms in their most honest forms . . . The early artists have had integrity in the way they deliver their art forms, and there is a lot of commercialism and influences that change it from its core. BET will protect its core."[25] Yet, suggests Whitworth-Barner, "In the case of music videos on BET, black star power is associated with fantasy, nostalgia, sex, and euphoria. The illusion of reality poses blackness as something or someone sexual, primitive or exotic and culturally rich and extremely desired."[26] This problematic construction of blackness in a vehicle financed by and at least covertly marketed to whites creates a scenario ripe for African-American feelings of resentment, disillusionment, and disempowerment to fester and grow.

A duplicitous relationship exists between the BET/hip-hop mogul and the audience/consumer and resonates throughout hip hop. BET and hip hop share the same impetus not only because, like BET, hip hop promises more but actually produces less—only a few Puffys, Jay-Zs, and Master Ps —but more critically, because of the plantation-like structuring of the music business itself. BET harvests the fruits of the mostly disenfranchised artists-laborers, without compensating them. Moreover, the BET network itself suffers from similar sorts of master-slave accusations in its practices of paying low wages, maintaining low budgets, and generally maintaining low expectations beyond accruing profits. This scenario is largely lost on its consumer-audiences.

BET taps into whatever the current popular thing is in African-American culture. In this time frame, it is hip hop. The network is certainly not alone in this. The spread of hip hop has been aided by corporations hawking their wares on the bodies of black and brown poor to gain white and suburban youth consumers. In *No Logo,* Naomi Klein argues that in the world of super brands, including Nike, Hilfiger, and DKNY, "cool hunting simply mean Black-culture hunting."[27] She cites Nike's aggressive strategy to illustrate this point: "The company has its own word for [observing and copying Black style]: bro-ing. That's when Nike marketers and designers bring their prototypes to inner-city neighborhoods in New York, Philadelphia, or Chicago and say, 'Hey, Bro, check out this shoe,' to gauge the reaction to new styles and to build a buzz."[28] Journalist Marc Graser adds, "Advertisers are only eager to leverage the power of hip-hop as a marketing tool and generate exposure for their brand among the music genre's young urban consumers."[29] This word-of-mouth, man-on-the-street phenomenon forwarded hip hop in the same way that BET spread and sought to position itself in the lives of African Americans.

The paradox of BET is that, like the music industry, BET has perpetuated both the making of a singular black market and a system that disregards those whom it attempts to address. Like hip hop, BET is now ubiquitous. In gyms, university cafeterias, clubs, barbershops, and malls, BET can be viewed writ large. In the most unlikely of places, you may encounter the bounce and the bust of a BET music video.[30] Yet as Gray reminds us, "Whites are the ideal subjects of consumerism and representation, while people of color are simply political subjects on whose behalf civil rights advocates must make special appeals for recognition and representation."[31] Or as a *Los Angeles Times* article surmises, "Networks can't afford to alienate whites, who make up the vast majority of potential viewers, and remain the ones advertisers privately concede they want most."[32] This desire, this longing applies equally to a network that targets non-whites.

Conclusion

BET's initial reluctance even to show music videos and then its slow movement into hip hop suggests that it capitalized on something different than what the larger corporate cable culture taps. In the twenty-first-century cable universe, we see more people of color, hear more disparate voices, and experience more global places than at any moment previously. Yet the mechanics of representation remain the same—capitalism that privileges individualism and white audiences who remain the ideal target. Increasingly, as media scholars deconstruct images and assign meaning to discrete phenomenon or programs, they move further away from interrogating an economic system that disempowers most individuals, including networks that replicate sameness and encourage stasis, and thus away from inciting activism against such systems and corporations.

By embracing and offering up hip hop and its populace as of the people and for the people, the BET network has effectively merged competing ideas of capitalism and social responsibility. Yet despite acknowledged representational limitations and problems, the value of Johnson's mainstream capitalist success outweighs any protests. Many believe still that salvation for African Americans and other people of color comes through their ability to produce and exhibit programming that reflects the concerns, issues, and arts of black American culture in its multiplicity. In this way, equality will become manifest. Others, however, such as television scholar Deborah Jaramillo, argue that people mistakenly assume that "economic viability equals cultural force" or conversely, that cultural acknowledgment accords power. As she contends, "Unfortunately, this conflation of economics and culture has been naturalized in the discourse surrounding greater visibility for ethnic minorities on television."[33] Making a broader point in "The Culture Industry: Enlightenment as Mass Deception," cultural critics Max Horkheimer and Theodor W. Adorno argue that the "truth that [media interests] are just business is made into an ideology in order to justify the rubbish they deliberately produce. They call themselves industries; and when their directors' incomes are published, any doubt about the social utility of the finished products is removed."[34] This reality speaks directly to BET—to the impetus for its founding and to the ways it continues to appeal to and through African-American audiences.

While some maintain that Johnson has been consistent about his focus on building a business rather than addressing social and cultural concerns, many others insist that he was duplicitous at best. For example, in the pages of *The Black World Today,* social and political activist Ron Daniels seethes: In "Youngstown, Ohio, some years ago, I distinctly remember being called by representatives of a start up venture called BET to mobilize Black people to demand that the local cable company carry the programming of this embryonic network. Out of a sense of racial pride and social commitment, we did just that, and I am certain that scores of communities across the country did likewise."[35]

Yet African-American feelings of betrayal and distrust have seemingly not hampered BET's success or changed its trajectory. For example, a testament to how white America regards BET's reach comes through two high-profile interviews, one with O. J. Simpson (adored by white media before the murder of his ex-wife), who granted the only interview about his acquittal to BET's Ed Gordon in 1995, and the other with Senator Trent Lott, who in January 2003 attempted to retain his majority speaker seat by also talking with Ed Gordon. Both hoped that by appearing on the channel that presumably *all* black people watch, they would find redemption.

Communication scholars Robert M. Entman and Andrew Rojecki's study of advertisements in *The Black Image in the White Mind* provides a key framework for imagining the future of Black Entertainment Television and upstarts modeling themselves on it. In it they show that even on a network dedicated to African-American representation and audiences, the preponderance of the advertisements, the real goal of television networks, feature white actors. This is a shift from what advertisers produced for black audiences in the past two decades. Since its move to Viacom, BET looks different—glossier and better produced. Many mainstream black actors, who in the past failed to appear there, have begun doing so. However for the most part, BET maintains a unified and definitive notion of blackness dependent primarily on music videos. As for developing original programming, the content is simply painted black. Its business plan seems set on assuming that the ways of whiteness are the keys to economic viability and growth. This model may prove to be an effective one in creating successful cable businesses focused on identity, but in terms of those audiences desiring alternative visioning spaces, the future appears extremely white-washed.

NOTES

The author thanks Kristin Baranack and Ayanna Whitworth-Barner for their research efforts and transcription prowess. She also thanks Salmon A. Shomade for his help in thinking through ideas developed here.

1. In its May 2003 issue, *Sports Illustrated* named Johnson the number one most influential person in sports—only two months after he purchased the team.

2. As found in Alice A. Tait and John T. Barber, "Black Entertainment Televi-

sion: Breaking New Ground and Accepting New Responsibility?" in Venise T. Berry and Carmen L. Manning-Miller, eds., *Mediated Messages and African American Culture* (Thousand Oaks, CA: Sage, 1996), 187.

3. Tariq K. Muhammad, "The Branding of BET," *Black Enterprise,* June 1997, 156–58.

4. Infomercials played a critical part of BET's programming as well—airing from 1:00 a.m. to 6:00 a.m. CST at least until 2002, when the network announced it was dropping them.

5. Networks sample their upcoming programming season for the advertisers, agency executives, and the media at presentations called "upfronts."

6. Robert L. Johnson, quoted in "Viacom-BET," *All Things Considered,* narr. Snigdha Prakash, National Public Radio, November 3, 2000. Johnson's dream came true with the sale of *Essence* magazine to Time, Inc., in 2005.

7. T. C. Melewar and Christopher M. Walker, "Global Corporate Branding: Guidelines and Case Studies," *Journal of Brand Management* (November 2003): 157; Al Ries and Jack Trout, *Positioning: The Battle for Your Mind* (New York: McGraw-Hill, 1986), 2.

8. See Dan Herman's "Al Ries Might Be Dangerous to Your Brand," where he takes Ries to task for outmoded thinking, http://www.themanager.org/Marketing/Al_Ries.htm (accessed April 18, 2005).

9. As found in Christine B. Hickman, "The Devil and the One-Drop Rule: Racial Categories, African-Americans, and the U.S. Census," *Michigan Law Review* (March 1, 1997): 1265.

10. In the "not" category, for example, although BET promoted and aired music videos of singer Mariah Carey since her emergence in the early 1990s, as a black woman, Carey herself claimed a sort of ambiguity of black heritage until the early twenty-first century. At that moment, she found herself on the heels of declining popularity and could make a comeback with the support of black audiences.

11. Arlene Dávila, *Latinos Inc.: The Marketing and Making of a People* (Berkeley: University of California Press, 2001), 238.

12. Allan Leigh, "All BET's Are On," *Billboard* online EBSCOhost (April 22, 2000), accessed September 25, 2004.

13. E. Franklin Frazier, *Black Bourgeoisie: The Rise of a New Middle Class in the United States* (New York: Free Press, 1957), 109.

14. A little clarification may be needed here. While MTV in particular airs many black performers and performances, these moments exist solely in the context of hip-hop. Thus, unless white audiences are already conversant with a particular popular and contemporary black musical artist, actor, writer, etc., such individuals will not be seen on MTV. So with MTV perhaps invisibility is not the case; rather, it is more like limited visioning or, as writer Ralph Ellison calls it, seeing with an "inner eye."

15. *Uncut,* a late-night music video program featuring explicit music videos, was canceled in July 2006 after a six-year run and extensive negative press.

16. Herman S. Gray, *Cultural Moves: African Americans and the Politics of Representation* (Berkeley: University of California Press, 2005), 30.

17. Laurie Ouellette, *Viewers Like You? How Public TV Failed the People* (New York: Columbia University Press, 2002), 146, 174.

18. See Beretta E. Smith-Shomade, *Shaded Lives: African-American Women and Television* (New Brunswick, NJ: Rutgers University Press, 2002).

19. Stuart Hall discusses these ideas in several spaces including the video *Stuart Hall: Representation and the Media,* Media Education Foundation, 1997.

20. I raise this conundrum in the article "Narrowcasting in the New World Information Order: A Space for the Audience?" *Television & New Media* 5, no. 1 (February 2004): 69–81.

21. Ayanna Whitworth-Barner, *The BET Super-text: An Analysis of Economics, Imagination and Identity of BET,* unpublished master's report, University of Arizona (Spring 2004): 12.

22. Nelson George, *Hip Hop America* (New York: Viking, 1998), 59.

23. Brett Pulley, "The Cable Capitalist," *Forbes,* October 8, 2001.

24. Denzel Washington, quoted in *BET: Celebrating 20 Years* (Washington, D.C.: BET Holdings, 2000), 7.

25. Russell Simmons, quoted in ibid., 26.

26. Whitworth-Barner, *The BET Super-text,* 15.

27. Naomi Klein, *No Logo* (New York: Picador, 2000): 74.

28. Ibid., 75.

29. Marc Graser, "McDonald's Buying Way Into Hip-Hop Song Lyrics," *Advertising Age* (March 23, 2005).

30. To think about the different spaces where television is now viewed, see Anna McCarthy, *Ambient Television* (Durham: Duke University Press, 2001).

31. Gray, *Cultural Moves,* 94.

32. Brian Lowery, Elizabeth Jensen, and Greg Braxton, "Networks Decide Diversity Doesn't Pay," *Los Angeles Times* (July 30, 1999).

33. Deborah Jaramillo, "Opening Up the Representation Discussion: Pay Cable, Minority Programming and the Capitalist Paradigm," unpublished manuscript, 5.

34. Max Horkheimer and Theodor W. Adorno, "The Culture Industry: Enlightenment as Mass Deception," in Meenakshi Gigi Durham and Douglas M. Kellner, eds., *Media and Cultural Studies: KeyWorks* (Oxford: Blackwell Publishers, 2001): 71–72.

35. Ron Daniels, "The Demise of Emerge and the Ethics of Black Capitalism," *The Black World Today* (June 29, 2000), http://www.tbwt.com/views/rd/rd_06-29-00.asp (accessed November 21, 2002).

Chapter 9

Monolingualism, Biculturalism, and Cable TV

HBO Latino and the Promise of the Multiplex

Katynka Z. Martínez

Home Box Office (HBO) is often described as a pioneer, a risk-taker, and even as a channel to which the title of "auteur" can be applied.[1] Such titles were originally given to HBO because it was one of the first premium cable networks to offer original programming. It also drew from genres that were not initially staples of other premium cable networks, such as live sporting events and comedy shows. While such nontraditional approaches to programming characterize the channel's early efforts at establishing an audience, HBO's relationship with its Latino audience is more rooted in traditional Hispanic marketing conventions. This essay provides a brief history of HBO's efforts to reach the U.S. Latino population through HBO Selecciones En Español, HBO En Español, and, most recently, through HBO Latino.

HBO Latino was launched in November 2000 while marketers were capitalizing on the popularity of Latino entertainers like Jennifer Lopez and Ricky Martin. During this time the marketing and entertainment industries were typically recognizing and promoting Latino pop artists at the expense of artists who did not fall under the pop format. Focusing on these artists and continuously invoking the younger generation of Latinos who consumed their music served as a shorthand device for referencing upwardly mobile class aspirations and ignoring large segments of the Latino population that were older, recent immigrants, non-middle class, and/or monolingual Spanish speakers. This essay positions HBO Latino within the context of the "Latin explosion" and discusses the extent to which HBO Latino approached and imagined its audience differently than did mainstream marketers. I also focus on how HBO's Latino-oriented programming, and HBO Latino in particular, challenge some of the maxims of Spanish-language television, such as the reliance on a non-nation specific construct of the Spanish language. However, an analysis of the few examples of HBO Latino programming for bilingual, bicultural viewers points to the limitations of the digital multiplex cable package, which claims to offer diverse programming that reflects contemporary U.S. demographics and addresses the interests of an ever-growing cable television subscriber base. HBO's expectation is that U.S. Latinos will patch together programs from HBO Latino and the main HBO channel and come away with an entertainment experience that reflects their bicultural experiences and speaks to their bicultural sensibilities. This essay will point to the limitations of this promise and also highlight the fallacy of equating consumer choice with control, especially as it relates to the current growth of "Latino-oriented" media.

A Channel for Latin America, SAP Instructions for U.S. Latinos

On September 14, 1988, HBO announced that it was on the verge of becoming the first television channel to offer a Spanish-language enhancement of selected programs. By January 1989, HBO had introduced Selecciones En Español to both its HBO and its Cinemax channels. Subscribers could tune in to Selecciones En Español and watch selected programs in Spanish via secondary audio programming (SAP). During this time, efforts were underway to expand HBO into Latin America and also to create an HBO channel for Latinos. HBO executives did not yet view U.S. Latinos as a profitable market and were therefore hesitant about creating an HBO channel that was directed at this demographic. Instead, they limited their courtship of this audience to providing instruction on how to use the SAP button. At the same time, HBO's foray into Latin America took off fairly quickly. In late October 1991, HBO announced plans for HBO Olé, Latin America's first 24-hour Spanish language pay television service.[2]

Megan Mullen has traced the ways in which media synergy assisted cable networks in gaining access to studio films while providing studios with a "massive distribution mechanism for its own productions."[3] This same structure applies to international cable ventures. In the late 1980s HBO's satellite transmission range, or footprint, covered the United States and regions south, up to and including Venezuela. However, HBO had film

rights to broadcast only within the United States. Most studios denied
HBO the rights to broadcast into Latin America. Nonetheless, HBO's sibling company, Warner Brothers—both companies are owned by Time
Warner, Inc.—agreed to give HBO international broadcasting rights to its
products. With access to this film library, HBO entered into a 50/50 venture with the Venezuelan media company, Omnivision. By 1991, the newly
created HBO Olé transmitted via the PanAmSat Satellite, PanAmSat 1,
with a footprint that extended from Mexico to Argentina.[4] Three years
later HBO segmented the Latin American market further and recognized
Brazilian culture and language as part of, but also distinct from, the Spanish-language Latin American experience. In 1994, Olé Communications,
HBO, Warner Bros., and Sony Pictures Entertainment joined forces with
Brazil's pay TV service, TVA, to launch HBO Brasil, a Portuguese-language
premium cable channel.

    While working as a consultant on HBO Olé, Concepción Lara began
creating proposals for an HBO project that would reach out to U.S. Latinos. She believed that the creation of a Latino channel was entirely possible given HBO's entry into Latin America and the emerging focus on multiplexing in the United States. Multiplexing allowed for many different
channels to be created from the same inventory of programs that had run
on the original HBO and Cinemax channels. While many new channels
were created through multiplexing during the 1990s, Lara recalls this period of media expansion as having been devoid of an ethnic-specific marketing orientation:

> This was in 1990, 1991, 1992. Before Ricky Martin. Cable operators in places
> like East L.A. or Bell Gardens or Pico Rivera would be pitched by the HBO
> people: "Put the multiplex." And the multiplex was very sexy so they would
> put five channels of HBO and that meant I could never get a channel on the
> air because they were all already running multiplexes. And I was like, man,
> these people should have a Latino channel.[5]

Lara's remark about a period "Before Ricky Martin" points to a distinct
epoch in the history of U.S. Latino participation in the mainstream entertainment and marketing industries. Martin's performance at the 1999
Grammy Awards ceremony is often viewed as the most recent juncture at
which Latinos began to receive recognition as producers of broadly popular (read: Anglo-friendly) entertainment. The English-language popular
press in the United States marks the "explosion" of Latin music as having

been set off by Martin's performance. However, the Recording Industry Association of America reports that U.S. Latin music sales grew 12 percent to 49.3 million units in 1998, compared with 6 percent for all music sales. I draw attention to this particular statistic because it refers to music sales that occurred before the popularity of Latin pop music presumably "exploded" in the United States (i.e., Ricky Martin and Enrique Iglesias). In 1999, the year of the "explosion" of Latin pop, Mexican regional music still accounted for approximately 60 percent of the U.S. Latin music market.

Although these statistics show that Mexican regional music was the top-selling Latino genre in the United States during the years of the "Latin explosion," Regional Mexican artists were rarely, if ever, highlighted in the U.S. English-language popular press. Instead, the focus was on the growing number of English-language singles that were making their way onto Spanish-language radio playlists. Lara's reference to East L.A., Bell Gardens, and Pico Rivera, three working-class Mexican immigrant and Chicano neighborhoods of Los Angeles, draws our attention back to specific geographic locales in which Regional Mexican music often fills the soundscape. Lara refers to these neighborhoods to emphasize that Latino audiences did, in fact, hold cable subscriptions in the early 1990s. This reality points to the existence of an audience and suggested to her that the creation of a channel that spoke to this audience was long overdue. At the time, however, HBO executives were not quite ready to accept the idea of U.S. Latinos (especially the fans of Regional Mexican music) being the "wealthiest Hispanics in the world."[6] Nor were they ready to make the programming investment that would recognize the presumed economic clout of this population.

Lara responded by focusing her efforts on contacting the cable operators with the highest Latino subscription rates. She pitched the idea of a Latino-oriented HBO channel to sixteen cable operators and asked them to submit letters stating whether they would be interested in such a channel. The letters included estimations made by the operators regarding the degree to which they would expect to see their business grow within one year of offering such a Latino-oriented service. Not surprisingly, given U.S. demographics, the cable operators with the highest Latino penetration were all in urban areas. They stated that rates of subscription would grow exponentially if they carried a Latino-oriented HBO channel. Lara presented this information to HBO executives and explained that the revenue generated by these sixteen cable operators alone would pay for the technology needed to launch a Latino-oriented HBO channel.

Eventually Selecciones En Español became HBO En Español, and, by 1993, all of HBO's programming was available to subscribers via SAP. However, Lara's grand vision of a Latino-oriented HBO channel still remained limited to the audio track option. Thus, Latino audiences were not provided with Latino-specific programming. Instead, they were simply witnessing how the Spanish language could be tacked on to HBO's already-existing programming. However, despite HBO's reluctance to provide Latino audiences with programming specific to their cultural interests and demographics, Lara was able to implement original interstitial programming in Spanish. Although this was an important achievement for the channel, as will be discussed below, it was not until November 2000 that HBO finally launched a Latino-oriented channel that stood apart from the HBO main channel. Selecciones En Español and HBO En Español existed prior to this, but HBO Latino was the first multiplex channel that was specifically designed to provide programming for a Latino audience rather than simply providing an audio feed of already existing HBO programming.

By the time HBO Latino was added to HBO's multiplex package, most cable subscribers were quite familiar with the other branded channels and knew how to navigate through the digital feeds. The multiplex package, "HBO The Works," already included HBO, HBO Family, HBO Comedy, HBO Zone, HBO Plus (formerly HBO2), and HBO Signature (formerly HBO3). The launch of HBO Latino was introduced in the pages of *Latina* magazine through an advertisement that instructed readers to prepare themselves for the new channel. In the ad a sepia-toned image of a single eye stares out at the viewer. The eye is framed by four fingers while a bottle of Tabasco sauce enters slightly into the frame. The red and green colors of the bottle, together with the small red scripted word "Latino" in the corner of the advertisement, add brightness to an otherwise grainy illustration. The assumption is that HBO Latino will add color, flavor, and perhaps even spiciness to the otherwise bland offerings of cable television.

According to Bernadette Aulestia, HBO Vice President of Affiliate Marketing and Brand Development, Latinos and African Americans had consistently over-indexed, or had a strongly noticeable presence, as HBO subscribers. Aulestia and others at HBO were expecting Latinos to be drawn to the new HBO channel by recalling HBO's earlier association with Latino-oriented programming. Examples of such Latino programming include the biopics *The Burning Season* (1994) about union organizer and Brazilian rainforest activist Chico Mendes, played by Raul Julia, and *For*

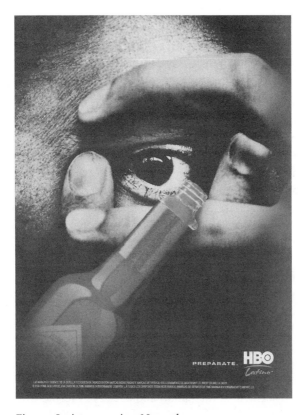

Fig. 9.1. *Latina* magazine, November 2000.

Love or Country (2000), starring Andy Garcia as the Cuban trumpeter Arturo Sandoval. In addition, HBO had aired the stand-up comedy shows of Paul Rodriguez (1987, 1995, 1998) and John Leguizamo (1991, 1993, 1998), as well as concert specials featuring Julio Iglesias (1990), Gloria Estefan (1996), and Marc Anthony (2000).

Reaching an Audience with Roots in U.S. Cities

While the dramas that aired as part of HBO's Latino-targeted programming often centered on Latin American figures, the comedy shows continuously spoke to the life experiences of Latinos living in the United States. HBO Latino's first promotional campaign for the channel directed attention to these experiences rather than to Latinos' home countries or to the Latin American films that would eventually be aired on HBO Latino. The "Ciudad Caliente" campaign consisted of interviews with Latinos living in Los Angeles, New York, Miami, Chicago, San Diego, Phoenix, San Antonio, Houston, and El Paso. Latinos from these cities spoke to the camera and identified their favorite local restaurant, described their city's entertainment hotspots, and offered speculations as to why their city's Latinos and Latinas were the most attractive in the country. These playful interviews were aired as interstitial programming segments broadcast on the channel's national feed. Viewers were also able to enter a contest to travel to one of these "hot cities" as part of the promotional campaign.

The "Ciudad Caliente" campaign suggested that HBO was trying to distance itself from solely associating with Latin America and instead choosing to focus on the diversity of Latinos that are solidly rooted in U.S. metropolitan cities. In her introduction of the launch of the campaign, Aulestia said: "People are proud not only of their Latino identities, but also of their cities."[7] This view of U.S. Latinos is quite different than the one usually held by marketers. Arlene Dávila's research on Hispanic advertising suggests that "it is Latin America rather than the deterritorialized U.S.-Latino culture that has traditionally been valorized as the source of cultural authenticity in Latino/Hispanic culture."[8] The "Ciudad Caliente" campaign, on the other hand, recognized the "deterritorialized U.S.-Latino culture" and segmented the Latino audience even further by acknowledging that the experience of a Latino living in New York is quite different than that of a Latino living in San Diego, Phoenix, or Houston.

HBO Latino's orientation toward U.S. Latino culture can be seen in the channel's second *Latina* magazine advertisement. This advertisement showcases Latinos who are easily recognizable among the magazine's bilingual, bicultural readers. Jennifer Lopez has graced the cover of *Latina* a countless number of times, while Andy Garcia, John Leguizamo, Marc Anthony, and Oscar de la Hoya have all been featured in the magazine's monthly *papi chulo* profiles of male celebrities.[9] The celebrities are all Latinos who have attained success in U.S. film, stand-up comedy, music, and sports. However, with the exception of Marc Anthony, all of the celebrities are best known for their performances in the English language. This puts into question the promise presented in the advertisement. HBO Latino claims to be in "*tu idioma*" or in "your language." But what *is* the language of this imagined community of panethnic Latinos?

Fig. 9.2. *Latina* magazine, February 2001.

The type of Latinidad that is presented in the advertisements for HBO Latino suggests that the Spanish language is what marks HBO Latino as the channel for Latinos. The inclusion of John Leguizamo's comedy show, *Freak,* which is rooted in his reflections on growing up as a confused bicultural kid in New York City, does not necessarily mark the channel as "Latino"—for this comedy special was first aired on HBO in its original English-language format. Instead, the channel is "Latino" because it airs *dubbed versions* of the programming that is aired on the main HBO channel. This suggests that the vision of Latinidad pursued by HBO Latino is one that holds the Spanish language as its guiding principle. However, by including the artists Jennifer Lopez and Marc Anthony within its advertisement, HBO Latino complicates this construct of Latinidad and alludes to the nuanced role that Latino music plays in the construction of Latino panethnic identity.

HBO Latino did not follow the lead of HBO Olé when programming its music videos. HBO Olé had teamed up with BMG Entertainment and the Warner Music Group in 1995 to form YA TV, a 24-hour Spanish-language music channel. YA TV used two satellite signals that enabled the channel to create distinct programming for Central and South America, which would reflect the music that was popular in these broad geographic locales. On the other hand, HBO Latino did not pursue a narrowcasting approach to the musical selections that aired on its 30-minute music video spots. Rather, it brought together music that is popular in different parts of Latin America and within the United States. For example, it is currently not uncommon to see a video by the Mexican pop singer Paulina Rubio alongside a *grupero* song by Alicia Villareal, which is followed by the *reggaetón* sounds of Noelia. Aulestia explains that this approach is representative of HBO Latino's continuing attempts to provide Latinos with an experience of Latinidad that goes beyond that which is already familiar in their local media. She speaks of HBO Latino's ability to "offer the broadest variety of music and . . . expose people to things that they may not necessarily always be exposed to just by nature of the fact that your radio station may not play that kind of music." She believes that this broad cross-section of music "represent[s] diversity in who we are."[10]

Advertising companies have suggested that Latinos in different parts of the country will respond more positively to an ad if it contains the style of music that is typically aired over the radio stations of their hometowns. As a result, advertisers have produced regionally specific advertising campaigns for everything from beer to luxury cars. Aulestia seems to acknowledge such marketing practices and radio station realities, but she views them as a challenge and an educational opportunity rather than as a mandate on how to schedule programming for specific parts of the country. Some Hispanic advertising agencies have resisted regionally specific marketing by targeting an imagined generic Hispanic nation that is presumably united by ethnic pride, nationalism, and a strict maintenance of traditions.[11] Within the world of television, the grand unifier has traditionally been the Spanish language. Thus, recognition as an audience and the subsequent granting of cultural citizenship that is offered by Univision and Telemundo rest on the maintenance of a grammatically correct and non-nation-specific Spanish language.[12] This approach to establishing a panethnic audience obscures the diversity of ways in which Latinos play with language both within the United States and throughout Latin America. In addition, positioning the Spanish language as the decisive factor that marks one's identity as "Latino" inevitably serves to exclude monolingual English speakers, or those who speak Portuguese, use an indigenous language, or communicate using any mixture of the above.

Although the vast majority of HBO Latino's programming is in Spanish, including the music videos alluded to above, the channel perceives its audience and their needs in a manner that is quite different than that taken by U.S. Spanish-language broadcast channels. Channels like Univision and Telemundo hold the Spanish language as the critical marker of Latinidad and of their programming. For example, it is not uncommon for a Univision spokesperson to invoke language when describing consumer choice and audience preferences: "Hispanics choose Univision because they want Spanish."[13] On the other hand, Aulestia differentiates HBO Latino from channels like Univision by drawing attention to the fact that "we never defined ourselves as a Spanish-language channel."[14] She goes on to describe the strategy behind the naming of the channel, noting that

> One of the reasons we avoided [the word "Spanish" in the title] is because HBO Latino is very different today than it was when it launched and I'm sure it will be very different five years from now. And I think what we wanted to do was to keep it as flexible as possible. The goal of the channel clearly was just always to continue to emerge and evolve as our business needs emerged and evolved and as our audience emerged and evolved.[15]

The strategy described by Aulestia is appropriate for a channel striving for longevity in the face of continuing immigration and varying rates of assimilation. However the reality is that HBO Latino does not appear to be that different than its predecessor, HBO En Español. Now, well into the first decade of 2000, HBO Latino programming continues to consist primarily of Spanish dubs of the English-language programming that airs on the main HBO channel. For example, during the month of November 2004 the vast majority of HBO Latino programming consisted of Hollywood films dubbed into Spanish. Only seventy-one hours of programming during this one-month period were devoted to non-dubbed feature films, documentaries, boxing, and music specials. The channel's Latino-oriented programming that *does* exist is often highlighted under the HBO

Latino banner, "Estrenos Latinos." And even here, the focus is on the Spanish language and on feature films that originate outside of the United States. To be sure, many U.S. Latinos could feel an affinity for this type of programming. However, the fact that only minimal reflections of U.S. Latino experiences are included within "Estrenos Latinos" exposes the actual rigidity of HBO's vision of Latinidad.

The Chicano Presence and Absence on HBO Latino

HBO's "Estrenos Latinos" lineup usually consists of films that have not debuted in the United States. For example, Guillermo del Toro's *El Espinazo del Diablo* (2001) made its premiere on HBO Latino, as did the documentary *The Bronze Screen: 100 Years of the Latino Image in Hollywood Cinema* (2002), which was screened as part of HBO Latino's celebration of Hispanic Heritage Month.[16] Airing independent films throughout the month of September as part of a program titled "El Otro Cine," the channel showed such films as *La Ciénaga* (2001) from Argentina, *La Comunidad* (2000) from Spain, and *Luminarias* (2000), an English-language Chicana film from the United States.

The presence of *Luminarias* on HBO Latino was somewhat of an anomaly. The film's storyline addresses homophobia, cross-cultural relationships, and the consequences of assimilation—Chicana/Chicano points of view on these issues are rarely represented on HBO. Indeed, while the cast of *Luminarias* features familiar Chicano actors such as Evelina Fernandez, Lupe Ontiveros, Cheech Marin, and Sal Lopez, the faces of these actors are even more of a rarity on HBO. This may be due largely to the fact that there are few feature-length Chicano films that are financed and available to be programmed by HBO. However, HBO has begun to make inroads in this area. For example, HBO financed the film *Real Women Have Curves* (2003) and subsequently released it with Newmarket Films as the first theatrical title under the HBO banner. HBO decided to release the film theatrically after witnessing its warm reception at the 2002 Sundance Film Festival, where it won the Audience Award for Best Dramatic Feature and the Special Jury Award for Acting. The film, based on Josefina Lopez's play of the same name, focuses on the college ambitions of a young Chicana high-school graduate who challenges her mother's traditional perspectives on sexuality and body image while working with her at a downtown Los Angeles sewing factory. Newmarket Films anticipated that these themes would resonate with audiences that had made a box office hit out of *My Big Fat Greek Wedding* (2002), also released by Newmarket.[17]

Although *Real Women Have Curves* performed modestly in the domestic box office, HBO has continued its support of Chicano filmmaking. The co-writer of the *Real Women* script, Lopez, is currently working on a new script for HBO about the murder of over 350 women in the Mexican border town of Ciudad Juarez. HBO also produced the film *Walkout* (2006), a historical drama about the 1968 Chicano student walkout at five East Los Angeles schools, which premiered on HBO and HBO Latino in February 2006.

Early on, HBO appeared to be looking toward Chicano culture as the means through which to generate revenue and gain viewership for HBO Latino. The first original programming to air on HBO Latino was *Oscar de la Hoya Presenta Boxeo de Oro*, which was produced by HBO Sports and first shown in January 2003. The premise of the show is de la Hoya's search for the next great boxer. Promotional spots for the program feature the Olympic gold winner and multi-weight champion stating: *"En algún lugar están los campeones de mañana. Boxeadores con corazón, velocidad, poder, y disciplina—están en algún lugar esperando por la oportunidad de brillar—y yo los voy a encontrar."* ("The champions of tomorrow are out there. Boxers with heart, velocity, power, and discipline—they are out there waiting for the opportunity to shine—and I am going to find them").[18] The show is filmed in boxing rings across the United States, including cities with large Latino populations such as New York, Miami, Chicago, Houston, San Antonio, and El Paso.[19]

Boxing has been a central part of HBO's programming since the early 1970s when it carried live boxing matches from Madison Square Garden. In fact, HBO initiated satellite use for regular transmission of programming by airing the 1975 heavyweight championship fight between Muhammad Ali and Joe Frazier live from the Philippines.[20] The focus on heavyweight fights continued through the 1990s, culminating in the largest pay-per-view event in television history, the 1997 rematch between Evander Holyfield and Mike Tyson. This pay-per-view event was bought by 1.99 million homes and grossed $100 million. But the boxing match is also noteworthy because it marked HBO's transition to focusing on lower weight classes since many of the star heavyweight fighters were nearing retirement. The inadvertent result of this reorientation was the inclusion of more Latino boxers on HBO pay-per-view events and on programs like "HBO Boxing After Dark."[21]

One of the biggest stars of the pay-per-view events was Oscar de la Hoya, who fought in the four most lucrative non-heavyweight pay-per-view events. The 1999 de la Hoya–Felix Trinidad fight was purchased by 1.4 million homes and grossed $71 million. In 2004 one million buyers purchased the de la Hoya–Bernard Hopkins fight that grossed $56 million. The 2003 rematch between de La Hoya and Shane Mosley resulted in 950,000 buys, and the 2002 de la Hoya–Fernando Vargas fight generated 935,000 buys.[22]

Clearly the decision to pursue *Boxeo de Oro* as the first HBO Latino original programming makes market sense. De la Hoya has a strong Latino fan base, especially among Chicanos living in the Southwest, and boxing has consistently performed extremely well on HBO. Yet one can't help but also hope that this decision may be signaling a change in television's treatment and understanding of the Latino audience. The program suggests that HBO may be challenging the common Spanish-language television practice of relying heavily on Latin American actors and programming. *Boxeo de Oro* is hosted by ringside commentators who speak in Spanish and also easily transfer to English when interviewing a non-Spanish-speaking boxer. De la Hoya himself also appears at the end of the program to offer commentary on the night's matches. He and the hosts end the show by telling viewers which city will be the venue for the next match. When appropriate, the hosts comment that they are looking forward to broadcasting the next fight from their home state or hometown.

Dávila has shown that the "close language and programming synergy between the U.S. Hispanic and Latin American media markets" usually results in the privileging of Latin American media over local U.S. Latino productions.[23] HBO Latino's decision to feature a Chicano icon and opt for bilingual, bicultural hosts challenges the practice of relying on economies of scale that do not differentiate between Latin American and U.S. Latino markets. However, the next original programming offered by HBO Latino was an Argentinean murder mystery dramatic series produced by the HBO Latin America service. *Epitafios,* the first Latino-oriented series to air on HBO Latino, made its debut in September 2005 as part of the channel's Hispanic Heritage Month celebration. The series was featured along with three other programs: *Torremolinos 73,* a Spanish film making its U.S. television premiere; *Oscar de la Hoya Presenta Boxeo de Oro;* and *Habla y Habla,* a series of short vignettes on Latino life in the United States. Although these shows represent strides forward, it is uncertain how long viewers will have to wait until they are presented with U.S.-based

HBO Latino original dramas that are as innovative and character driven as *The Sopranos* or *Six Feet Under*. For now, they must content themselves with the short interstitial programming, like *Habla y Habla,* that airs between dubbed versions of standard Hollywood fare.

Interstitial Programming and U.S. Latino Voices

As was discussed earlier, the inclusion of interstitial programming on HBO En Español was a landmark achievement for the channel. The presence of such programming suggested that the channel was not simply a dubbed version of the mainstream HBO. The interstitial programming represented moments in which viewers could recognize the Latino-specificity of the channel, even if this was limited to a colorful "ethnic" background behind text introducing an upcoming romantic comedy starring Julia Roberts.

Interstitial programming, such as bumpers and ID spots, are used to emphasize the brand of a channel. For example, beginning in the 1990s, channels like MTV and Nickelodeon started engaging in the extensive use of creative bumpers and IDs as a way to emphasize the uniqueness of their programming.[24] The most recent HBO Latino campaign emphasizes the uniqueness of the channel by presenting the experiences of second- and third-generation bicultural Latinos. However, in doing so, the campaign ultimately raises questions as to whether this audience is truly being served by HBO's multiplex system of programming.

The "Habla" (2003) and "Habla Again" (2004) campaigns were composed of short (30-second to 2-minute) vignettes by Latino New York and Los Angeles–based poets, actors, comedians and ordinary people. These vignettes presented mostly humorous stories related to being Latino in the U.S. They aired individually as station IDs and were also grouped together and aired as two 25-minute pieces. "Habla" was also aired as a half-hour show in celebration of Hispanic Heritage Month on HBO Latino.[25] The casting call for the "Habla" campaigns did not specify what kind of performers HBO Latino was looking for. The postings simply stated that the channel was "looking for people who had opinions."[26] The result was an array of predominantly 18–34-year-old second- and third-generation Latinos using Spanish and English, often in the same vignette, to refer to the culture clashes that they experience among family members outside of their age group. Each vignette begins with a clacker board that introduces

the performers by their first name. In the vignette "*Abuelita* (Again)," Eric
describes his grandmother's affinity for her bathrobe: "When she goes to
church she just puts a belt around it so it becomes a *bata*-slash-dress"
(*abuelita*: grandma; *bata*: bathrobe). In another one of the "Habla" vignettes, "*Pelea*," we are introduced to Caridad and her pre-teen son, Pedro
(*pelea*: fight). Caridad asks Pedro, in Spanish, why they fight. He says that
she gets mad when he doesn't do what she asks him to do. Caridad cor-
rects him and says that they fight when he makes fun of her English. He
giggles, then she tells the camera that he refuses to learn how to dance
salsa because he is more immersed in hip-hop culture. The vignette ends
with Caridad asking Pedro to dance and talk like "*los cubanos.*" Pedro engages in exaggerated forms of dance and holds his nose while he speaks.
Both Pedro and Caridad laugh at his antics.

Elaine Brown, HBO's Vice President of Special Markets, oversees on-air
promotion for all of the multiplex channels, including HBO Latino. She
explains the significance of vignettes such as "*Pelea*" by suggesting that
they speak to multiple members of the Latino household. "Earlier research
indicated that a lot of Hispanic households tend to watch TV together. I
mean, you'll have families in the household, which range from the young-
est people to the *abuelos* living together and so what we wanted to do is we
tried to capture that same spirit and target that same audience and be as
broad as possible."[27]

At times, this attempt to reach the multi-generational Latino household
simply draws from clichés that do not necessarily reflect a U.S. Latino-
specific experience. For example, with minor language adjustments, Eric's
"*bata*-slash-dress" story could serve just as effectively if told by a non-
Latino comedian. Nevertheless, other vignettes go beyond identifying simple differences between old-school parents and their assimilated offspring.
"Padres" is a very nuanced story told by Sandra in which she explains that
her parents represent "two different Mexicans." Her mother longs to return to the country in which she was born and sings praises of "*México,
pero que lindo México*" ("México, but how lovely is México"). On the other
hand, Sandra's cantankerous father responds, "*¡Esa gente de México, pura
gente chueca!*" ("Those people from Mexico are nothing but crooks!").

While the culture clashes referred to in the above vignettes highlight
the difficulties in communicating across generations, other vignettes draw
attention to the ways in which Latinos are different from Anglos. There
are stories about being forced to pronounce Spanish words like "*los ánge-
les*" with an English accent; having to schedule play dates with Anglo kids

rather than being able to engage in the spontaneity one experiences with Latino family members; and learning that when people in the United States ask "How are you?" they usually do not want a sincere answer. These are all familiar jokes to those who have attended Latino comedy shows, and when told on HBO Latino they serve as a reminder of how different the channel's audience is from that of its mainstream sibling station. However, the vignettes do not limit themselves to pointing out superficial differences. Some of the stories told amongst these Latino/Anglo culture-clash vignettes focus more squarely on the confrontations that occur when Latinos enter previously Anglo-only environments. In these interesting moments of self-reflection, the Latino performers gain an opportunity to speak from the television and address the entertainment industry directly by telling horror stories about casting calls and demeaning "Latino" television shows.

Jon uses his vignette, "Lotería," to define what he calls "*gringo*-friendly Spanish" (*lotería*: lottery). He explains that TV shows often use this "cheap knockoff of Spanglish except they use only easy words that everyone can understand" because they "don't want to confuse the *gringo* viewers when the Latins are speaking their native tongues." In "Sufrida" Marcia explains that "*Americano*" directors love it when she comes to auditions wearing a *sarape* and a wig with long braids (*sufrida*: suffering woman; *sarape*: shawl). In the vignette, "Asian," casting directors tell Nicole, a Puerto Rican Jew, that she is "not exactly ethnic-specific enough." The casting directors attempt to dismantle Nicole's identity at the same time that they comment on her facial features. They ultimately make the recommendation that the young actress respond only to casting calls for Asian characters. On the other hand, Marilyn has mastered the art of performing her ethnicity at auditions. In "Audition" she re-enacts the experience of running into a Latina friend at a casting call: "Oh my God, you look beautiful! No, no, no, no, the miniskirt works. The hoops, they're perfect. You're going in for The Ho, right? Ho One or Ho Two? Oh, you're going in for Two, *bendito* [God bless], I'm going in for One. Yeah, she has a line. No just a line. One." Marilyn turns to look off camera, suggesting that an actor's name has been called. She tells her friend, "That's you. You break a leg, girl. Remember— Latino power. Latino power. No, no, no, no—*con su cabeza arriba* [with your head held high]. You do your thing, girl!" Then she whispers, "*Pobrecita* [poor thing]."

The director of the "Habla" and "Habla Again" campaigns, Alberto Ferreras, says that some of the most intriguing stories came out when the actors let go of the scripts that they had brought with them and instead spoke about their own lived experiences. The stories that the actors told reinforced his view of the current state of Latinos in the media:

> I have this whole theory about how Latinos live. I say that as Latinos in the nation, we live in a house without mirrors. Because if television is the mirror of society we as Latinos look at TV—we look at Anglo television and we're not there. We look at Latino television and it's all South American stuff—these people don't talk like we do, they don't have the same experiences that we do so what it does to your self-esteem to never be able to see yourself in a mirror is bad. What I wanted people to do is to look in the television that almost looks like a mirror because we do talk like that. And also to realize the beauty of our experience and our language.[28]

By contrast, John Sinclair proposes that "watching an Argentinean or Mexican telenovela reminds [Peruvian viewers] of the similarities they share with neighboring countries in their region (and perhaps also the differences), while flipping over to CBS Telenoticias or a Hollywood film dubbed into Spanish might make them feel more like privileged citizens of the globe."[29] Suggesting that this is not the case for U.S. Latinos who are constantly receiving South American programming on U.S. Spanish-language television, the Spanish-Venezuelan Ferreras is not attempting to disparage South American programming or accents; in fact, one of the vignettes that he repeatedly praises, "*Patria*," features the 1940s Argentinean actress Graciela Lecube (*patria*: native land, home country). Rather, Ferreras is lamenting the lack of U.S.-Latino accents, mannerisms, attitudes, and stories on television that is broadcast within the United States.

Writing in a different context, Moya Luckett explains that the success of the BBC's Asian sketch comedy, *Goodness Gracious Me*, was due largely to its strong Britasian sensibility. That is, the sketches resonated with audience members who recognized the specificity of the home furnishings, characters' catchphrases, and humor.[30] Ferreras hopes that the "Habla" performers' non-self-conscious use of Spanglish when telling their personal stories will serve to bring a strong U.S. Latino sensibility to HBO Latino: "When you're a Latino in the U.S., we—anyone who speaks Spanglish has a license to play with the language . . . When you're a Latino in the U.S. and you look at Telemundo or Univision you never see that playfulness of the way we speak both languages—you hardly ever see that reflected and that's something else that we wanted to show."[31]

Ferreras' comments echo the findings in Ana Celia Zentella's study of language in New York City's East Harlem Puerto Rican community. Zentella's study includes many examples of how children engage in code switching to honor community norms and to meet social and communicative goals. She argues that the code switching does not position New York Puerto Rican children as "semi- or a-lingual hodge-podgers," but as "adept bilingual jugglers."[32]

It remains to be seen whether cable channels will also recognize U.S. Latinos as "adept bilingual jugglers" who want more programming that reflects their lived experiences. Recently multiple-system operators (MSOs) have begun integrating more English-language channels into packages targeting Latinos. In 2001, DirecTV added ten new English channels to its extended Latino package, DirecTV Para Todos Opción Extra Especial. Around this time EchoStar Communications also initiated programming packages that included equal numbers of English and Spanish-language channels.[33] However, code switching between English and Spanish on one channel continues to remain virtually absent.

Latinos and Cable Television: A Future of Multiplex Patchwork Viewing Habits?

In 1977, an article published by the National Council of La Raza declared that cable represented "the only avenue available for Hispanics to gain substantial control over a communications medium."[34] Securing cable ownership was obviously not an outcome of the so-called 1980s "Decade of the Hispanic." Nor did cable ownership occur as a consequence of the late-1990s "Latin explosion." Instead, consumer choice has been championed, and HBO has made efforts to reach out to the Latino community via dubbed HBO Latino programming and the advocacy of multiplex patchwork viewing habits.

Bernadette Aulestia suggests that HBO offers the type of consumer choice that MSOs are now pursuing: "When you get HBO you get multiplex channels, six or seven of them, including HBO Latino. The nice thing about that was that we weren't asking anybody to give up their English-language HBO. We were saying, 'Listen, if there's someone in your house that—'; like me, I would never watch *The Sopranos* in Spanish."[35] Aulestia's reference to *The Sopranos* points to an unfortunate reality of HBO Latino. For the most part, when a bicultural, bilingual Latino tunes in to

HBO Latino he or she finds *The Sopranos, The Wire,* or the latest Holly-wood blockbuster dubbed into Spanish. Finding no significant difference between HBO Latino and the main HBO channel, this Latino often ends up back at the English-language HBO channel.

In his article on Showtime's No Limits programming and the origi-nal program *Resurrection Boulevard,* Scott Wible invokes Chon Noriega's history of early Chicano advocacy efforts in creating Chicano television. Wible looks to the current state of the cable industry and finds that "The act of production, then, no longer entails owning stations and writing, producing, and acting in television programmes but instead involves 'cre-ating' a product by piecing together different programmes of the viewer's choice."[36] This is exactly the kind of piecing together that Aulestia cele-brates. Yet she herself admits that she would not watch certain program-ming that is dubbed into Spanish. Why *would* an English-speaking Latino watch *The Sopranos* in Spanish? The reality of bicultural, bilingual Lati-nos' viewing experiences points to the fallacy of consumer choice being equated with citizen control.

HBO Latino is, indeed, challenging the narrow definitions of Latinidad that have been offered by Spanish-language television. However, U.S. Lati-nos deserve more from cable television's recent expansion. We should have enough dignity to know that we can—and we must—make demands for more than just a multiplex patchwork of television programming. Do we really need to hear Tony Soprano and Carrie Bradshaw in Spanish? What if this were to be replaced by Luis Montez and Lauren Fernández speaking English, Spanish, and Spanglish?[37] I, for one, would put down the remote and watch the show. Who knows—maybe Marilyn (from the "Audition" Habla vignette) would have a lead role.

NOTES

1. See Al Auster, "HBO Movies: Has Risk-Taking Made the Cable Giant the 'Auteur' of the New Century?" *Television Quarterly* 31 (2000): 75–83; and John Horn, "HBO Emerges as a Mecca for Maverick Filmmakers," *Los Angeles Times,* September 19, 2004, A1.

2. HBO's first operation outside the United States was the launch of television service in Hungary in September 1991.

3. Megan Mullen, *The Rise of Cable Programming in the United States: Revolu-tion or Evolution?* (Austin: University of Texas Press, 2003), 145.

4. Janet Stilson, "HBO Olé to Launch in Latin America," *Multichannel News,* January 21, 1991, 3.

5. Concepción Lara, telephone conversation with the author, October 16, 2004.

6. John Sinclair, *Latin American Television: A Global View* (Oxford: Oxford University Press, 1999).

7. Monica Hogan, "Hot Promos Back HBO Latino Launches," *Multichannel News,* September 18, 2000, 42.

8. Arlene Dávila, "Mapping Latinidad: Language and Culture in the Spanish TV Battlefront," *Television and New Media* 1, no. 1 (2000): 79.

9. Katynka Z. Martínez, "*Latina* Magazine and the Invocation of a Panethnic Family: Latino Identity as It Is Informed by Celebrities and *Papis Chulos,*" *Communication Review* 7, no. 2 (2004): 155–74.

10. Bernadette Aulestia, telephone conversation with the author, October 19, 2004.

11. Arlene Dávila, *Latinos, Inc.: The Marketing and Making of a People* (Berkeley: University of California Press, 2001).

12. Ibid.

13. Elana Levine, "Constructing a Market, Constructing an Ethnicity: U.S. Spanish-Language Media and the Formation of a Syncretic Latino/a Identity," *Studies in Latin American Popular Culture* 20 (2001): 33–50.

14. Aulestia, telephone conversation with the author.

15. Ibid.

16. Elaine Brown, telephone conversation with the author, October 20, 2004.

17. Martin A. Grove, " 'Women' Could Show Real B.O. Legs for Newmarket, HBO," *Hollywood Reporter,* October 9, 2002, http://www.hollywoodreporter.com/.

18. Unless otherwise stated, all translations are my own.

19. Dan Rafael, "De la Hoya Juggles Training, New Cable Series," *USA Today,* January 13, 2003, 10C.

20. Mullen, *The Rise of Cable Programming in the United States,* 107.

21. Jennifer Reingold, "I Can Lift the Name of Boxing," *Business Week,* July 7, 1997, 115.

22. Richard Sandomir, "De la Hoya's a Champ to Paying Viewers," *New York Times,* September 24, 2004, D1.

23. Davila, "Mapping Latinidad," 78.

24. Mullen, *The Rise of Cable Programming in the United States.*

25. Alberto Ferreras, telephone conversation with the author, October 20, 2004.

26. Ibid.

27. Brown, telephone conversation with the author.

28. Ferreras, telephone conversation with the author.

29. Sinclair, *Latin American Television,* 133.

30. Moya Luckett, "Postnational Television? Goodness Gracious Me and the Britasian Diaspora," in Lisa Parks and Shanti Kumar, eds., *Planet TV* (New York: New York University Press, 2003), 402–22.

31. Ferreras, telephone conversation with the author.

32. Ana Celia Zentella, *Growing Up Bilingual* (New York: Blackwell, 1997).

33. Andrea Figler, "More English for Spanish Subscribers," *Cable World*, May 7, 2001, 22.

34. Chon Noriega, *Shot in America: Television, the State, and the Rise of Chicano Cinema* (Minneapolis: University of Minnesota Press, 2000), 88.

35. Aulestia, telephone conversation with the author.

36. Scott Wible, "Media Advocates, Latino Citizens and Nice Cable," *Cultural Studies* 18, no. 1 (2004): 39.

37. The character of Luis Montez is the middle-aged lawyer and former Chicano movement activist who serves as the protagonist in the detective novel *The Ballad of Rocky Ruiz* (1993) by Manuel Ramos. Lauren Fernández is one of six Latina characters that sound off on homophobia, the glass ceiling, sex, and intra-Latino discrimination in Alisa Valdes-Rodriguez's debut novel, *The Dirty Girls Social Club* (2003).

Chapter 10

Gay Programming, Gay Publics
Public and Private Tensions in Lesbian and Gay Cable Channels

Anthony Freitas

Between 2001 and 2005, several media firms introduced commercial channels aimed at lesbian and gay viewers. These very narrowly targeted channels employed innovative funding and programming strategies to overcome financial and regulatory barriers as well as anticipated negative reactions from both anti-gay groups and lesbian and gay communities.[1] Like other media outlets targeted toward particular demographics or interests, these channels necessarily imagine an audience or public and hope to attract them to their channels. This chapter examines the creation of some of the channels targeted at gay audiences and the publics that they imagine.[2]

Conceived of and built by established media interests in consultation with regulators or lesbian and gay media groups, these channels have been based largely on a series of assumptions about the lesbian and gay community. These assumptions represent the lesbian and gay community as relatively uniform, able and willing to pay for these channels, numerous enough to support channels this narrowly cast, loyal enough to keep ratings steady, and interested in programming separated out from non-gay media content. A result of these assumptions is the refining of this community as a market. Indeed, this community is understood as media-savvy, brand-loyal, trendsetting with large disposable incomes. It is often touted as the last under-served market niche and is seen by many in television and advertising as the ideal market: a seemingly perfect audience waiting to be addressed.[3] However, these assumptions have proved vexing for the owners of these channels. The imagined audience has not materialized. Carriers have been reluctant to adopt the channels. Initial business plans have been radically restructured. PrideVision and Q Television struggled to stay afloat, Logo debuted two years late, and Here! remains available only as a pay-per-view selection.

This essay examines the difficulties in the launch and survival of these channels in an effort to better understand some of the aims and pitfalls of targeting entire, private channels to a lesbian and gay public. A primary concern is the envisioning and creation of a gay audience that is distinct from a general audience. The very act of defining an audience from any community is difficult. Like many other modern communities, the lesbian and gay community is dispersed and diverse. There are varying degrees of contact between individuals and people come from an array of political, economic, educational, racial, religious, and ethnic backgrounds. Because of this diversity, the channels will necessarily attempt to speak to a wide segment of lesbians and gay men. However, in calling out to this perceived audience, these channels also help to constitute a community. This act of interpellation—the simultaneous calling out to and formation of an identity or subject position and its related publics—is one of my central interests in this essay.[4]

As Louis Althusser describes, interpellation is a key element in the formation of subjecthood within a culture's ideology. Althusser employs the term to describe how "ideological state apparatuses" operate to constitute the subject within the ideology of the state. These apparatuses include governmental bodies such as courts and police, but they also include private or non-governmental institutions such as businesses, media, churches, and unions. Through interpellation, we come to recognize ourselves and our roles within the organizing ideas and structures of a society. Althusser likens interpellation to a situation in which a police officer yells, "Hey, you!" into a crowd, and the guilty party recognizes that the comment is directed at him. This both identifies the guilty individual for the police and invests the one attempting to flee with the identity of a suspect. Another analogy would be the experience of being in a large crowd and, above all the din, discerning someone calling your name; you immediately turn toward the voice. In hearing your name through the noise, you locate not only your hailer, but also yourself. You are that name, you are the subject that is called forth with that hail.

This analogy can be applied to media and consumer culture. When browsing selections on TV, in other media, or in a store we come across thousands of items that we pass up, some that we look at fleetingly, some that hold our attention, and some that even compel us—hail us. Our recognition of ourselves as the audience is often overt but sometimes less than conscious. In a split second, we decide to watch the program, read the article, or buy the product. In doing so, we not only reaffirm that the product is for us, but that we are for the product. We formulate a bit of our identity around that product. This might be a very small part of our identity, but, sometimes at least, many of us do identify with the brands we buy, the shows we watch, and the media we consume.

Niche marketing to a community works in a similar way. Ads and programming use subtle or overt cues to hail viewers—"Hey, queers, watch me!"—not only as individuals—the hail was not "Hey, queer!"—but as a group: not only does the product hail me, it hails others like me. As the marketers of the product seek a queer market, they also produce that market; they call it into being. The individuals and their attractions to others of the same sex existed before the hail, as did the community itself. But the hailing pulls them together again as queers, if for a moment, and it ties this product into that pre-existing notion of community while simultaneously reinforcing that the community exists.

These channels, and the identities, audiences and communities they create, probably have many positive effects for lesbians and gay men—including, for example, recognition of oneself in the media, connection to a larger national media community, and discussion of issues and concerns that are often ignored in the mainstream but that directly impinge on our lives. However, the dynamics of these channels also involve some more troubling aspects of identity formation. They operate to privatize the community, separating it and its interests and issues from those of the general public. They further link the lesbian and gay community and access to its culture to commercial interests. Moreover, these channels may also give strength to accusations by religious conservatives that lesbians and gay men are privileged and need no legal protections. Finally, although these channels may be problematic, if they fail, will others be willing to take risks on gay and lesbian media and content? Is that a problem?

The Channels

PrideVision

In September 2001, Headline Media launched PrideVision in the top four Canadian cable markets. Founded and run by heterosexual Canadian cable entrepreneurs John Levy and Frank Mersch in 2000, Headline Media was started in part with the $160 million Levy made from selling his family-owned cable system, Cableworks, to Cogeco Cable.[5] The initial business plan for PrideVision was a combination of subscription and advertising. Headline established that the channel would be available for about $7 Canadian a month to be subsidized by selling advertising spots on the channel. The Canadian Radio-Television and Telecommunications Commission (CRTC) awarded Headline Media a license to develop a gay- and lesbian-themed cable channel in late 2000 as part of an initiative to develop 15 new digital channels.

Headline Media used a combination of standard and novel approaches in its attempts to draw and maintain interest in PrideVision. Hoping to develop a loyal base of viewers that it would retain after its switch to a pay-for-service channel, PrideVision was initially offered to digital-cable subscribers for no additional fee. The more novel aspect of PrideVision's business plan—similar to the strategy initially proposed for Logo—established revenue for the channel as a combination of pay-for-service and sponsored programming. However, because PrideVision was stand-alone —that is, not bundled with other digital network channels—it had a relatively higher price than other digital channels introduced at the same time in Canada.[6] Some cable providers, most notably in Calgary, declined to carry the channel because they perceived only a small market for it in their conservative areas, and they feared that offering the channel would offend their customer base. Eventually, the CTRC informed Canadian cable companies that in order to get any of the new digital cable channels, they must carry all of the channels.

Initial interest in the channel appeared to be great, and early subscription sales were brisk. However, by the fall of 2002, PrideVision was mired in debt and having trouble recruiting enough subscribers—only 35,00 of the 240,000 needed—to meet its costs. Ultimately, the channel failed to attract enough subscribers and advertisers to be self-sustaining. In an attempt to boost subscriptions, PrideVision began a new, innuendo-heavy ad campaign imploring viewers to recognize that "With only 20,000 subscribers we are impotent! Help PrideVision TV get it up!" The channel also began showing more, and more explicit, gay-male erotica. Neither strategy succeeded. Additionally, the low subscription numbers and the addition of erotic programming hurt attempts to gain advertisers for the channel, limiting the other projected source of revenue.[7]

Lacking the anticipated revenue, PrideVision closed its storefront studios on Toronto's King Street, eliminated all original programming—a condition of its license—and cut its staff from twenty to ten and later to two, neither of whom identified as members of the LGBTQ community. Straight folks owning, operating, and programming a gay channel is not a problem in and of itself. However, it is not in keeping with the spirit of the Canadian regulation that seeks to bring previously under-represented minorities and views, and presumably bodies, into the public discourse.

Beginning in December 2002, Headline Media attempted to sell, merge, or partner PrideVision with other companies. Eventually, in late 2004, Headline Media worked out a deal with web-caster Bill Craig's company iCrave to buy a 90 percent stake in PrideVision—once valued at $30 million Canadian—for $2.3 million Canadian.[8] In 2005, Craig spun off PrideVision's lifestyle and non-erotic entertainment onto a new channel branded OUTtv. Advertiser-supported OUTtv is carried on the basic tier reaching nearly 700,000 households.[9] PrideVision was renamed Hard on PrideVision and became a 24-hour, subscription-only gay and lesbian adult-entertainment channel.[10]

In August 2005, iCrave settled a discrimination complaint against Shaw Communications and BellExpressVu. In this first-ever programming discrimination complaint, iCrave contended that Shaw's and Bell's refusals first to carry PrideVision and later only if offered as a stand-alone subscription channel had cost it Can$2.3 million and denied lesbians and gay men in western Canada access to the channel. OUTtv is now available on both Shaw Cable and Bell's system. Craig attributed the change at Shaw to audience demand, commenting that "[t]here was a lot of pressure coming out of Vancouver and Calgary [subscribers], and there are gay cowboys."[11]

Logo and U.S. Competitors

In January of 2002, Viacom subsidiaries MTV Networks and Showtime announced a joint project to develop a cable channel aimed at lesbian and gay viewers in the United States. Originally named Outlet, the channel was set to debut sometime in mid-2003. Developers hoped to capitalize on the niche-audience development expertise of MTV and the gay-friendly reputation of Showtime. Like PrideVision, Outlet was set to be supported by both subscription fees and advertising.

The producers at Outlet also worked closely with groups such as the Gay and Lesbian Alliance Against Defamation (GLAAD). Founded in the mid-1980s as a response to negative portrayals of lesbians and gay men in the media, GLAAD has historically served as a watchdog, and it has often worked with entertainment producers to ensure sensitive portrayals of lesbian and gay characters and issues. Outlet enlisted GLAAD's input, and by association its imprimatur, to "make sure the channel is sensitive and representative." While the cooperation of a media watchdog in the formation of a network and its featured programming—the GLAAD Media Awards were one of the first televised events on Logo—ensures positive portrayals, such an arrangement could undermine the tenacity of GLAAD's oversight of the channel.

From the outset, a number of cable operators had expressed doubts about the channel's likelihood to succeed in their markets, and some further voiced concern over boycotts by conservatives. In July 2003, citing the flagging economy and difficult ad market, Viacom postponed the debut of Outlet indefinitely. In February of 2004, MTV announced that it was no longer teaming with Showtime but would, on its own, launch Outlet as an entirely ad-supported channel as part of a package that includes MTV, MTV2, VH1, Nickelodeon, and other channels. Changing the launch plan for the planned gay channel from a partial pay-for-service to a fully ad-supported and bundled channel meant that the channel would be available to all subscribers at a particular tier, not just those who opt in. This, in turn, led to a shift in content away from the originally planned edgier series and movies to more lifestyle-based programming.[12]

On June 30, 2005, MTV launched the renamed channel—Logo—on cable systems reaching 11 million households.[13] Negotiations with Comcast, the largest multiple-system operator (MSO) in the United States, soon brought total receiving households to 18 million. The channel runs a combination of off-network syndicated series, movies with gay themes or gay appeal, and some original programming—primarily biographies of celebrity and "ordinary" lesbians and gay men—and reality and talk shows. Additionally, the channel partnered with PlanetOut.com's subsidiary Liberation Publications, Inc. (LPI) to develop lifestyle programming related to travel, home décor, and fashion. Logo debuted with a handful of charter sponsors: Subaru, MillerLite, Tylenol, Key West, and Orbitz.com. These companies expressed faith in the ability of the channel to deliver a well-heeled, trend-setting, and brand-loyal audience and dismissed concerns of boycotts spearheaded by conservatives.[14]

Logo was not the first gay channel available in the United States. The now defunct Q Television Network (QTN) debuted in 2000. The channel was a fully-owned subsidiary of the Palm Springs–based, publicly held

Fig. 10.1. An advertisement for Here! on a bus stop in Chelsea in New York City. The text above the two men's heads reads "Possessed and Undressed."

company Triangle Multi-Media Limited, Inc.[15] In September 2005, QTN had 22,000 subscribers, produced much of its own content, and had rights to independent productions and to broadcast the Gay Games VII. QTN management firmly believed in the premium model for gay channels and, fearing a backlash, derided Logo for moving too fast too soon and for not following an opt-in strategy. Frank Olsen, then-president of QTN, commented that "QTN wants to be invited into your home." Logo, he continued, "is pushing the gay agenda too far on straight people"—borrowing, perhaps unintentionally, a phrase from those opposing gay rights. Olsen also claimed that providing the channel through subscription allowed QTN to offer edgier and live programming.[16] However, by May 2006, QTN was closed by its parent company for lack of funding.[17]

Another channel on offer in the United States is Here! Founders Paul Colichman and Stephen P. Jarchow, who are also founders and principals of the film production company Regent Entertainment, launched the channel in 2003. While Here! began as a pay-per-view gay- and lesbian-themed movie channel available over DIRECTV, a year later, the channel added options for monthly subscriptions or daily blocks of program-ming.[18] The linear-programming channel shows classic and recent-release films and a small amount of original content—primarily dramatic and comedy series—and is available to about 25 million subscribers. In 2004, Here! reported nearly $12 million annual revenue.[19] President Paul Colich-man says that Here!'s offerings are "not adult content . . . not erotica, but [are] designed for a mature audience that pays for it and wants it."[20] Like Olsen of QTN, Colichman asserts that pay-per-view or subscription ser-vices are the only ways such channels can survive. "Advertisers were inter-ested in the audience but they were concerned about the type of program-ming we'd have. You might say they were hyperconcerned."[21]

Benefits and Costs of Gay Networks

In an effort to secure a profitable niche in the competitive cable market, these television channels are following a number of trends within both ca-ble television and lesbian and gay media. Chief among the trends in cable television is the development and availability of virtually unlimited space for on-demand viewing. When discussing the reasons for the emergence of three gay networks in the United States in such a short period of time, Page Thompson, General Manager of Comcast's On-Demand, said, "I think it's primarily a function of technology." Digital cable and satellite services, he states, allow programmers "to reach an audience that might be underserved."[22] Although new technologies are crucial in enabling the creation of a number of narrowly cast channels, these channels do not emerge only because the technology is available. They must also be of in-terest to programmers, carriers, and, in some cases, advertisers. They must be targeted to an audience, and they must connect with that audience in order to remain viable.

The recognition of the gay and lesbian audience by larger corporations is one of the principle trends within lesbian and gay media in the past dec-ade or so. This recognition can be seen in the increase in lesbian and gay content and outlets within broadcast and other media. Clearly, these chan-nels were only imaginable with the ratings, syndication, and re-distribu-tion successes of programs such as *Will & Grace, Queer as Folk, Absolutely*

Fabulous, and *Six Feet Under,* as well as the enormous growth of commercial and independent lesbian and gay films.

Other trends these channels follow are the increase in corporate, non-gay (or not-necessarily-gay) ownership and the consolidation of gay film, Internet, and print media, in addition to a blending or overlapping of what were once thought of as gay concerns into wider genres such as men's fashion and fitness magazines. For example, in the 1990s, Liberation Press Incorporated (LPI), parent of the longest continuously published gay magazine, *The Advocate,* bought Alyson Books, a gay and lesbian publisher, and *The Advocate*'s chief competitor, *Out.* Soon after, PlanetOut bought LPI. PlanetOut, the world's largest gay web portal thanks to its acquisition of Gay.com, is now partnered with Logo and operates as its main web presence.

Apart from following such trends, many of these television channels initially pursued a relatively uncharted approach to their financing by combining subscription fees and advertising.[23] Most networks sell their commercial programming to the cable or satellite operators for a minimal charge per household, and these charges are then rolled into the consumer's monthly bill without itemization. The original business plans for PrideVision and Logo, however, called for them to be unbundled—that is, offered as stand-alone channels available for separate subscription fees. While both Logo and OUTtv are part of the basic digital tier in their respective countries, Logo is now offered as part of a bundle, in part because MTV Networks believed this would ensure wider distribution of the channel.

One argument, often deployed by conservatives, takes the position against bundling because it reduces parental control. Such control, accordingly, is said to be better served by "à la carte" methods in which the subscriber chooses from a menu of channels. However, there are forceful objections to the very real financial, social, and political implications of this argument. One is that unbundling could raise the cost to subscribers, as MSOs would no longer be able to aggregate costs. It would also be likely to make entry into the cable market even more onerous for new channels such as Logo, which are launched by companies that can offer the new channel bundled with its current offerings at low or no cost. In addition, unbundling could make survival difficult for channels with small but devoted audiences. Many of these channels are subsidized by their larger corporate siblings. Socially, unbundling continues the partitioning of audiences, especially in the United States, into increasingly narrower and potentially more isolated media clusters. Furthermore, where such a scenario reinforces an existing tendency of viewers to choose media outlets that agree with their philosophies, politically, this means that we are less likely to encounter ideas that may challenge our ways of thinking or expand our understanding of the world, our nation, our fellow citizens, and ourselves.[24]

A number of factors motivated the original two-branched approach to revenue generation. First among these was the fear that cable companies, out of concern of offending and possibly losing some customers, would not carry the channel if access was not somehow restricted. This concern played itself out in Canada when Shaw refused to carry even the subscription-based PrideVision. It has also been a concern in the United States among both supporters and detractors of gay media, as the observation, cited earlier, concerning the "gay agenda" by Olsen of QTN illustrates.

A second factor is that a channel dedicated to lesbian and gay programming is unlikely to survive on advertising alone. Here owners are responding to two potential concerns among advertisers. The first of these concerns is that any association with lesbian and gay programming will result in negative publicity and possible boycotts by anti-gay groups. This concern is not unfounded. In 2003, Bravo accidentally aired an Applebee's commercial during its reality show on gay weddings. In response to Applebee's perceived support of the program, and by association gay weddings, a number of anti-gay groups threatened a boycott of the restaurant chain. Instead of asserting a right to advertise where it thought appropriate, Applebee's threatened to pull its entire account from Bravo. Bravo issued a public apology to Applebee's.[25] The second concern is that the channel cannot deliver enough viewers to make advertising cost-effective. Since ad prices for television are based largely on audience share, if these channels cannot guarantee an adequate audience, large national advertisers are not likely to buy time on the channel or sponsor particular shows.

In spite of the perceived need for combining subscription and advertising in gay and lesbian channels, the combination did not work as well as the owners had hoped. As PrideVision's failure may demonstrate, it is questionable whether this strategy is viable for such a narrowly cast channel. According to industry sources, gay men and lesbians make up 6.5 percent of the television market in the United States.[26] But the issue is not just one of numbers. In counting on lesbians and gay men to be willing to pay for a channel directed at them even though it has advertising, owners of these channels were assuming more than is usually asked of premium-channel customers. Additionally, unless the owners can guarantee a certain number of viewers (or subscribers), it is unlikely that they will be able to attract the advertising revenue needed to keep subscription costs low. In the PrideVision case, subscription levels were less than 15 percent of what was needed to begin turning a profit, even with advertising included!

The possibility of combining subscription and commercial television stems in part from a third factor: the combination of the construction of a gay market niche and the idea of a gay audience or public. What makes these channels and these financial arrangements appear possible and appealing is the circulation of marketing data that represents lesbians and especially gay men as an affluent and easily targeted market. These representations typically portray gay men as representing 10 percent of the male population, as having large disposable incomes, few dependents, and more leisure time, and as being more interested in entertainment issues than the general population. Additionally, lesbians and gay men are said to be very brand loyal to companies that reach out to them through community services or advertising. These representations of a gay market were developed and deployed by a couple of marketing firms and have attracted a great deal of attention from media corporations and advertisers. In press releases and interviews, the purported strength of the gay market was cited by all of the media outlets as one of the driving factors behind these channels.[27]

The formation of and the assumptions about a gay and lesbian market niche are problematic, as are some of the consequences, and they have been critiqued by a number of authors, including myself.[28] As I have discussed elsewhere, the representation of a unitary lesbian and gay market niche erases much of the variation within and between these communities. Most lesbians and gay men live and consume much as their straight educational, age, and occupational peers do. As Lee Badgett details, while some gay men do have incomes above the national average, most of those live in urban areas where there is likely to be more tolerance and where costs of living are higher. Further, most lesbians, like most women, earn less than the national average. In a two-woman household the gap between men's and women's salaries becomes more pronounced. Lesbians are more likely to have lower disposable incomes than their straight contemporaries.[29]

However, the representation of lesbians and gay men as a "well-heeled minority," as *The Economist* put it, extends not only to, and from, marketers, but also to the larger public and to lesbians and gay men as well.[30] It is impossible to predict what impacts these channels will have on representations of lesbians, gay men, and issues that concern them. And, while it is possible that less stereotypical portrayals will appear, the demands of advertisers may make this less likely than on purely subscription television. If current programming and magazines can be taken as predictors, it is likely that at least on Logo, lesbians and gay men will continue to be portrayed in the same way that their target market is: as white, middle-class or wealthy, urban, and young—not unlike other people on commercial television, and very much like Will Truman of *Will & Grace.*

It is also likely that these channels will mirror other commercial television stations by taking a status quo and "centrist" perspective on social and political issues. Although MTV is often thought of as pushing boundaries, it is unlikely that its offspring Logo will push the envelope of sexuality, relationships, and what constitutes family as far or in as many ways as the gay community does. For example, whereas MTV could take up a progressive and politically charged stance in favor of same-sex marriage, it is less likely that Logo will engage in a well-rounded debate on whether marriage itself should be entirely rethought or abolished as many in the gay community advocate. It is also possible that more radical and left-progressive opinions, including longstanding lesbian and gay critiques of heteronormativity and commercialism, will emerge. However, it is also possible that the centrist voices will share space with, be "balanced by," or even be eclipsed by the voices of gay neo-cons such as Andrew Sullivan or Bruce Bawer. Progressive, radical, and politically controversial topics are often not easily addressed within short time slots. Further, as Cheri Ketchum discusses elsewhere in this volume, programming that challenges commercialism may not be scheduled or survive as concerns over advertiser sentiment and appealing to the widest possible audience overshadow concerns about representing the wider community and its diverse interests.

On the other hand, gay and lesbian channels may be a place where some under-represented perspectives can be voiced. Sirius, the satellite radio service, includes John McMullen's progressive talk show on its lesbian and gay channel, having moved him from its political channel. While such channels may become *the* place for gay programming and information, this seems unlikely. After all, programming devoted to sports, nature, and cooking has not significantly decreased on other channels since the advent of channels devoted to these interests.

Certainly, sections of the lesbian and gay community do fit or aspire to the contours of the imagined niche market, and many in the larger queer community would watch such channels. However, insofar as lesbians and gay men make up an already political and politicized public as well as an audience, the linking of market niche and public is fraught with contradictions, not the least of which are the demands and restrictions that commercial interests place on a public. This is not to argue that in the West a public can and must emerge separate from or outside of larger more established, commercialized publics. That is not possible. Rather, it is to underscore that most publics in the West bear some mark of capital. As Oskar Negt and Alexander Kluge put it, the horizon of publics within capitalism are always hemmed in by commercial interests.[31] Gay and lesbian communities are certainly no different. Indeed, from early on, lesbian and gay communities and identities were closely intertwined with commercial interests.[32] Channels devoted to lesbian and gay programming represent both a continuation of this path and a slight deviation from it.

Further, these gay-themed channels constitute what Kluge refers to as partial publics—publics that are constituted through commercially mediated forms such as television rather than through face-to-face interaction.[33] Unlike the bar, the bookstore, the café, the gym, or the bathhouse, gay-themed channels constitute, and are constituted by, a different and differently intimate, differently sexualized, and differently political public. For, the existence of an already politicized public to which these channels can appeal is not lost on the founders of these channels. In fact, it is partly to this idea of a public that executives and other representatives from the channels are appealing. Spokespersons and marketing material often evoke the notion that the channels serve or help constitute some form of community, safe space, or public. Thus, for example, PrideVision appeals to the gay and lesbian community for more subscribers, Logo teams up with GLAAD, and Viacom executive Brian Graden comments that Logo seeks to offer its viewers the "notion of a 'home.'"[34]

Each of these examples illustrates the very complex ways in which these for-profit ventures interpellate gay subjects and in doing so call them both to commune (share a commonality, or a "home") and to consume. In so doing, although these channels help to construct and fortify a gay public, at the same time they also play on the larger societal opprobrium that haunts most gay men and lesbians. That is, inasmuch as these stations portray themselves as ameliorating the isolation of, say, the closet or rural or suburban lesbians or gay men who may live miles from an accepting queer community, this very idea of refuge can—although not always expressly—conjure the specter of homophobia. Moreover, even as these

channels offer a "home" to a community that is often vilified and perpetu-
ally in a struggle to assert its rights, this "home" is not free. Either through
subscription or the expectation of supporting the advertiser, or both, this
community is expected to make payments for its "home" to people, such
as shareholders of large corporations, who are outside the community.

Further, the community that television channels create is itself limited.
As Miriam Hansen notes, television watching in the West largely involves a
privatized participation in a public sphere.[35] Subscription-based channels
further privatize and limit this television public by making their programming available only to those who have or are willing to devote the money
for it. To be sure, conversations in this public could potentially be more
explicit and more expressly political because they would not have to appeal to a broader audience or worry as much about offending advertisers.
At the same time, however, such a service does not afford the "privacy" of
the closet, as one must cop to a desire to have this channel. On the other
hand, the business model that does away with the subscription portion in
favor of a more broadly based, advertiser-sponsored, unrestricted channel
can open up this television public to more potential viewers/participants.
It can overcome the hurdles of cost as well as those of the closet, shame,
and reluctance to sign-up for a gay channel. But then there are disadvan-
tages here, too. Such a business model works at the expense of its ability to
address many potentially controversial or sexually explicit issues and to
provide more explicit content that advertisers, the larger public, and FCC
regulators may deem offensive.

Yet another complication arises from the ways in which gay and lesbian
channels deterritorialize these publics. For even as these channels allow a
linking of interests across state or national boundaries and foster national
and even international gay identities, these gains can come at the cost of
ignoring specific publics that are confronting issues of discrimination or
hardship in their immediate locale. The national may come to eclipse the
local. As mentioned above, this is a trend in lesbian and gay media generally. It can also be seen in the political mobilization of the lesbian and gay
movement in the last fifteen years, which has seen the rise of advocacy
groups such as the Human Rights Campaign that focus heavily on federal
issues. More troubling, though, is that channels this narrowly cast may operate to privatize an already privatized public even more, separating it and
its interests and issues from those of the general public. At the same time,
the existence of such channels may give strength to accusations by reli-

gious conservatives that lesbians and gay men are privileged and need no legal protections.

Finally, these channels appeal to and rely on an already troubled notion of a gay public or publics. As Michael Warner and others have argued, understanding lesbian and gay communities within a framework of "publics" is problematic within a homophobic culture.[36] Historically, American liberalism has held that a public is predicated on the existence of a "private." Entrance into the public sphere—into the conversations and debates over matters of the state and society—has meant bracketing the private and putting aside specific particularities. As Nancy Fraser has illustrated, sexuality, as one of these particularities, has long been considered one of the most private.[37] Because of the centrality of sexuality to gay identity, Warner suggests that one of the problems of conceptualizing a gay public is that a gay private has not been allowed to exist. In other words, because lesbians and gay men were not given a sanctioned space to place their private particularities no gay public could exist. Until June of 2003, American sodomites were not even guaranteed a "zone of privacy." In numerous ways, lesbians and gay men continue to be denied both private and public spheres. And, as the ongoing showdown over marriage underscores, much of official America remains hostile to public acknowledgement of a queer private sphere. Even after *Lawrence v. Texas*, pervasive homophobia and the persistent necessity of the closet, even if only to get home safely or to avoid loosing your job, still troubles the existence of the gay and lesbian community.

Viacom's long-delayed release of Logo, Logo's and PrideVision's restructurings of their revenue plans, the failure of QTN, and the fear that if these channels fail no one will again try such an endeavor, call attention to some of the more specific barriers to launching a channel so narrowly cast to such a politically charged group. While it may in fact be true that future entrepreneurs would be far more hesitant to launch a commercial channel directed at lesbians and gay men than they are currently, the assumption is that the failure of such a service would be due to a lack of sufficient audience support and commercialization. What if, instead, such a failure were a result of the limitation of commercialization itself? Certainly, there are demonstrable tensions between a commercial audience and a gay public. What if these were more widely perceived? What if it were recognized that a commercial channel may not be meeting the needs or addressing the issues that are relevant to lesbian and gay publics? Maybe these constituencies do not feel compelled to pay for access to a cordoned-off zone? Perhaps the lackluster response to PrideVision and the hesitancy of advertisers and cable carriers for it and Logo illustrate a limit of commercialization and the conflation of audiences with publics.

NOTES

1. The term "lesbian and gay community" is an imperfect shorthand for describing a group of people loosely allied on the basis of shared attraction to members of the same sex. "The community" is by no means as coherent as the term implies. With often deep divisions along gender, economic, racial, political, geographic, and other lines, it is part of a much larger grouping frequently including those who identify as bisexual, transgender, queer, and/or intersexed.

2. There are other channels targeted to lesbians and gay men that have emerged in the past few years—for example, Gay.tv in Italy, GayDate and GayTV in Britain, and PinkTV in France. I restrict my discussion here to North American channels. The other channels elsewhere have followed many of the same strategies and have met similar obstacles, but emerged in their own unique and interesting environments.

3. "The Agony and the Ecstasy: TV for Gays," *Economist,* July 2, 2005, http://www.economist.com/displaystory.cfm?story_id=4135336 (accessed January 10, 2007); "Advertisers Drool Over Gay Market," *The Record,* November 15, 2003, C6.

4. On interpellation see Louis Althusser, "Ideology and Ideological State Apparatuses (Notes Towards an Investigation)," in Slavoj Žižek, ed., *Mapping Ideology* (London: Verso, 1994), 100–141.

5. Matthew Fraser, "Bad Deals, Bungling Hurt PrideVision: Money-Losing TV Channel for Gays on Auction Block," *Financial Post,* December 23, 2002, FP3.

6. Tamsen Tillson, "Struggling PrideVision Pinkslips Half of Staff," *Daily Variety,* December 23, 2002, 5.

7. See Tony Atherton, "PrideVision & Prejudice: A Gay Specialty Channel Struggles with Suspicions of Its Target Viewers, and Their Fears of Being Outed," *Ottawa Citizen,* January 23, 2003, F1; and Gillian Livingston, "Loss Widens at Headline Media," *Toronto Star,* January 10, 2003, E4.

8. John Dempsy, "MTV Revs Up Gay-Themed Outlet," *Variety,* February 10, 2004; Fraser, "Bad Deals"; Etan Vlessing, "iCrave's Craig Buys PrideVision," *Hollywood Reporter,* December 9, 2003, http://www.hollywoodreporter.com (accessed September 25, 2005).

9. Tamsen Tilson, "PrideVision Doubles Up," *Variety,* April 4–10, 2005, 23.

10. "Gay Specialty Channel PrideVision Goes Hardcore, Spins Off New Outlet," *Canadian Newswire* (retrieved from Lexis-Nexis).

11. Laura Bracken, "OUTtv Settles with Shaw," *Playback,* August 1, 2005, 2.

12. Dempsy, "MTV Revs Up Gay-Themed Outlet."

13. The channel was renamed in 2003 in response to a complaint filed with the U.S. Patent and Trademark Office. A web-casting radio service already owned rights to the name. Christian Grantham, "Outlet Radio Looks Ahead," Outletradio.com, October 19, 2003, www.outletradio.com/grantham/archives/2003_10.php (accessed October 16, 2005).

14. "Miller, Orbitz Among Logo Advertisers," *Advertising Age,* July 4, 2005, 10.

15. Anthony Crupi, "Logo Debuts with Multi-Carriage," Mediaweek.com, July 5, 2005 (accessed October 12, 2005).

16. Linda Moss, "Pushing the Boundaries: Fledgling Nets Will Supply a Groundswell for Gay Fare," *Multichannel News,* October 18, 2004, http://www.multichannel.com/.

17. "Q Television Network Is No More," Outnowconsulting.com, May 26, 2006 (accessed July 17, 2006).

18. Linda Moss, "Gay-Aimed 'Here! TV' Goes Premium," *Multichannel News,* October 4, 2004, 10.

19. Moss, "Pushing the Boundaries."

20. Moss, "Gay-Aimed 'Here! TV' Goes Premium."

21. Tony Gnoffo, "3 New Premium Channels Focused on Gay Viewers," *Philadelphia Inquirer,* April 6, 2005, Financial News.

22. Gnoffo, "3 New Premium Channels."

23. The Golf Channel attempted this model in 1995. Within a year the channel converted to an all-advertiser supported channel offered as part of premium-tier packages. Several people have compared the two channels, asking, as Katherine Sender did, "If there's a golf channel, why not a gay one?" (cited in Michael Willoughby, "Gay Broadcasting: Pink TV Hopes for a Rosy Future," *Guardian,* May 23, 2005, 4). Sender probably did not intend to equate being gay with an avocation for a particular sport. Instead, she and others who make such claims are probably trying to convey that if there are enough golf devotees to support an entire channel on golf, then it is likely that there are enough lesbians and gay men interested in entertainment directed toward them to support such a channel. However, the quotation does underscore the ease with which communities and identities, especially those based on sexuality, overlap and are intertwined with commodity culture and the language of choice with which it is associated. An appreciation for or skill at golf may be fundamental to one's identity but it is a conscious choice. Sexuality is more deeply rooted and emerges as a far less conscious choice. Further, sexual identity is fundamentally less reliant on purchase even if its contours are necessarily shaped by it. However, comparing a channel for lesbians and gay men to a channel for people sharing an immutable genetic trait as Adam Sternbergh has done is also off the mark. Sternbergh writes: "programming a niche channel around gayness is less like targeting, say, golf lovers than it is like targeting, say,

people with red hair" (see Adam Sternbergh, "I Want My Gay TV," *New York Magazine*, July 27, 2005, http://newyorkmetro.com/nymetro/arts/tv/12045/ [accessed October 17, 2005]). Although the origins of our sexuality are heavily debated, sexuality, even if it is largely laid down through genes and biology, manifests itself through particular social, economic, and political circumstances. It is neither as genetically immutable as hair color nor as adoptable as a sporting trend.

24. The National Cable & Telecommunications Association (NCTA), the General Accounting Office (GAO), and the Federal Communications Commission (FCC) have commissioned reports concluding that à la carte pricing could raise the barriers to market-entry for niche channels, especially those targeting underrepresented groups. See GAO, Report to the Chairman, Committee on Commerce, Science, and Transportation, U.S. Senate (October 2003, GAO-04–08) and the FCC's Media Bureau "Report on the Packaging and Sale of Video Programming Services to the Public" (November 18, 2004), esp. Section IIF: "Feasibility of A La Carte and Themed Tiers," 26–59. A more recent report argues against the Commissions earlier findings and concludes that à la carte pricing would reduce costs to consumers. "FCC Media Bureau Report Finds Substantial Consumer Benefits in A La Carte Model of Delivering Video Programming," *All Business* (February 9, 2006), http://www.allbusiness.com/government/3528377-1.html (accessed January 7, 2007).

25. "Applebee's Says Ad on 'Gay Weddings' Ran by Mistake," BGILT-News (October 2, 2002) http://lists.uua.org/pipermail/bgilt-news/2002-October/000716.html (accessed September 26, 2005).

26. This figure is derived primarily from Viacom's internal surveys. The figure has been reported and repeated frequently in the press in discussions of gay viewership. Among its earliest appearances is in Bill Carter's "MTV and Showtime Plan Cable Channel for Gay Viewers," *New York Times,* January 10, 2002, C1.

27. "The Agony and the Ecstasy"; "Advertisers Drool Over Gay Market," *The Record,* November 15, 2003, C6; Moss, "Pushing the Boundaries"; Joe Mandese, "The Rainbow Connection: The Gay Community Has Money to Burn, But Few Marketers Know How to Reach It," *Broadcasting & Cable* 55, no. 30 (July 25, 2005): 22.

28. Anthony Freitas, "Belongings: Citizenship, Sexuality and the Market," in Jodie O'Brien and Judy Howard, eds., *Everyday Inequalities: Critical Inquiries* (Malden, MA: Blackwell, 1998), 362–84; M. V. Lee Badgett, "Beyond Biased Samples: Challenging the Myths on the Economic Status of Lesbians and Gay Men," in Amy Gluckman and Betsy Reed, eds., *Homo Economics: Capitalism, Community, and Lesbian and Gay Life* (New York: Routledge, 1997), 65–72.

29. M. V. Lee Badgett, *Money, Myths, and Change: The Economic Lives of Lesbians and Gay Men* (Chicago: University of Chicago Press, 2001).

30. "The Agony and the Ecstasy."

31. Oskar Negt and Alexander Kluge, *Public Sphere and Experience: Toward an*

Analysis of the Bourgeois and Proletarian Public Sphere, trans. P. Labanyi, J. Daniel, and A. Oksiloff (Minneapolis: University of Minnesota Press, 1993).

32. John D'Emilio, "Capitalism and Gay Identity," in Ann Snitow, Christine Stansell, and Sharon Thompson, eds., *Powers of Desire: The Politics of Sexuality* (New York: Monthly Review Press, 1983), 100–123.

33. Negt and Kluge, *Public Sphere.*

34. Geraldine Fabrikant, "Two New Cable Ventures Seek to Tap the Market," *New York Times,* April 11, 2005, C1.

35. Miriam Hansen, "Foreword," in Negt and Kluge, *Public Sphere and Experience,* vii–xiii.

36. Michael Warner, *Publics and Counterpublics* (New York: Zone Books, 2002).

37. Nancy Fraser, "Rethinking the Public Sphere: A Contribution to the Critique of Actually Existing Democracy," in Nancy Fraser, ed., *Habermas and the Public Sphere* (Cambridge, MA: MIT Press, 1992), 109–42.

Chapter 11

The Nickelodeon Brand

Buying and Selling the Audience

Sarah Banet-Weiser

As is well-known in the media industry, five multimedia conglomerates—Viacom, Disney, News Corporation, NBC Universal, and Time Warner—exert unprecedented power in marketing messages and products to young people, capitalizing on the lifestyle culture of "cool" and incorporating what historically have been subversive and anti-establishment ideologies as the very center of their marketing strategies. This marketing trend has been a major impetus for the development of brand culture, where the brand matters more than the product, and corporations sell an experience or a lifestyle more than a thing. As cultural theorist Naomi Klein points out, "What these companies produced primarily were not things, they said, but *images* of their brands. Their real work lay not in manufacturing, but in marketing."[1] The relationship between commercial culture and youth has become one based on brand bonding, where differences between "authentic" youth experiences and the experiences sold to youth through corporate branding are no longer (if they ever were) distinct. The identity of the brand is maintained by personal narratives—lifestyle, identity, empowerment—more than by a more historical language of advertising, which relied heavily on a product's efficiency in a competitive market.

In this essay, I will discuss the ways in which the children's cable channel Nickelodeon developed its distinct brand identity as a consequence of recent cultural and economic developments within the media landscape. These developments include the emergence of the cable industry, the subsequent development of niche networks, and the increasingly formidable presence of the children's market in the media economy. Within this context, I examine the ways that contemporary strategies of branding work to commodify an *experience*—not just a product—for audiences. For Nickelodeon, which claims to "respect" kids by creating a network just for them, this experience is about kids' empowerment and a particular kind of citizenship. Within the Nickelodeon context, though, empowerment is not a discrete political function or privilege or action taken by the child, but rather a product itself, a crucial element in the brand identity of the channel. The network addresses its child audience as empowered citizens who are able to make decisions about politics, culture, and relevant social issues by virtue of their membership in the brand community. Brand loyalty thus becomes much more than an inclination to buy a particular product; it further involves a kind of cultural affiliation, where being a "Nick kid" means experiencing a shared community and common values about current youth culture. The specific development of the Nickelodeon brand is thus an ambitious market strategy that appropriates political (and personal) rhetoric about empowerment and agency as a way to promote the network. As the Nickelodeon employee handbook states, "Nick is more than just another kids' entertainment outlet; our big orange splot [the network's logo] stands for a set of ideas kids can understand and trust."[2]

The early loyalty of a child audience is important for television networks, where children are understood in marketing terms as a "Three-in-One Market." Representing potential buying power on three different levels, they are a primary market for goods and services; they are an "influence market," directing parents' spending toward their own wants and needs; and they are a future market for all goods and services.[3] The construction of a "children's market" has been met with hyperbolic and inflated claims from industry executives about its lucrative potential—claims that function to solidify the children's market as a legitimate one to be exploited. According to children's marketing expert James McNeal, the kids' market has been growing at a rate of 10 to 20 percent a year since the mid-1980s (a claim which is clearly exaggerated, but serves to emphasize the increasing power of this market). Within this kind of economic and cultural environment, where kids are thought to represent "more market potential than any other demographic group," competition for the attention and loyalty of children within media companies has become even more intense.[4]

Combined with the continued reach of the cable industry into more and more niche channels, including children's channels, the brand identity of channels and specifically the marketing of brands as particular experiences have achieved a new economic significance. Effective branding strategies that result in attracting both narrowly specific audiences and advertisers concerned with reaching those same specific audiences have become the norm for transnational media conglomerates such as Viacom.[5] Unlike some of the other brands discussed in this volume, such as Discovery Channel and HBO, Nickelodeon is often touted as a particularly successful brand for kids and has clearly provided an economic model (in terms of both brand success and advertising revenue) in the media landscape for channels such as Cartoon Network and Fox Kids. In order for a network's brand campaign to make sense to a child audience, however, the child needs to be cultivated as a consumer, and in the present media context, as a cultural citizen—a member of the brand community—as well.

Children as Consumer Citizens

Consumer citizenship represents part of a larger dynamic of citizenship that moves between agency and conformity in media culture. I contrast consumer citizenship with a more traditional notion of citizenship, which has historically been shaped by different understandings of rights and responsibilities of citizens and the roles and responsibilities of governments. While consumer citizenship indicates a certain willingness to participate in consumer culture through the purchase of goods as well as a more general affirmation of consumption habits, it also points to something broader, where the distinctions between cultural and social practice and consumption are not so finely drawn. This ambiguity allows Nickelodeon to claim, not disingenuously, to "empower" kids within the cultural context of consumer citizenship. Empowerment, then, in the context of Nickelodeon designates the power that is attached to a particular consumer demographic, such as the child or "tween" audience. However, empowerment means not only the power to make consumer choices. The branding strategies of Nickelodeon also frame the language of empowerment and kids' rebellion in a way so that the very successful channel ironically adopts a kind of counter-hegemonic, "under-dog" identity. This identity is then marketed to children as a means of empowerment, specifically in relationship to adults.[6]

It is within this cultural economy that Nickelodeon imagines its audience as a group of active consumer-citizens. The fact that this imagined audience is similar for both Nickelodeon and advertisers does not automatically mean that the network is merely giving lip-service to their claims that they are on the "kids' side." What it actually means, however, to be on the "kids' side," and what the consequences are for the child audience when an empowered identity is marketed as a kind of product, needs to be critically explored.

For Nickelodeon, the emphasis on generation (suggested by its being on the "kids' side") is framed in terms of "respect," wherein the network positions itself as a rebellious entity that respects children and is capable of liberating them from the stifling world of adults. But given the facts and imperatives of marketing, this "respect" is *already* a kind of product for Nickelodeon. Consequently, the network's signature brand of "kids rule!" is not only about recognizing that kids influence consumer habits and behaviors in the household; it also offers up the very notion of the empowered, respected kid as a kind of product. The economic imperative and the political message of the network thus become indistinguishable through the careful construction of the Nickelodeon brand as a discrete place, where "Nick kids" are created and marketed by the network as empowered citizens.

Community, Belonging, and the Brand: Nickelodeon's Identity

Naomi Klein argues that branding has become the key symbolic frame of reference for contemporary identity.[7] Put simply, the advertising trend in the 1990s was that the brand mattered more than the actual product. Industry executive Donna Mitroff confirms this notion in regard to Nickelodeon, arguing that the network focuses less on individual shows and more on the brand itself. She states:

> What Nick did was so systematic about this whole branding thing that they were doing, so the programming and the thing were all woven together. And you didn't just pick shows in isolation because someone liked them. It all had to merge to create that concept which was Nickelodeon. And they spent a lot of time with kids, trying to figure out what that would be. I mean, that's how the big orange blob came to be, the splat . . . That was all by spending time not just with your branding people, but with kids.[8]

The corporate employee handbook, titled "How to Nickelodeon," discusses the imperative for Nickelodeon employees to "find the kid" in themselves in order to really fit in at the company. There is a clear mission statement and a "Bill of Rights" for kids displayed in the offices and within the employee handbook.[9] The brand loyalty of Nickelodeon is not simply about the Nickelodeon products that are consumed by the public, but also about what one former employee calls "the Nick way of life" or "the Nick voice."[10] This notion, that Nickelodeon is a "way of life," or has a distinct voice, or can function as a kind of verb, for that matter, as something that one *does,* or that one *is,* rather than as a product that one consumes or uses, is part of a relatively new culture of branding. Watching Nickelodeon programs thus becomes a cultural practice, akin to other citizenship practices, such as demonstrating national affiliation and loyalty.

In an article aptly titled "The Nickelodeon Experience," the then-president of the network, Geraldine Laybourne, details the way in which the concept for the channel developed:

> Nickelodeon decided to do what nobody else was doing—raise a banner for kids and give them a place on television that they could call their own. In doing this, we reevaluated everything from the program lineup to the logo. We replaced Nick's original inflexible silver ball logo with a bold, brash and ever-changing orange loop that can take as many shapes as kids can imagine. This became the symbol of Nickelodeon's new identity and mission, and in January 1985, we relaunched as a network dedicated to empowering kids, a place where kids could take a break and get a break.[11]

Thus, from the time that Laybourne envisioned a new image for the network that erased the trait of "green vegetable television" from the audience mind, the channel was conceived as a place, not a product, a set of ideas, not as merchandise. Nickelodeon subsequently became many different things to its audience. As Laybourne puts it, the channel is not simply something a child turns on and off at whim; rather, it is a kind of "understanding friend" who "is always there, from breakfast to bedtime, everyday, whenever kids want to watch. Nick is their home base, a place kids can count on and trust."[12] But is this kind of media interaction a means to empowerment? And what are the consequences of considering a television channel one's most trusted friend? The costs involved when children adopt the Nickelodeon ideas that they are victims in an adult world and can become empowered by watching a television station are, needless to say, rarely examined by the network. The ubiquity of the channel (at least for families who can afford cable), which is shown 24 hours a day, airing programs that are either designed for children or, in the case of Nick at

Nite, appropriate for children, means that the channel *is always on*—and hence can position itself as something (or someone) that children can trust, even more than their own parents.[13]

The Nickelodeon viewer is the savvy consumer—one who knows brands, has strong name recognition, and can easily move between the television set, the computer, and the mall. As children themselves become more confident in their use of the media, the advertising and branding campaigns that attempt to cultivate them as loyal customers also become more sophisticated and hipper so that the distinctions between popular culture and consumer culture are increasingly difficult to discern. Even more significant, the cultural definition of what it means to be a savvy, empowered kid is *created* by networks like Nickelodeon. Consequently, the network's claim to empower kids becomes a kind of self-fulfilling prophecy: as Nickelodeon crafts the definition of a contemporary empowered kid, what it means to be empowered comes to take the form of being a "Nickelodeon kid." In the following sections, I discuss two interrelated themes that are crucial to Nickelodeon's brand identity: the Us vs. Them philosophy, wherein kids are "oppressed" in an adult world; and the Nickelodeon Nation, in which the network can be a "home" or neighborhood.

Us vs. Them: Kids Rule!

Nickelodeon has built its brand identity around a primary core idea: "Us vs. Them." This idea developed in part as a response to the cultural debates over educational and entertainment programming for children that were coming to a head in the mid-1980s. Advocates of educational programming, such as educators, legislators, and media watchdog groups like Action for Children's Television (ACT), recognized the symbolic power of television and saw the potential of television to be a kind of educator. Obviously, *Sesame Street* and the Children's Television Workshop capitalized on this notion early on and enjoyed great success.[14] From the mid-1980s, Nickelodeon very purposefully constructed its brand identity in opposition to the educational potential of television.[15] Specifically, for the network, education equaled adults, and therefore educational television was television that *adults* thought children would like. Framing this kind of television as patronizing to kids, Nickelodeon shaped the network's identity around the idea of Us vs. Them. As the employee handbook states about the network's identity: "It's an adult world out there where kids get talked down to and everybody older has authority over them. For kids, it's 'Us vs. Them' in the grown-up world: you're either for kids or against them . . . we were on the kids' side, and we wanted them to know it."[16] In a context of media advocacy on behalf of children, Nickelodeon insisted that children needed to be protected from precisely those who claimed to want to protect them. Nickelodeon would not represent the authoritative adult world, but would rather identify with kids and "empower" them.

For Nickelodeon, empowerment means, in part, a particular tone when "speaking" to a child through the network—a tone that apparently refers to a media-defined notion of respect. As former Nickelodeon employee Linda Simensky comments about this rhetorical strategy:

> I always thought it was interesting—and right on—that you could never say something was "fun" or "cool." You couldn't tell kids that it *was* something—you just had to present it. And kids could decide if it was fun or cool, but you just had to present it in a fun or cool way, then you didn't have to say it, you didn't have to qualify it. I think that is really smart. You watch other networks—you watch the Disney channel and they say "watch our show—it's cool, you'll have fun!" That feels wrong to me.[17]

Simensky's observations point to a tension between speaking for kids, or on behalf of kids, and allowing kids to speak for themselves, or make an independent judgment or decision about what they like and what they do not like. While it is clear that Nickelodeon *is* speaking for kids—it is the network, after all, that defines what empowerment means for its child audience—it does so in a way that capitalizes on contemporary notions of "cool": cool means anti-establishment, alternative, the under-dog. Nickelodeon re-shapes these ideas within the brand context and positions itself and its audience as victims in an on-going power struggle. Most significantly, liberation from this kind of oppression comes not from traditional political action, but rather from choosing the right television network. This is one way in which Nickelodeon conflates a residual notion of political citizenship with marketing.

Nickelodeon smoothly adopts political rhetoric in a way that indicates that it is through watching television that children can become empowered. What kids "need," according to the network, is simply better TV. This idea of "better TV" for kids was a crucial element of the re-launch of the network in 1985, which involved the airing of a continuous set of promotional spots called "promise spots"—commercials that highlighted the new claims and "promises" of the network to its child audience. Through the kids-only rhetoric of the Nickelodeon promises, these promos seamlessly incorporated the Us vs. Them philosophy as part the network's core ideology, rather than as an efficient marketing tactic. In an apparently unselfconscious manner, these promises "define Nickelodeon and ensure that everyone talks and thinks about Nick in the same way. They serve as tools for thinking about the network and describing our differences from the competition."[18] Above all else, the "promise spots" "turn Nickelodeon, the TV network, into Nick—a place where kids know they can relax and be themselves." Some of these promises include: "Nickelodeon Is the Only Network for You," "Nick Is Kids," and "Nickelodeon Is Every Day."[19] As Simensky comments, "[These 30-second promos were] just drilled into kids in those early years, these promises of what they were going to get. It was just branding in every possible way—it was smart to do that because kids would hear that and they would believe it . . . and it was true."[20] And indeed, because each of the network's promises had to do with Nickelodeon being the only network for kids, and because in this branding environment it was necessary for Nickelodeon to actively define a contemporary empowered media citizen, the promises *were* true. In other words, Nickelodeon's promise spots helped assure the network dominance in the children's television market by defining who and what the audience was in an appropriated language of political empowerment.

A "promise spot" called "Nickelodeon Is the Only Network for You" asserts: "Nick talks directly to kids and appeals to their sense of humor. Nick is the place where kids come to take a break and get a break. Nick empowers kids by saying to them, 'You're important—important enough to have a network of your own.'"[21] In a commercial context of media visibility, having a "network of your own" has an interesting twist on the meaning that Virginia Woolf's "room of one's own" had in the early twentieth century: it inspires independence, freedom of thought, and ownership over one's body and thoughts, but in the specific milieu of a commercial media network. This tactic is similar to the one used by Nickelodeon's corporate sibling, MTV, in its early slogan "I want my MTV"—a strategy that constructed the channel as both a space for the viewer and a link to ownership or entitlement. In fact, in the Nickelodeon universe, the concepts of independence, freedom of thought, and ownership over one's body *have meaning only* within the context of a commercial media environment—these definitions of identity and subjectivity are pre-packaged products that the channel encourages its audience to buy. Consequently, empowerment from having "a network of your own" has meaning only within the landscape of commercial television—outside of this environment, ownership of a network, even if illusory, has no power and indeed no significant meaning. Not only does this imply that "real" power is consumer power, it also renders cultural realms outside the commercial media as invisible or at the least unimportant for citizenship.

Nickelodeon taps into another aspect of consumer empowerment by emphasizing the idea that, within the world of consumption, kids are oppressed by adults. The channel strategically uses the concept of oppression as part of its self-identity; specifically, Nickelodeon claims to alleviate the oppression kids apparently feel simply by living in today's culture. Kids are different from adults, Nickelodeon insists; they have different needs, different desires, and require different kinds of entertainment. This binary opposition between adults and children, like many other binary relationships that Nickelodeon exploits as part of its brand identity, is difficult to maintain. Nickelodeon's strategy in this regard was to find an effective logo and mascot.[22] When Nickelodeon was re-launched in the mid-1980s, the channel wanted to symbolize its new irreverent, playful attitude through its constantly-changing, flexible orange logo. Ironically, in order to achieve the apparent flexibility of the "orange splat," Nickelodeon designers must follow explicit, precise rules and regulations concerning its design and display. The color must never vary (so much for the "irreverence" of the color orange), and the word "Nickelodeon" must always be easily decipherable. The shape itself changes depending on the context, so that form follows function: a turkey shape for a Thanksgiving promo, a flag shape for a Fourth of July promo, and so on.

The paradox between the network's claim of flexibility in its ideas and its insistence on conformity is evident, as Simensky comments, in the way that Nickelodeon employees were talked about in the industry as

> bleeding orange . . . People joke now that it was a cult . . . you know, do you drink the Kool-Aid. You believe Nickelodeon's the only way to be but there was just something about it that was right and part of it was that there was never any sense of exploiting kids—that was never part of it. If anything, there was this sort of idealizing kids—it was about respecting them, almost "over-respecting" them—people really felt like they were working in service to kids.[23]

The idea of "over-respecting" kids in such a commercial environment is ironic—if the network was indeed created to serve children, then *doing that* would hardly constitute "over-respecting" them. More to the point, what does it mean, in this context, to be "working in service to kids"? This is not to say that Simensky's claim that there was "never any sense of exploiting kids" is insincere, but the idea of creating a commercial network that is intended to serve kids begs the question of *how* they are being served. In other words, the logo, like the network itself, capitalizes on the political valence of irreverence and flexibility, and transforms the logic of these concepts to better fit the commercial environment.

Alongside the logo, Nickelodeon also used a material artifact to represent itself: green slime. Green slime is part of Nickelodeon promos, programming, and magazine articles, as well as a product in and of itself. The presence of slime (or "gak" which is a similarly gooey, messy substance) on Nickelodeon translates as a symbol of kids' oppression—it is anti-adult, anti-authority, and kid-centered. As the corporate handbook puts it: "Mess, slime and gak are more than a gag or theme, they are Nickelodeon's mascots. Pretty much by accident, on shows like *You Can't Do That on Television* and *Double Dare*, we discovered that green slime and mess excite kids emotionally, symbolize their rebellion and totally gross out adults."[24] Slime is clearly much more than simply a licensed product for the network (although not insignificantly, Green Slime Shampoo was the first licensed product to emerge from Nickelodeon). It is an important symbolic object that, in its messiness, its refusal to stay within conventional spatial borders, its sheer disgustingness and audacity, characterizes not only the mission of the network, but also the personal identities of its audience.

Slime was first used on *You Can't Do That on Television,* in which every time an adult or a child on the show used the words "I don't know" he or she was "slimed"—a huge amount of green slime was poured on his or her head. Slime, because it is understood to be fundamentally anti-adult, is construed by the network as a part of its commitment to respect children rather than condescend to them. The copyright-protected toy product, for the Nickelodeon universe, becomes a particular kind of unifying symbol, one which signifies a specific meaning that makes sense in a contemporary context where market forces and cultural politics have encouraged a generational divide. Slime is precisely what "adults don't get" on Nickelodeon—and the network works to keep it that way. As the handbook states:

Kids watching at home identify with the poor kids on TV who gets slimed. They say to themselves: "He's like me, he's just trying to get through his show and he's getting dumped on. I'm just trying to do my best to get through life and I get dumped on." Green slime becomes a symbol for kid oppression and solidarity. It's like a great common denominator. A slimed kid says to all kids: "You're not stupid and you're not alone! We all get dumped on!"[25]

To be "dumped on" obviously has at least two meanings in this context: the literal dumping of slime on the child in the game show, as well as the metaphorical meaning of being "dumped on"—kids are victims here, oppressed by adults, by network television fare, by life in general.

Indeed, in this context, childhood itself is a state of victimization and can be thus considered part of a contemporary construct of liberalism, wherein liberalism (and by extension, citizenship) is now understood primarily in terms of identity positions or identity politics. Within this context of liberal citizenship, Nickelodeon provides a refuge, by not only creating a material artifact that is a symbol "for kid oppression and solidarity" but also "standing up" for kids in this kind of oppressive environment. This, of course, has been an effective strategy within liberal politics: create the problem, thus creating an opportunity to resolve it. Nickelodeon generates an imagined space of oppression through its relentless focus on the cultural divides between adults and children. Because the channel is dedicated to "kids only," it—rather than parents—can then liberate the child through the use of an anti-adult artifact. This "liberation" is contained and commodified, encouraging the members of the audience to feel solidarity with each other. Consumer citizenship is built around this kind of identification, where shared consumption of products provides for a kind of safe liberal practice. The representation of oppression symbolized by consumer artifacts such as slime thus suggests an underlying dynamic: liberation is also gained by consumption. Consumption as a kind of liberation shapes the Nickelodeon brand identity, and the question of whom or what one is liberated from is answered by the constant presence of the Us vs. Them philosophy. In this sense, Nickelodeon's branding establishes a particular kind of social identity, where loyalty to the channel offers a kind of temporary or provisional liberation from adults. The network then can claim to be for "kids' rights" that are constituted in opposition to those of adults, or to the way in which adults treat them.

Slime and mess on Nickelodeon represent a new order for children, one in keeping with the overt philosophy of the network. It is not so much— or even especially—that children understand the significance of slime in Nickelodeon programming, but that *adults don't*—that is the crucial importance of the symbol. Because of the extreme difference in how adults and children define slime, slime becomes *the* symbol for Us vs. Them— it embodies the playful, flexible logo of Nickelodeon itself. And importantly, slime is not just an anti-adult statement, but it is also an affirmation of kid-hood. Thus, slime is not about humiliating the kid, but rather is framed as a "triumphant event" that symbolizes kids' rebellion. Constructed specifically to alienate adults, slime is a symbol of not only the cultural definition of "play" that Nickelodeon maintains, but is also symbolic of a kind of liberation. Nickelodeon is set up as a "place" where kids can confront other parts of their lives and identities: rules, adult authority, schoolwork. Nickelodeon is the neighborhood for hanging out, the cool place to be.

Citizens of the Nickelodeon Nation

Establishing the network's brand identity as specific place—the transformation of Nickelodeon, the network, into "Nick," the space—was a crucial precedent for the network's 1999–2000 brand campaign, "Nickelodeon Nation." This campaign follows an even more contemporary brand design than the "promise spots," although the "Nickelodeon Nation" campaign also followed the mission of Nickelodeon carefully. The appropriation of nationalist rhetoric by a children's cable channel seemed seamless at that historical moment. The design of the network's brand identity as a "nation" was multi-dimensional and tapped into several different cultural conversations circulating at the time—consumer patriotism, branding loyalty, the transformation of products into generalized brand ideas, such as nationalism—all the while maintaining the Us vs. Them ideology.[26] As one reporter described the campaign: "Nickelodeon wants world domination. Or at the very least, its own nation."[27] The network launched the Nickelodeon Nation campaign in October 1999, on network television, during what is widely considered a favorite American national event, the World Series. Veering from its "kids only" rhetoric at least for this campaign, the network courted a more transgenerational audience with Nickelodeon Nation: the campaign was aired in primetime and late-night programming, and during the early morning news shows *Today* and *Good*

Morning America. Print ads to advertisers were run in trade magazines such as *Advertising Age* and *Adweek* and ads aimed at gaining (and keeping) subscribers were directed to audience members in magazines such as *TV Guide.* In effect, the campaign for the "Nickelodeon Nation" amounted to a campaign for a network for the "nation."

This theme was particularly effective in the context of intense competition for brand recognition within the television landscape. Nickelodeon Nation, like other successful brands, has become synonymous with the name of the company, but more than that, the hyperbolic claims of the campaign—the network *as* nation—captures both the way in which Nickelodeon dominates the children's television landscape and the rhetorical insistence that the network is more than just TV—it is a place, a nation, a feeling of belonging, a stand-in for parental authority.

The ads that formed the heart of the Nickelodeon Nation campaign always featured several children, both boys and girls, of a variety of ethnic and racial backgrounds, dancing in the street, in playgrounds, and in neighborhoods. These ads became immediately recognizable because of their snappy do-wop sound: a medley of hip-hop and nostalgic pop music. The "Nickelodeon Anthem" was a revision of the Dixie Cups' 1965 hit tune "Iko Iko":[28]

> 2, 4, 6, 8—SPLAT!!
> Me and my generation
> Don't you like kids who look like that
> Nickelodeon Nation
> Talk about Hey now (hey now), Hey now (hey now)
> Going to Celebration
> I believe in Nick 'cause Nick believes in me
> Nickelodeon Nation!

The color print ads usually featured only one child, caught in a candid moment of laughing or playing, and included large orange print that said "Nickelodeon Nation" or "We Pledge Allegiance to Kids."

The versions of the Nickelodeon Nation campaign directed at advertisers stressed the importance of kids as consumers, as Cindy White and Elizabeth Preston point out, labeling kids as "amazingly powerful consumers" and promoting the idea that "They may be small. But if you're a marketer, kids can be positively superheroic."[29] As White and Preston argue, with these kinds of ads, "Nickelodeon tries to capitalize on its undisputed brand identification as a service singularly devoted to attracting child consumers and uniting them with appropriate advertisers."[30] More than simply attracting the child consumer, however, the use of national rhetoric in these appeals allows the network to *create* its audience as particular kind of consumer-citizens. White and Preston point out that "The young inhabitants of this community, this nation—Nick-branded kids—are not simply citizens: they are powerful consumers of the products that litter its landscape."[31] In fact, these "Nick-branded kids" are citizens in this context precisely *because* they are powerful consumers—it is their consumer behavior that authorizes them as citizens of the Nickelodeon Nation. The Nickelodeon Nation promotional strategies signal the various ways in which Nickelodeon understands both itself and its audience as an empowered force in the consumer market. This kind of empowerment, created by Nickelodeon as a particular kind of product, is not one to dismiss lightly. It is the central feature of the present-day child consumer-citizen.

It is not simply a hyperbolic assumption that Nickelodeon forms a nation. The children's market that Nickelodeon has both created and captured does form the network's national borders, with the other "states" being other channels and other market shares. The consequences of framing a television station as a nation are important: this construction is more than simply a fan relationship, but indeed encourages children to identify with and personally invest in Nickelodeon as citizens of a nation. Created by the appeal of consumer participation and commercial belonging, this kind of child citizen is encouraged to actively participate in Nickelodeon's definition of political action: "voting" on the *Kids' Choice Awards,* where kids determine favorite pop stars, music, actors, and so forth, or choosing a candidate on *Kids Pick the President,* where during an election year children "vote" for president. This kind of consumer-driven political action, where election coverage is used as a kind of interstitial between programming, in short sound bites for easy understanding, helps the notion of media nationhood have resonance with audiences. In this scenario, divisions between a kind of "real" national identity (ostensibly built upon political action and civic agency) and a consumer national identity, one that connects empowerment with the purchase of products, are illusory at best.[32] In particular, the collapse of the network with nation establishes Nickelodeon's superiority in the landscape of children's television, confirming the network as the key competitor in terms of programming content, style, and promotional campaigns.

This strategy is evident in the history of Nickelodeon in relation to that of other cable networks. While initially Nickelodeon enjoyed success in part because it had no other competitors, by 1999, there were four full-time children's cable networks: Nickelodeon, Disney, the Cartoon Network, and Fox Family Channel. The Nickelodeon Nation campaign needs to be interpreted within this context: the campaign can be read as a familiar, continuous reminder that Nickelodeon was (and is) the standard against which these other channels could be measured. Given the intensely competitive commercial environment of cable television, all niche networks attempt to define themselves as being different from—and superior to—their competition, and Nickelodeon was certainly no exception. The competition between Nickelodeon and other children's networks is often noted in the media, in part because Nickelodeon had so thoroughly captured the market. Thus, when Disney entered into the fray, the struggle for territory itself became a media event. As reporter Josh Young describes the competition between the networks, it was "the battle of enchantment versus empowerment: . . . in this schoolyard showdown, the Big Orange Bully (the industry nickname for Nick's citrus-colored logo) has already thrown the first punch."[33] In its branding strategy and self-definition, Nickelodeon implicitly mocked Disney as the goody-two-shoes of children's television, again capitalizing on the commercial potential of being the cool channel. Construing Disney as the patronizing (and infantilizing) parent, Nickelodeon presented itself as the hip best friend and thereby enhanced its brand tactic of having a renegade, and indeed counter-hegemonic, identity. In the battle between enchantment and empowerment, Nickelodeon's tactics clearly worked in its favor. Even Cartoon Network, which has a more irreverent style than the Disney Channel, often seems to be no more than a thinner imitation of Nickelodeon. As Simensky commented about the competition between Nickelodeon and Cartoon, "If you look at [Cartoon's] Friday night packaging, they might as well run a line that says 'we'd like to be Nick—how can we do it? Email your suggestions to us.' Because they are trying but they're not pulling it off. They have a bad case of Nick envy. Disney also has Nick envy."[34] Nickelodeon's dominance in the kids' television market has allowed the channel to take the moral high ground; as Cyma Zarghami, then Senior Vice President of Programming, commented when asked about Nickelodeon's competition, "There's the Superman quote that with great success comes a great responsibility. We will remain responsible, but there's a lot of irresponsible ways that will be used to get ratings. That's what alerted the government to begin with. Kids' programming is like food. If you give them candy, they'll eat it. We try to make our programs more nourishing. We won't do anything gratuitously in a show."[35]

Yet, despite Nickelodeon's self-conscious claim to be the "alternative" choice in the television landscape, by the mid-1990s, the rhetoric of the network had even less of the feel of an upstart and more of that of a major global corporate player. In fact, the current president of Nickelodeon, Herb Scannell, used marketing rhetoric when speaking about Nickelodeon's attempts at capturing new audiences with different media venues: "The battleground for tomorrow is not just for share of TV audience, but for share of mind . . . With Nickelodeon now in magazines, online and in movies, we have three more opportunities to reinforce the Nickelodeon brand."[36] And the multimedia reach of Nickelodeon is indeed enormous in scope. In addition to Nickelodeon magazine, the channel is a corporate sibling of Paramount Pictures, which released *The RugRats Movie* in 1999, the first non-Disney animated film to gross more than $100 million. Paramount is an important part of Viacom's empire-building, as well as an effective way to edge out competition such as Disney or The Cartoon Network. Reporter Marc Gunther claims, "The game they're playing at Nickelodeon these days isn't *Guts* or *Double Dare*. It's Extend That Brand. Name a category that touches kids, from toys to macaroni and cheese, and there's a product out there with Nick's name on it. The cute little children's channel that once ran no ads at all now looms over the kids' market for just about anything."[37]

The Nickelodeon Nation campaign helped confirm the network's standing as a dominating force not only in the area of television programming, but in all areas of children's consumer culture, from film to toys and games, to basic merchandise. Because the network owns so much of the commercial media that are directed at children, and because commercial media are an important element in the cultural definitions of both childhood and consumer citizenship, Nickelodeon contributes significantly to these definitions. At the same time, the network's identity remains that of a rebel. This dynamic reflects the contemporary pose of branding, where the enormously successful Nickelodeon can ironically continue to claim an alternative subjectivity. The network as "alternative" is precisely its most effective branding strategy—it is by virtue of Nickelodeon's claim about victimization, where kids are oppressed and Nickelodeon is the savior, that the network enjoys its commercial success.

The rhetoric that structures the branding strategies of Nickelodeon refers to a politics that is fundamentally about empowerment: rights, re- spect, security, and liberty. Through this kind of brand development, Nickelodeons presents to its audience a network—and an identity—that seems non-commercial. Cultural anxieties about advertising to children and training them to be "little consumers" were, and continue to be, based on an exploding kids market, more and more segmented, age-specific ca- ble programming and advertising, and the increasing purchasing power of children, especially tweens. Nickelodeon responded to these anxieties in a way commensurate with the contemporary branding environment: by positioning itself as the renegade, the cool network, the outsider. The "re- spect" Nickelodeon has for its child audience is not necessarily insincere, but it is based on commodifying and packaging trends and values impor- tant to particular age groups—and the network certainly profits from the revenue. Importantly, this construction of empowerment *becomes* empow- erment for a child media audience—empowerment exists for this genera- tion within the bounds of consumer culture. This is not to assume a dis- tinct opposition between "real" culture and marketing culture, but it is to argue that Nickelodeon's claim that the network empowers kids makes sense only within the culture of the market.

NOTES

1. Naomi Klein, *No Logo: No Space, No Choice, No Jobs* (New York: Picador, 2002), 4.

2. *How to Nickelodeon,* Corporate Employee Handbook (New York: MTV Net- works, 1992), unpaginated.

3. James McNeal, *The Kids Market: Myths and Realities* (New York: Paramount Market Publishing, 1999), 30.

4. Ibid.

5. Elizabeth Preston and Amy White, "Commodifying Kids: Branded Identities and the Selling of Adspace on Kids Networks," manuscript submitted to *Commu- nication Quarterly,* Spring 2004.

6. The child consumer is an identity blurred with other social categories such as love, parental obligation, and education, and is a commonplace way of under- standing children in global capitalist societies. Unlike the earlier part of the twen- tieth century, where arguably the U.S. public was "training" to be a nation of con- sumers, in the contemporary context consumption is as much a part of life for U.S. children (albeit in different ways for different socioeconomic groups) as for-

mal education. In fact, cultural definitions of who—and what—a child is interconnects with consumer behavior. As Stephen Kline, Dan Cook, and others note, the actual physical and intellectual development of children—the definition of childhood—is charted, mapped, and in some ways defined by how the market characterizes this development. In the current brand environment, the early cultivation of children as loyal consumers is particularly important to corporate culture. However, this strategic, aggressive cultivation intervenes in the dominant discourse about children as inherently innocent and in need of protection. Indeed, it is precisely the sophistication of children in terms of their real and potential income, their influence with parents, and their potential as a future market that advertisers rely upon.

7. Klein, *No Logo*, 21.

8. Donna Mitroff, interview with author, March 4, 2004.

9. This is how the Nickelodeon "Bill of Rights" is described in the employee handbook:

In the course of history, it has become pretty clear that all people are born with certain inalienable rights; among them life, liberty and the pursuit of happiness. But these rights haven't always applied to kids. And that stinks!

Now, 200 years after the creation of America's Bill of Rights, this declaration proclaims to the world that you have rights too:

. . . You have the right to make mistakes without someone making you feel like a jerkhead.

You have the right to be protected from harm, injustice and hatred.

You have the right to an education that prepares you to run the world when it's your turn.

You have the right to your opinions and feelings, even if others don't agree with them. So there!

10. Interview with author, January, 2004.

11. Geraldine Laybourne, "The Nickelodeon Experience" in Gordon L. Berry and Joy Keiko Asamen, eds., *Children and Television: Images in a Changing Sociocultural World* (London: Sage, 1993), 304.

12. Ibid., 305.

13. Nick at Nite is a separate programming block on Nickelodeon that airs every evening, presenting television sitcoms from the 1960s through the early 1990s. It airs until the early morning, when Nickelodeon begins again.

14. For more on this, see Shalom M. Fisch and Rosemarie T. Truglio, "G" *Is for Growing: Thirty Years of Research on Children and Sesame Street* (New Jersey: Lawrence Erlbaum Associates, 2001).

15. Nickelodeon's stance against educational television is ironic, given the channel's later production of the extremely successful educational programs *Blues Clues* and *Dora the Explorer*.

16. *How to Nickelodeon.*

17. Linda Simensky, interview with author, February 2004.

18. *How to Nickelodeon.*

19. Ibid.

20. Linda Simensky, interview with author, February 2004.

21. *How to Nickelodeon.*

22. Joseph Turow, *Breaking Up America: Advertisers and the New Media World* (Chicago: University of Chicago Press, 1997).

23. Linda Simensky, interview with author, February 2004.

24. *How to Nickelodeon.*

25. Ibid.

26. Marita Sturken, *Tourists of History: Memory, Consumerism and Kitsch in American Culture* (Durham: Duke University Press, forthcoming).

27. Phillips Business Information, Inc., *Reaching a Nation Nickelodeon-Style* 5, no. 6 (April 5, 2000).

28. I am grateful to Ken Wissoker for this reference.

29. Cited in Preston and White, "Commodifying Kids," 10.

30. Ibid., 24.

31. Ibid., 25.

32. Consumer nationalism also illuminates the ways in which candidates and issues are packaged as products. In other words, the vote-purchase analogy works both ways.

33. Josh Young, "A Better Mousetrap?" *Entertainment Weekly,* December 4, 1998.

34. Linda Simensky, interview with author, February, 2004.

35. Cyma Zarghami, interviewed by Vicki Mayer, August 8, 1997.

36. Donna Petrozzello, "Nickelodeon's Herb Scannell: In Toon with Kids," *Broadcasting & Cable* 128 (February 2, 1998): 28–31.

37. Marc Gunther, "This Gang Controls Your Kids' Brains," *Fortune* 136 (October 27, 1997).

Part III

Cable Programs
The Platinum Age of Television?

Introduction

In recent years, cable programs have achieved the reputation of "quality TV"—expensively produced, intelligently written, they utilize edgy and graphic themes and are often humorous in a dark, ironic sort of way. The flexibility of the cable industry is largely responsible for these quality programs: the relatively light FCC regulation in terms of language, action, and sexuality, the ability to air episodes multiple times during the week, and the freedom from the traditional broadcast confines of "least objectionable programming" have all helped create an environment that allows for interesting, edgy shows to air. Home Box Office (HBO), said by one reporter to have set off the "Platinum Age of Television," has used the freedom of cable to produce highly-acclaimed adult series, and programs such as *The Sopranos* (discussed by Dana Polan later in this volume), *Six Feet Under,* and *Oz* have become the benchmark for original cable programming.[1] That was not always the case, of course—cable programs in the 1980s were more known for schlocky re-runs of tired sit-coms like *Full House* and *Growing Pains* than for cinematic-quality television series. But in 2005, cable channels spent $13.4 billion on their own shows, up 65 percent in four years,[2] and with their novel programming, they have been recently said to "remake television." The question is: Remake it how? Are cable programs truly different from broadcast television, and if so, what do we make of this difference?

In the last several decades, television, media, and cultural studies scholars have examined both televisual texts and audience reception as a way to determine the cultural, political, and economic significance of television in people's lives. Depending on one's theoretical position, this significance is read as corporate capitulation, an effort to homogenize the masses, a useful function for individuals, or a kind of pleasurable activity. New changes in technology—"new media"—have spurred another generation of television studies, which often follow critical traditions by understanding new media either as more oppressive forms of capitalism or as an expression of liberalizing politics. Between these two poles lie issues of representation—a topic with which this section of the book deals centrally. The essays that follow engage, in a variety of ways, the question of whether or not the format and function of cable enable television producers to create a *different* sort of representation from that created by the programs on broadcast television.

While many television studies scholars theorize the production of particular representations and how audiences interpret and make meaning out of those representations, much of this scholarship has investigated the various ways in which groups with the least social and political power develop readings of and uses for cultural products like televisual representation—in fun, in resistance, to articulate their own identity. In the last decade or so, such studies have profited greatly from Stuart Hall's groundbreaking essay, "Encoding, Decoding," which challenged in important ways a linear model of production/message/reception that previously characterized studies of television representation[3] and enabled TV Studies scholars to reframe audiences as actively "decoding" a televisual message, rather than passively receiving it. There have been, of course, many important works that have since offered an increasingly sophisticated understanding of the complex issues surrounding representation and how viewers make meaning. Moreover, on the industry side, television producers have also increasingly pushed the envelope, recognizing, with the escalating presence of niche audiences, that advertisers were not as insistent on providing bland, "non-offensive" television that would appeal to a large mass of consumers. Based on narrow niche audiences, cable channels have also been able to tailor programming even more tightly to particular demographic groups.

Regardless of one's theoretical perspective on the nature and effects of representation, it is clear that the television landscape is changing and that cable programs are an important part of this change. Because of the intense competition for audiences in the cable industry, wherein hundreds of channels struggle for a share of the pie, it seems inevitable that new program formats should emerge. While certainly there are cable channels that repackage broadcast television programs in order to capture already loyal audiences, there are others that make use of the flexibility of the industry to create new, more innovative shows. In this section of the volume, we examine both of these groups. Treating the differences within cable programs, the essays that follow also demonstrate how both kinds of programs follow a more general post-industrial, late-capitalist logic whereby media become more centralized on the one hand and audiences become more fragmented and niche marketing becomes narrower on the other.[4]

Inasmuch as it is clear that cable programs have expanded the scope of the representational landscape on television, it is not so clear whether the increase in kinds of representation lives up to the utopian claims of the early cable industry. As Lynn Spigel has argued, "the current multi-channel landscape is not a world of infinite diversity but rather a sophisti-cated marketplace that aims to attract demographic groups with spending power."[5] Nonetheless, the slick packaging and expensive production of original series on cable channels such as HBO and Showtime have altered traditional understanding of television as being "low culture." High qual-ity programs, and the increasing number of hours U.S. consumers spend watching these programs, have forced scholars' attention to the fact that television programming cannot be simply characterized as homogenized, mass audience fare. On the contrary, it is apparent that there are different kinds of TV: while each is geared towards particular markets, some qualify as "quality," others simply reiterate and confirm a formulaic broadcast tel-evision model, and still others are innovative even as they do not necessar-ily strive for or achieve "quality." Thus, while HBO is often recognized as the benchmark in quality television series, other cable channels, such as Showtime, FX, USA Network, and Cartoon Network have recently pro-duced critically acclaimed series such as *Weeds, The Shield, Monk,* and *Boondocks,* all programs that explore new themes, plots, and characters in ways that have come to be recognized as "quality" television. At the same time, though, these cable channels also air programming such as re-runs of *Beverly Hills, 90210,* and *Nash Bridges,* alongside other shows that may or may not be considered "quality" such as Comedy Central's *South Park* and *Chappelle's Show.*

In the following section, two of the essays offer a critique of the ways in which television scholars have examined and analyzed recent cable pro-gramming. Dana Polan's essay focuses on HBO's *The Sopranos,* which seems to be what established HBO firmly as a brand—*Sopranos* viewers are HBO viewers, consumers of an expensive, high-end production, origi-nal cable series—and to have reinforced the slogan that "It's Not TV, it's HBO." Whether or not this is so, television scholars and critics do seem quite ready to label *The Sopranos* "quality TV," and in his essay, Polan asks why this appears to be the case, as well as why some read the show as a transparent display of meaning or a clear and simple morality play, rather than delving deeper into the constructs of meaning that make this series unique within a particular historical moment in cable television. Tapping into the theme of "quality TV," Polan offers reflection on some of the ways "one might think about cable television's performance of distinction," focusing on how in *The Sopranos*, the representations of ethnicity and class do not so much increase an awareness of class and ethnic politics as they allow middle-class white viewers to "slum" in the underworld of a generalized New Jersey mafia.

Toby Miller also critiques television studies scholars, but rather than take them to task for a kind of naïve acceptance of the "meaning" of quality TV, he argues that the lack of scholarly analysis of cable television news within the disciplines of television and cultural studies has privileged entertainment programming as a more important cultural site. Given the historical importance of the news in constructing U.S. audiences as particular kinds of citizens, the dearth of academic studies of the news—especially cable news networks such as CNN and Fox News—poses a real problem. Positioning cable television as both a "bank" and a "flag," Miller finds that cable news channels function as agents for multi-media conglomerates and as uncritical expressions of U.S. nationalism in the context of international turmoil and strife. Further focusing on how and in what ways cable television news has failed to live up to its potential as an alternative to broadcast network news, Miller takes us through several recent world events, including the World Trade Center attacks on September 11, 2001, and the second war with Iraq, and argues that the U.S. cable news primarily reiterated the government party line. Indeed, he sees cable news as "a form of uncompromising nationalism [that] supplements the ordinary business of U.S. journalism in a neoliberal age." Moreover, in pointing out the connection between the kind of news U.S. audiences watch and the ways in which these same audiences constitute a nationalist identity, Miller argues that we need to be much more vigilant about our expectations for a kind of "quality news."

In her essay on the cable programming aired on Bravo, Katherine Sender also theorizes the connections between actual programs and what the programmers envision their audience to be. Examining the Bravo original program *Queer Eye for the Straight Guy*, Sender argues that the positioning of Bravo as a "gay channel" is not so much about a loosening of homophobic norms and practices in the United States as it is a marketing strategy intended to make the channel appeal not only to high-income gay audiences, but also to women. The channel thus "dual-casts" by offering the same shows to two different audiences for the purpose of advertisers. Situating this kind of marketing strategy within the cable industry, Sender argues that the production of gay-themed programming on Bravo is an important example of cable's approach to attracting a fragmented and somewhat unpredictable audience. In so doing, Sender richly theorizes the connections between cable programmers, audiences, and advertisers and sees the political-economic context of cable to be the perfect grounds for a convergence between all three: "Bravo's gay-themed programming of 2003 thus reflects not a brave attempt by a cable renegade to bring gays to basic cable, but a confluence of existing industry, marketing, and representational trends that made the channel ideally placed to develop gay television."

Christine Acham also discusses the unique capabilities of cable television to produce programs that could not air on broadcast TV, but focuses her essay on what has now become one of the most-talked-about shows on television, *Chappelle's Show*. Positioning *Chappelle's Show* as the catalyst for intersections between network television, cable, and blackness, Acham concludes that the kind of blackness that is performed on *Chappelle's Show* transgresses the bounds of representation on network television and is thus only possible in the cable venue. Moreover, in specifically connecting representations of blackness with the material realities of African-American life, she asks, "Does cable television facilitate or deter the black performer, and in what ways? . . . [W]hat has Chappelle's use of cable and Comedy Central specifically envisioned about black people?" Acham sees promise in cable as the most viable media arena for "black humor," but ultimately asks what it means for ideas of community that black humor can only be shown on broadcast television in a domesticated, diluted form.

The final essay in this section looks at the ways in which cable programs have extended beyond the boundaries of the actual television venue into new media forms, such as the Internet and video games. Examining one of the most successful cable television programs/brands, World Wrestling Entertainment (WWE), Ellen Seiter analyzes the various marketing strategies of this franchise among broadcast, cable, and pay-per-view events and on-line content and product merchandising. She also looks specifically at the ways in which WWE has cultivated a Latino audience through targeted marketing strategies, including sensationalistic personal stories about the wrestlers themselves, as well as exploitation of racial and ethnic tensions between wrestlers. Seiter's essay provides an incisive examination of the ways a struggling franchise, the WWE, "used cable, satellite, pay-per-view, the Internet, and home video sales to push aggressively into an international market and cope with domestic declines."

NOTES

1. Sarah Lacy, "I Want My Flexible TV," *Business Week,* June 21, 2005.

2. Scott Woolley, "Rebranding The Boob Tube," *Forbes,* December 12, 2005, http://www.forbes.com/business/forbes/2005/1212/052.html.

3. Stuart Hall, "Encoding, Decoding," in Simon During, ed., *The Cultural Studies Reader* (New York: Routledge, 1993), 90–103.

4. Lynn Spigel, "Introduction," in Lynn Spigel and Jan Olsson, eds., *Television After TV: Essays on a Medium in Transition* (Durham: Duke University Press), 16–17.

5. Ibid.

Chapter 12

Cable Watching
HBO, The Sopranos, *and Discourses of Distinction*

Dana Polan

[C]ontent is a great business.
—Comcast Cable CEO Brian Roberts in an interview
with *The Wall Street Journal*, September 24, 2004[1]

How might we write about the ways in which, in the complicated land-scape of today's big business media production, certain cultural products seem to stand out and gain distinction? To offer some reflections on cable television's performance of distinction, I divide the following essay into two parts. The first starts with the point of reception—in large part be-cause we often imagine, in our neo-liberal moment, that the exercise of taste happens in a free and spontaneous gravitation of audiences to just those sorts of cultural work that are ready-made for them. I concentrate here on one such mythology of the immediate (i.e., direct and unmedi-ated) meeting up of culture and appropriate audience, a mythology on exhibit in academic *interpretative* discourse on popular culture. Here, it is imagined that gifted readers, outside any history other than the eternal history of "great thought," can find in works of culture, themselves outside any history, deep meanings rich with wide resonance and timeless applica-bility. In popular culture, the *ne plus ultra* of this epiphanous encounter of discerning spectator and deeply meaningful culture takes the form of what has come to be known as "quality TV," and it is clear that both the astute viewer and the television product benefit from the process by which some shows are singled out as "quality." My goal here will be to place discourses of reception, such as the interpretative, back into history—to suggest that reception is precisely a sociological phenomenon coming from a specific social stratum and practiced according to well-ingrained teachings of the members of the community.

The discourse of free and spontaneous reception as an unfettered, supra-historical exercise of taste and interpretation has its role to play in the flow of cultural work from production to consumption. In this respect, the second part of the essay moves back from reception to examine how one producer of cable programming, HBO, aims at, or benefits from, particular kinds of viewing, academic and not. I will concentrate on HBO's most noted show, *The Sopranos*, the acclaimed gangster-meets-the-family-sitcom original series and offer some thoughts toward a political economy of its production and a sociology of its reception. Here, the radiant meeting up of culture and consumer appears less a "free" flow of Art to its rightful audience than a set of calculated and calculable business strategies.[2]

Academia as Taste Community

Strikingly, there are now a half dozen or so *academic* books on HBO's *The Sopranos*. With some scattered exceptions (a few isolated essays in several anthologies), the bulk of these writings come from humanities scholars outside of television and media studies and set out inevitably to engage in thematic analysis.

Specifically, such writings treat *The Sopranos* as a simulacrum of real-life issues the raising of which supposedly accounts for the show's emotional and intellectual appeal. Popular culture here is a virtually transparent approximation to everyday existence: the television show represents life quandaries we all face. The title of one such book, *The Sopranos and Philosophy: I Kill, Therefore I Am*, says it all: the show is examined as a parable that stages metaphysical and ethical questions. And the essays in the volume become predictable and obvious both in their treatment of television characters as veritable real-world beings and in their assumption that such characters could virtually be given expert advice on how to handle their lives. Jennifer Baker's essay on "The Unhappiness of Tony Soprano: An Ancient Analysis" is typical in this regard:

> When it comes to what he wants, Tony has both aim and reach. . . . But Tony is not happy. . . . In this chapter, we'll look to the diagnosis ancient philos-

ophers would offer. That's right, what would Plato, Aristotle, the Stoics, and the Epicureans have to say about the happiness of a self-described 'fat crook' from Jersey? They have plenty to say. And I would like to suggest that Tony Soprano, who has hardly changed at all from his psychotherapy, could benefit from a little philosophy.[3]

It has long been the proclivity of a humanist and humanizing brand of popular culture criticism, when faced especially with examples that it can deem to be "quality," to concentrate only on message and theme. For instance, when, in the so-called Golden Age of 1950s television, writers such as Paddy Chayefsky, Rod Serling, and Reginald Rose offered up stirring narratives of little people who fought against the system, critics singled out these works—often against the seeming formulaicness of genre television (the Western, especially)—for their ostensible accomplishments in rich and rounded character depiction and moral suasion. Although the 1950s were certainly a period marked by the "failure" of encounter, as Andrew Ross puts it, between American intellectuals and their country's popular culture,[4] and although such intellectuals often took television to represent the extreme of what they saw as popular culture's own failures of morality and creativity, Golden Age drama could in the best of cases serve as the exception to the rule. As William Boddy strongly suggests, both critics in the budding profession of television criticism and the prestige shows they wrote about benefited from the process: each received cultural legitimation and each took on the aura of moral emissary bringing enlightenment into a crass culture.[5]

Indeed, a discourse of cultural mission and mass salvation was key to much of American intellectual life in the 1950s. Representative was Perry Miller's influential book on American letters from the middle of the decade, the aptly named *Errand into the Wilderness,* which studied the morally inflected discourse of the first European settlers in North America.[6] Miller's book captured a characteristic 1950s sense of American arts as being born from, and gaining their ongoing relevance in, a sense of life-as-mission—an idea that was symptomatic of a 1950s America caught in Cold War struggle and seeing culture as one of the fronts in that war. Many 1950s American intellectuals delighted in imagining that from within the seeming morass of American popular culture, there could be signs of a resonant indigenous art, and the Golden Age television writers served the cause amply. For example, it was easy for practitioners of the budding art of television criticism to applaud Rod Serling—a convert to the strongly social-reformist world of Unitarianism—who saw himself on a dual errand into the wilderness: on the one hand, to fight against prejudice and pettiness in everyday American life and, on the other, to fight to make television become more than cheap entertainment.

That the *mise en scène* of Golden Age dramas either tended toward a zero-degree style of unobtrusive transparency or, at most, manifested moments of stylistic flourish only when this could serve to express inner psychology or dramatic tension further aided critics in their quest to treat such televisual offerings again in humanist terms as parables with real-life purchase. And in this respect, it may well be the case that *The Sopranos* benefits from the potential that today's critics and interpretive commentators possess to inscribe it within the legacy of a "quality television" whose lineaments were first established in the legend of a 1950s Golden Age.

As a growing library of books and essays attests, Quality has a complicated history in television criticism. If one brand of Quality came in the form of elegant adaptation from canonic Western literature—the *Masterpiece Theatre* tradition of *Upstairs/Downstairs* or *The Forsythe Saga*—where the model of uplift was British in nature (including the Sirs and Dames, such as Diana Rigg, who served as cultural mediators to introduce the fictions), another quality tradition was more indigenously American and focused on a supposed Everyman as he confronted life's quandaries. This working-class everyman appealed to the wizened liberal who could watch from a position of superiority but who could also appreciate those moments when the working-class stiff found insight from within the heart of his inarticulateness. From Paddy Chayefsky's *Marty* to Norman Lear's Archie Bunker to David Chase's Tony Soprano, there is a continued emphasis on a simple man who from time to time has flashes of enlightenment and visions of self-amelioration and who works to confront the limits of his own situation. Not for nothing does *The Sopranos* receive so many nominations for writing and acting since it reinvigorates a tradition that valorizes dramatic and narrative qualities and that sees these as central to attribution of high value to a television show. (It is perhaps revealing that *The Sopranos* has never won in the directing category but only in fact in acting and writing.) Award-winning Quality television is imagined here to revolve around rich and rounded figures who face moral dilemmas and either grow from the encounter or enable hip spectators to feel they have grown. Not for nothing did one of *The Sopranos*' first two Emmys go to the writing in the Season One stand-alone episode "College," in which

Tony Soprano, while taking his daughter on visits to prospective colleges, comes across a Mafia stoolie in witness-protection and sets out to kill him (the other Emmy to the show that year was an acting award to Edie Falco). "College" makes the moral questions explicit in a final scene where Tony, in the hall of one college, confronts an engraved motto from Nathaniel Hawthorne: "No man can wear one face to himself and another to the multitude without finally getting bewildered as to which may be true." Not for nothing do several essays in *The Sopranos and Philosophy* predictably hone in on this scene and argue about the ways it shows Tony grappling with his internal demons.

I need to be clear here. I am not saying that *The Sopranos* can't be read to offer values of drama and narrative and even moral interrogation. From the opening scene of its very first episode—in which Tony stares with wonder at an artsy statue of a woman outside the office of therapist Jennifer Melfi—*The Sopranos* clearly taps into an audience that has been trained (through, for example, years of high school and college courses in literary study as theme-hunting) to understand cultural work as hermeneutic—as meaning-making. The opening shots of Tony and the statue announce a show about interpretation: Tony will, among other things, endlessly have to confront the enigmas of femininity and of cultural hierarchy. The scene stages what literary critic Paul de Man termed an "allegory of reading," and it is worth adding that de Man himself had an infamous encounter with the tradition of quality television. In his 1973 essay "Semiology and Rhetoric," de Man set out to examine claims to meaning in linguistic declaration and, for his demonstration of the instability of meaning, chose an example, as he put it, "from the sub-literature of the mass media": a scene in *All in the Family* in which Archie Bunker posits against his wife's literalist uses of language a more open awareness of language's rhetorical qualities. Interestingly, de Man's point was to assert language's essential *ambiguity* of meaning but his own procedure was to take the, for him, low cultural form of the television show as a relatively easy dramatization of the issues. Once again, popular culture reveals deeper parables beneath its seeming innocuous surface.[7]

Again, my point is not to disavow that for select viewers an HBO show like *The Sopranos* can have thematic qualities of moral parable. In an interview with *Variety*, HBO head Chris Albrecht clarifies that the channel has absorbed a philosophy of meaningfulness as the path to distinction. Speaking of the revisionist Western show *Deadwood*, Albrecht promised, "This show is one that you haven't seen before. It is on HBO. It has a point of view. It is really about something."[8] That television is "really about something" becomes a way to market it: this is the language of relevance and meaningfulness, of theme and profound depths.

That *Deadwood* may also gain its appeal from a surface of raunchiness, Western revisionist wackiness, and randy wallowing in immorality—just as *The Sopranos* itself may also play on political incorrectness, risqué display of sex and violence, comic revisionism (of gangster film and domestic sitcom), postmodern allusion, and so on—is not something that is bragged about in the discourse of prestige. Such features are not supposed to lead to the winning of Emmys. But they are no less potentially a strong aspect of the channel's success with its audience. Similarly, by its setting in a funeral home, HBO's *Six Feet Under* can seem inevitably to raise big questions of life's finality, but it also clearly uses its opening sequence— where someone dies in some quirky way and ends up a cadaver at the home—to ludic ends: knowing the convention of the opening, the viewer participates in a game of who will die and how? what variation can be made in the formulae?

In a legendary assertion of the seductive values of artistic form for its own sake in ways that anticipated postmodernism's play with the surfaces of art, Susan Sontag (in)famously declared in the 1960s that "In place of a hermeneutics we need an erotics of art."[9] When Sontag wrote those words, they had a certain New York avant-gardist manifesto quality to them: against the middlebrow conventionalism that couldn't find a use for cultural production unless it could be converted into serious theme, Sontag's call for enlightened urban sophisticates to open themselves to art's potential for direct experience, a sensuousness of form, gave "Erotics," in that context, a revolutionary quality as a direct use-value that could not be exchanged for the values of meaningfulness. In our cultural moment, however—one in which, for example, the urban sophisticate comes increasingly to understand the "meaningfulness" of life as accomplishment of life-*style* (that is, surface look, commodified shows of self, personal performance, and so on)—erotics and sheer experience become themselves forms of capitalizable middlebrow activity. I'll return to this later in this essay when I suggest that HBO's appeal has as much to do with lifestyle qualities of urban sophisticate audiences as with their desire to find meaningfulness in their culture. For the moment, though, I want to say a little more about the discourse of hermeneutics and high seriousness.

There is a continuity from the Puritan "missionaries," whose New World exploits Perry Miller examined as central to American self-definition, to the television philosophers who wish to treat popular culture as vehicle for philosophic insight. That is, there is something puritanical about the approach to media culture on exhibit in the writings of the cultural mediators: they instruct the consumer that the need is to go past surface fun to life-lessons that are to be taken with deep seriousness.

The privileged spectator for Quality Television constitutes him- or herself within a hierarchical framework governed by a notion of moral pedagogy. That is, the spectator comes to assume that he or she has found something in a particular show that others have missed but that would make them better if they were to learn about it. Interestingly, in the case of the philosophers writing on *The Sopranos,* the pedagogy has two targets. On the one hand, the lessons target a larger group of viewers or potential viewers to whom awareness of the show's deeper philosophic value must be imparted: one might cite, again, not just the book that is titled *The Sopranos and Philosophy,* but also the fact that this book belongs to a series (with titles also on Philosophy and *Buffy the Vampire Slayer, The Simpsons,* and *Seinfeld*) that sees its goal as bringing philosophy to the masses by applying it to popular works they are already familiar with but didn't realize the ideational significance of. On the other hand, in a curious confusion of life and art, the philosophers take Tony Soprano as himself a veritable real-world (albeit low-life) figure whose moral failings come precisely from his ignorance of a philosophical tradition that the philosophers are just dying to have the chance to tell him about. Here it matters that Tony is a criminal because he can then stand in as the fallen and un-intellectual figure who is lacking in the higher learning that the philosophers themselves possess.

A class of viewers comes to constitute itself as veritable cultural mediator between the show itself and a broader public that, it is felt, needs to be instructed about the true—and deeper—meaning within that show. And this is not just an academic matter. When, for instance, in the presidential debates of 2004, John Kerry admonished that "Being lectured by the president on fiscal responsibility is a little bit like Tony Soprano talking to me about law and order," he played on and played into cultural hierarchies in America. Here was the East Coast liberal intellectual distancing himself from redneck (or red state) low-lifers and associating the ostensible cowboy man-of-the-people with failure in law, leadership, and economic wisdom alike. Tony might have things to say about law-and-order (and President Bush might have things to say about fiscal responsibility) but their "lessons" would have no real pedagogical value: they need themselves to be schooled in the right lessons by intellectually-inclined cultured mediators.

In effect, few of the scholarly writings deal then with the television show as being in fact a television show. On the one hand, they disavow stylistic analysis and offer little in the way of recognition of the show's specific formal practices. They read past form to imagine that the show enables one to peek in on real-life situations and raises moral or philosophical issues that come from the narrative's placing of characters in thematically resonant situations. On the other hand, and concurrently, they also tend, with a few exceptions, to eschew examination of the contextual situation of *The Sopranos* within a cultural industry and to imagine instead that the show carries its values in itself, in the themes it investigates —and not in its circulation as a commodity fabricated to flow from producer to consumer. Not unpredictably, again, the most extensive attempt to connect *The Sopranos* to questions of economy does so at the level of the show's "content": David Simon's *Tony Soprano's America: The Criminal Side of the American Dream* reads the show as a veritable investigative report on crime, corruption, and capital. Again, the critical discourse has to disavow the specific status of television-styled fiction in order to transform it into real-life parable. As the last lines of Simon's Preface declare,

> Tony is a fictional character, but he symbolizes deep yearnings on the part of a frustrated citizenry. . . . That he is in denial concerning all manner of contrary evidence [about the immorality of his actions] speaks to the deep contradictions that lie at both the heart of the American dream and social problems analyzed in this book.[10]

Moreover, while Simon argues that *The Sopranos'* disquisitions on crime find one strong sociological resonance in the ways the series' depictions of job advancement within the "Family" speak to the increasing corporatization of America (and of the corporation as a site of systemic malfeasance), one searches in vain in his book for any mention of Time Warner, the corporation that owns HBO, which is also unmentioned. For Simon, the show dramatizes the contradictions of industry in America today but he can only make it do so if he disavows its own nature as industrial product, its own participation in the nexus of culture industry practice.

And concomitantly, the interpretative critics must work to disavow their own participation in the production, distribution, and reception of cultural work. To note just one example, *Tony Soprano's America* takes pains on its front cover to issue the disclaimer that "This book was not prepared, licensed, approved, or endorsed by any entity involved in creating or producing *The Sopranos* television series." (Regina Barreca's anthology, *A Sitdown with the Sopranos,* in which specifically Italian-American university professors offer essays on Italian-American issues in the show, bears a comparable disclaimer.) No doubt one should take the publishers at their word and imagine that there is a relative autonomy of academic writing from the objects of media production it addresses. But it is important, I am arguing, to see how, despite its declarations of independence, critical discourse inevitably becomes part of the political economy of media flow. And I wouldn't want to claim greater "purity" of my academic discourse over that of the interpreters. At the very least, though, I'd like to imagine that an examination, even if itself academic, of conventional academic discourse on popular culture might take up a different relationship to that culture by seeking to understand what, as social practice, culture does and what, as social practice, academic writing does. The point is not that academics should not address the culture around them. Quite the contrary. But they need to be self-reflexive about their discourse and what it enables and what it disables.

The Sopranos *as a Television Practice*

It's interesting to note that the disclaimers of economic relationship to the show's production on the covers of several academic books on *The Sopranos*—which are also then declarations of purity from such relationship—bear a strong resemblance to the declaration of independence on the website for the Sopranos bus tour (tickets reservable at $40 plus $2 handling fee):

> *The Sopranos* and all related characters are the property of Brillstein-Grey Entertainment, Sopranos Productions, Inc., Chase Films and Home Box Office, a Division of Time Warner Entertainment Company, L.P. On Location Tours, Inc. is in no way associated with the aforementioned companies. The sceneontv.com content is assumed to be within the realm of the Public's Right of "Fair Use" and no copyright infringement is intended.

Each in its own way, the books on *The Sopranos* and the bus tour participate in an "informal economy" that erects itself around the authorized transactions the show engages in. In some cases, the informal economy can be quite profitable. An article in *The New York Times* notes that Open Court Publishing, the publisher of *The Sopranos and Philosophy: I Kill, Therefore I Am,* has hit it quite big with its "Popular Culture and Philosophy" series: *The Simpsons and Philosophy: The D'Oh of Homer* sold more than 200,000 copies, and *The Matrix and Philosophy: Welcome to the Desert of the Real* eventually hit the *Times* bestseller list (at the same time, the article notes that series editor William Irwin decided not to pursue volumes on *Friends* and *E.R.* "because they lacked the basic depth and literacy for a thorough philosophical discourse").[11]

The relations of this informal economy to the authorized activities of HBO are complicated. Certainly, it is to the advantage of HBO and Time Warner to keep as many revenue streams in-house as possible and not to allow "unauthorized" activities to proliferate. As a 2003 article in *Variety* on the economics of HBO explains, the company appears to account for 17 percent of Time Warner's annual sales (an impressive figure especially when one remembers that the Time Warner setup includes AOL, a profitable operation even if mention of it disappeared from the name of the conglomerate). The article goes on to note that currently 80 percent of HBO's sales come from "pay TV subscribers and affils" (that is, cable systems that carry the HBO channels), but it suggests strongly that HBO will need new revenue streams beyond cable subscriptions to continue to grow. In typical *Variety*-speak, the article explains,

> Company execs say that they saw subscriber growth stagnation in the increasingly competitive multichannel landscape sufficiently far ahead of time to devote resources to differentiating and building a brand around programs for which it owned 100% of the backend [that is, revenues beyond their original presentation on HBO channels]. Not only was groundbreaking original programming key to finding new subs and reducing churn [that is, customer cancellations of cable service], it formed the basis for a more diversified business and earnings stream.[12]

In other words, HBO needs to be more than just a set of shows on its channels; it needs to diversify its sources of revenue both by moving beyond show production (to, for example, the production of tie-ins, as we'll see in a bit) and by multiplying the media platforms in which its shows can appear (from cable channels to syndicated re-reruns to DVD boxed-sets).

The subscriber success of shows like *The Sopranos* may lead us to think of HBO as a stable operation (and thereby it may, as the first part of this essay suggested, encourage the academic viewer in imagining its productions as themselves stabilized works of art to which resonant interpretations may be affixed). But, from the start, HBO has had to be an operation of resilient re-fashioning, always seeking new means to generate cash flow from culture's flow.[13] HBO first achieved product differentiation of, and spectator fidelity to, the individual works that embody distinction through special offerings in cinema, sports, and sex (as also was the case, in the latter realm, of the early efforts of Showtime, its cable competitor); while HBO has not abandoned any of these, it has increasingly moved into upscale regions of culture through push-the-envelope comedy and artsy original programming in the form of home-produced feature films and episodic series.[14] Sports have come to matter less as sports-specialty channels like ESPN have come to target the niche audience. Cinema, while still quite central to HBO's schedule, has lost some importance as the VCR has made frequent replaying of a film title (an early staple of HBO) less attractive and as other distribution practices have provided alternative means by which films can enter domestic space (through, for example, video-on-demand and into-the-home mailing of DVDs through services such as Netflix, to which we might add Blockbuster's decision to get rid of late fees for DVD rentals). Sex remains a staple of HBO—especially now that the FCC has decided, post–Janet Jackson's "wardrobe malfunction," to crack down on nudity on network television—but increasingly it has been centered less in raunchy "reality shows" that purport to give voyeuristic and exoticizing glimpses at weird practices than in narrativizations of sexuality within ostensible "quality programming."

As it has moved through this history, HBO has tried to generate new audiences (as well as keep current audiences from churning) both by honing a particular quality of envelope-pushing entertainment and by seeking multiple venues in which that entertainment can gain ancillary afterlife. To take just one example of the attempt by HBO to recognize that its products can endlessly serve as new sources of revenue: just as I was finishing up this essay, NATPE (the National Association of Television Program Executives) was in Las Vegas for its annual convention—devoted primarily to the pick-up of shows for syndication—and all the buzz was about *The Sopranos*.[15] Whereas many program owners listed the shows they had come to hawk in the special NATPE issue of *Variety*, HBO could get away with simply providing its location in the convention hotel: it had no need to tell potential customers what its wares were since everyone knew, and everybody lusted for them. In the last days of the convention, the two big bidders for syndication rights to *The Sopranos* were two ca-blers (as *Variety* calls them), TNT and A&E. Going into the auction with the bid per episodes starting at $1.8 million (a little below the record price paid for episodes of a show—$1.92 million offered by USA and Bravo for *Law & Order: Criminal Intent*), HBO quickly got the auction up to $2.1 million per episode and awaited the cablers' next move. On the one hand, going with TNT would keep *The Sopranos* in-house since that channel is part of Time Warner. On the other hand, signing with TNT wouldn't nec-essarily bring with it a reputation for prestige (it is a channel of action and adventure, qualities which represent only one side of *The Sopranos*). A&E still has a patina of cultural cachet, and rumor had it that it planned, if it won the auction, to air a cut-down censor-friendly version of the show (what some people refer to as *Sopranos*-lite) early in the evening and an uncut version at 11 p.m. (with lots of parental-advisory warnings).

Ultimately, A&E won the bidding war. Press speculation has it that the final bid was $2.5 million per episode. Interestingly, as it moves from HBO showings to ones at A&E, *The Sopranos* will have new tasks to perform: as several reports on the deal clarify, A&E hopes that the show will bring an edginess to its programming and thereby bring in younger viewers, and it is further hoped specifically that the gangster aspects of the show will cross-promote A&E's in-house Mafia reality-show, *Growing Up Gotti*. At A&E, *The Sopranos* will serve then both as what the industry terms a "tent-pole" (that is, a cultural offering that is so strong and prestigious that it props up the rest of the operation) and as a ploy to attract viewers to its own in-house productions.

At the same time, it rebounds to the benefit of the cable channel to have critics, who declare themselves outside the market of culture, to wax eloquently about the show. The patina of purity that comes with the aca-demic critics' declarations of independence makes it appear as if academia has discovered the deep virtues of HBO on its own and has gravitated naturally to its cultural riches. While there is every rationale for media companies to carefully control the destiny of their products (as we see, for example, in Hollywood worries about film and DVD piracy), the com-panies discover increasingly that it is worthwhile to let there be some fuzziness around the full ownership of a product's cultural resonances. At the very least, from these companies' points of view, academic discourse is relatively harmless. At best, again from their point of view, it becomes itself a form of promotion in which scholarly propensities are taken as signs of a larger cultural disposition: that is, beyond the professors, there is certainly a milieu trained through years of classes in interpretation to want thematic resonance from its culture, and in this way academic buzz may point to larger interpretative communities ready to be flattered for their willingness to engage in hermeneutic work. Clearly, HBO cultivates reading protocols of this sort. Think, for instance, of "Employee of the Month," an acclaimed episode from Season Three of *The Sopranos*: Tony's therapist, Dr. Melfi, has been raped and seems to be thinking of going to Tony for vengeance. The end of the episode—where Tony inquires of a hesitant Melfi if she has something to ask him and, after a beat, she says "No"—appears designed to get people thinking and talking: What would they do in such a case? What would become of their own moral convictions? Undeniably, such a scene encourages a discussion of the show's content in terms of moral choice and taps into a market that is ready to approach culture in ethical terms. Scheduled for their first showing on Sunday nights, HBO prestige series are clearly intended to serve as what the biz terms "water cooler shows": that is, programs that so encourage reflection and discussion that the weekend viewing of them in the domestic context will be extended by fervent dialogue about them in the work sphere.[16]

For its part, however, HBO has run promotional spots on its channels that mock the very idea of "water cooler shows." In dry fashion, managers and employees from the water cooler industry speak to the camera to thank viewers for making their business so profitable, while a series of sketches shows such viewers offering pithy interpretative comments but delivered in deadpan monotone. HBO's own marketing of a show like *The Sopranos* and of various tie-ins around it suggests then that the channel also cultivates audience relationships that are not inevitably about morality, seriousness, depth of meaning, and so on. High seriousness is overlaid smugly with a knowing wink, a putting of deep purpose into quotations. Today's urban professional needs not only meaningfulness and substance but also hipness, newness, and cutting-edge innovation.

It is noteworthy that when pop sound-bite economist Richard Florida comes to announce the emergence of a new social stratum in America today—what he calls "the Rise of the Creative Class," geared to the generation of exciting ideas—there is little discussion of the content of creative notions: it is as if innovation in itself becomes value (virtually everything Florida says about the creative work of an artist or academic would apply as well to a high-level military researcher).[17] Florida's creative worker dwells in a vibrant city culture in which he or she needs endless visual stimuli and strong experiences, variety of social interaction, and malleable boundaries between work and play, as well as always wanting to be abreast of the new, next thing. Ethics is often little more than an after-thought. (Florida waxes eloquently over his moral wish that class divides could be overcome and everyone enabled to turn creative. But this quantitative expansion of creativity still brackets out questions of the qualitative ends of such creativity, however much it spreads to lots of people.)

At best, meaningfulness itself becomes a marketing strategy to be trotted out at the right moment (as, for example, with the responsible and respectable grown-up films that come out in the last months before Oscar-nomination season). Or it serves as a disposable element in an array of pragmatic considerations, one mere tactic among many. HBO's own marketing of Sopranos commodities targets a sophisticate audience that substitutes for reverent respect for ideas a push-the-envelope creativity for which morality and meaning matter only insofar as they remain saleable. The attractions here are precisely irreverence and playful amorality. For example, when the fourth season of The Sopranos comes out on DVD, the tone in the newspaper ads is sharply sardonic: in the center of a vast field of cute snowman, one stands with its head shot off, and the ad sports the slogan, "Bada Bing, Bada Boom. Your Shopping's Done." Whereas the discourse of interpretation has to find in popular culture a seriousness that makes this culture matter, media industries themselves take a more insouciant approach to their products and set out to sell the very fact that they don't matter—that they are without resonant moral consequence.

Revealingly, the most noted and most successful commodity tie-in with The Sopranos, other than the DVDs themselves, is a cookbook (jokingly attributed on the book's spine to a character from the show, chef Artie Bucco, but actually written by former experimental TV advocate Allen Rucker, who has also devised the Sopranos board game). With a print run of over half a million copies, the Sopranos Family Cookbook was a huge success and even reached the number one spot on the Miscellany bestsellers list of the New York Times (it held that title for six weeks and was in the top five for twenty weeks). It has been translated into eight languages: French, Hungarian, Swedish, Danish, Finnish, Dutch, Norwegian, and Korean.[18]

In several ways, the cookbook is revealing of HBO's own marketing approach to its premium show and of its image of its target audience. First, there is a deliberately self-deflating gag quality to the enterprise that indicates that none of this should be taken too seriously: starting from its first inside cover image of Tony's mother, Livia, confronting her stove on fire (from season one: she had left mushrooms on while talking to Tony on the phone), *The Sopranos Family Cookbook* reminds the reader how food is often a source of jokiness in the show. In Rucker's own words, the tome is a "funny book with recipes and a cookbook with jokes." Here, we might again note how the targeting of *The Sopranos* to the urban sophisticate audience taps into class values. Both the show and the cookbook mock a lower class world that tries to find "taste" in thick sauces, in abundant meals centered on fatty meats, in an entire panoply of the tacky, the excessive, and the unhealthy. (If the ethnic word once used to described the thickly laid-on emotions of popular culture was "schmaltz" [the coagulated fat in Jewish cuisine], it may be appropriate that today's word is "cheesy," although the reference is perhaps less to the goo on pizza than to the sense that in modern America, cheese often appears as processed, artificial, ersatz.) In contrast to an Italian-American cuisine that supposedly combines everything into messy amalgams of noodle, sausage, cheese and sauce and that serves it up in massive, overflowing bowls, yuppie cuisine has often been about separation of elements (the nouvelle cuisine which dots the plate with little designs of food and lightly traced patterns of sauce dancing around them) and about a lightness of effect (as in the current vogue for tapas-like small plate meals).

At the very same time, however, while food has long been a prime site for the articulation of class and status difference, it is also the case that in the hecticness of today's urban culture, there is a yuppie rediscovery of rich, even heavy food—a deliberate wallowing in the crossing of class boundaries of taste. We are still in a moment of ostensible refinement through exoticism (the vogue for zen Asian cuisine), but there is also a strong investment in comfort food which tends to center on the thick, the mushy, the saucy, the soft, the cheesy. Comfort food is about primal regression—we go back to the food we loved as kids, we go back to food that we've supposedly grown away from in developing good taste and in rising in the class hierarchy. And in this respect, it may well be that *The Sopranos Family Cookbook* attracts attention not only for its mockery of a world of pasta and sausage but for its embrace of them. Between its jokes, the volume also is a credible cookbook that luxuriates in abundant, thick, and comforting cuisine.

That the biggest selling commodity tie-in for the show is a cookbook is also in keeping with cuisine trends of the urban sophisticate in another way. While urban professionals have turned increasingly to upscale versions of fast-food to save themselves time but maintain self-image—as in the rise of so called "casual dining," supposedly a cut above the fast food options—it has also been the case that food preparation in the home has become one of the prime sites for invigoration of the urban professional life-style. After a day's labor in the information economy, after that conceptual work that Richard Florida sees as the province of the creative class, the urban professional can imagine a reconnection to physicality by playing hard in the kitchen. Like Food Network cabler's *The Iron Chef, The Sopranos* cookbook both jokingly turns everything into media game and yet promises a path through all simulacra to the reality of direct experience of a lost craft of cuisine. No doubt the gag tone of the book *and* its utility as a path to comfort food prepared emphatically in the home come together in the reported proliferation of Sunday-night *Sopranos* parties (usually around first or last episodes of a season), wherein the playful premise is to make baked ziti and other unrefined delights that one normally wouldn't deign to produce.

But the strongest way in which *The Sopranos Family Cookbook* reveals the cultivation of audience has to do with a conceit that is also in evidence in Rucker's previous tie-in book, *The Sopranos: A Family History.* Both volumes pretend that the Sopranos and their cohort really do exist and that the books come from people in their orbit (the cookbook, as noted, purports to come from Artie Bucco while the "Family History" is a blend of FBI reports and of materials, such as diary entries, that supposedly were written by various Sopranos themselves). Moreover, the volumes not only re-narrate information that was imparted in episodes of the show (in the cookbook, for instance, "Artie" retells how his restaurant Vesuvio went up in flames, an occurrence that viewers of the show witnessed in the first episode of the first season) but they also pretend to impart background history that goes beyond what the show itself offered (for example, "Artie" talks in the book of having gone to cooking school in London before becoming a chef in New Jersey). The fictional premise here is one that appears also on *The Sopranos* page at hbo.com where, as the *Rocky Mountain News* explains,

> [the] HBO site also contains an interesting conceit. The Wernick files, which offers FBI files and other inside stuff about the Soprano family, compiled by Jeffrey Wernick, one of America's most celebrated authorities on organized

crime. At least he would be if he existed. Wernick was a briefly-seen character in the series' first season. Incorporating him in the Web site, complete with a biography, is the kind of nudge-and-wink knowingness that connects with fans.[19]

The important point here is the nudge-and-wink aspect of the matter. By pretending to introduce the audience to a world that opens up into ever more complex bits of background information, the commodity tie-ins are not trying to make fans imagine that fictional characters are somehow actual and offer real life lessons. Rather, the expanding universe conceit turns the narrative space of the series into a game, a structure of endless expansion and inter-connection in which there are always new permutations of situation and story.

In an essay in Lynn Spigel and Jan Olssen's recent anthology on the nature of television today, *Television after TV: Essays on a Medium in Transition,* Michael Curtin argues how the use of clarification of narrative structure helps market television works in a complex and rich media landscape. Speaking of what he terms "Neo-network Hollywood," Curtin suggests that, despite claims that new technologies will open up new riches of media interaction, the very proliferation of media choices gives a new impetus to story-telling and character elaboration as strategies of branding. In Curtin's words,

> Today, prime-time shows on the terrestrial stations average only a three to five rating, while more than eighty cable channels vie for the remainder, of them averaging ratings of one or less. In this environment, media producers find that the branding of products is often more important than futile attempts to control the mode of distribution. . . . Given a greater range of choices, audiences are drawn to the products by textual elements—characters, story lines, special effects—rather than by the technological and regulatory constraints formerly imposed on the delivery systems.[20]

In this respect, the frequently pared down look of *The Sopranos*—a focusing in on characters with a concomitant avoidance of special effects (even dream sequence are filmed with an unmarked look)—works not so much in the cause of Golden Age realism as it does to establish a mark of distinction from the overly burdened rich image that is often in evidence on television today. In his classic study, *Televisuality: Style, Crisis, and Authority in American Television,* John Caldwell had shown trenchantly how contemporary television increasingly turned to a style of excess—graphics combined with images, screens within screens, videographic manipulation of the image's "reality," morphing and mutating of sights, frenetic montage, and so on.[21] While HBO dramas like *The Sopranos* or *Six Feet Under* haven't abandoned televisuality (*The Sopranos* frequently exhibits an extreme depth-of-field composition that belies the supposed shallowness of television's visual space to render it more seemingly cinematic), they mute style and stand out for that very reason.

Think again, for instance, of the opening sequence of Episode One of *The Sopranos*: Tony at Dr. Melfi's office, contemplating the statue of the woman and then going in for his first therapy session. Beyond the allegorizing of the act of interpretation that I have already discussed, the scene is noteworthy for its sheer distinction from televisuality. On the one hand, much of the scene plays out in silence as Tony contemplates the statue and then, once inside Melfi's office, he and the therapist contemplate each other but say nothing. On the other hand, that very silence goes on for a long time, and the scene turns into a veritable minimalist play with duration. Clearly, David Chase and his team have internalized the styles of European art cinema but here silence and pause have less to do with ineffable mysteries of existence than with a very concern to slow television down, to make it be a medium that can take its time. (The very structure of HBO programming allows *The Sopranos* to eschew a strict sixty minute format and extend temporality when so desired; of course, syndication may require the episodes to be re-purposed for the standard time slot.)

Through this expansive temporality, a show like *The Sopranos* can introduce its audience to an ever greater narrative universe that is always propelling audience interest forward and thereby solidifying audience fidelity. Indeed, in another essay in *Television after TV*, Jeff Sconce could be seen as giving fuller textual specificity to Michael Curtin's argument about text-as-brand when Sconce argues that central to the narrative structure of today's television dramatic series is a combination of older formats of episodic series and ongoing serial form. That is, television shows increasingly blur the boundaries between stand-alone episodes in which a narrative is begun and resolved in one single viewing time *and* a continuing plot that remains unresolved over a set of episodes. In Sconce's words,

> In the "postnetwork" era of increased competition for viewing audiences, "cumulative" narration provides distinct programming and demographic advantages. By combining the strengths of both the episodic and serials for-

mats, this narrative mode allows new and/or sporadic viewers to enjoy the stand-alone story of a particular episode while also rewarding more dedicated, long-term viewers for their sustained interest in the overall series.[22]

The Sopranos, for instance, will offer the veritably self-contained mini-narratives like that of the aforementioned Emmy-winning episode "College," but it will also offer narrative arcs that extend over a season (Tony's tensions with Ralphie and Ritchie, two key antagonists who die by season's end) or beyond (for example, the expectation that some day Tony will come to loggerheads with this or that rival Mafia boss). In addition, we might note that another tactic of *The Sopranos* has to do with blurring the boundaries between narrative closure and a simple disappearance or temporary suspension of a narrative line. Episodes and seasons may resolve certain narrative points but others simply are not returned to: there lingers the possibility that the suspended narrative will revive at some later date (probably the most famous case of such suspension is the "Pine Barren" episode which left viewers wondering if a tough and resilient Russian that Chris and Paulie, two members of Tony's gang, set out to whack is alive and will ever come back). The suspended narrative both builds a suspense that sustains audience fidelity *and* requires the audience to hone viewing skills around narrative memory (what happened when, what might still happen later on). Such complex narrativity helps explain the comparison of the show by *New York Times* critic Caryn James to ambitious and epic experimentation in the nineteenth-century novel: in James's words, "As no single film or ordinary television series could, *The Sopranos* has taken on the texture of epic fiction, a contemporary equivalent of a 19th-century sequence of novels. Like Zola's Rougon-Macquart series or Balzac's 'Comédie Humaine,' *The Sopranos* defines a particular culture (suburban New Jersey at the turn of the century) by using complex individuals."[23]

As Sconce notes, there are a number of specific strategies that television programs employ in the expansion of their narrative space. These can involve spin-offs of characters into new narrative worlds (so that, for example, *Friends* generates *Joey*), amnesia plots that mean that a narrative virtually needs to be started again, "evil twin" plots that create new dramatic resonances among characters who already have been fixed in place, "what if?" plots by which characters move into alternate universes that generate new narrative possibilities, "it was all a dream" plots that erase establish narrative and start things from scratch, "fish out of water" plots (in which the characters move to a new narrative world where interest is generated from the contrast between what they were and what they are now forced to be), and what Sconce terms the "meat locker" plot in which the characters are cut off from their typical narrative universe and new narrative interest is generated from their interiorized interactions with each other. The setting of *The Sopranos* in a somewhat realist narrative universe governed by rules of ordinary logic means that the show can't go off in more science-fictional directions in its search of narrative expansion—although it is striking to see how dream sequences do seem to open the show up to a certain degree of the fantastic (as in David Lynch's *Twin Peaks,* dreams do seem to have revelatory qualities). But it is noteworthy how the show uses other of the tactics that Sconce outlines. For example, the trip that Tony and his gang take to Italy in one of the seasons of the show allows them to be fish out of water. Even more, the aforementioned and famous "Pine Barrens" episode is a veritable "meat locker" narrative in which two characters (Paulie and Christopher) get cut off from civilization and drama is generated from the unforeseen conflict that develops between them (and that then generates suspense for the next season: will Paulie and Christopher come to blows?).

Typical HBO episodic series play to an intellectually savvy and culturally informed spectator who has been trained (or is being trained) to take cultural works to be enigmas or puzzles in which one goes beyond the text at hand to something else. This "something else" can be moral or philosophic meaning, but it can also be the less meaningful display of puzzles for their own sake (the game, for instance, as I've noted, in the first scene of episodes of *Six Feet Under* where one guesses about who will die and how). There is also the knowing game of *intertextual reference* where any one show will throw out mentions of culture-at-large to flatter the viewers who catch the reference. To the extent that urban professionals define themselves increasingly in cultural terms (and transform political and economic concerns into lifestyle issues), such references to a larger field of entertainment industry confirm the spectator in his or her "cultural citizenship."[24] To take a minor case, when, in the first season of *The Sopranos,* Tony's nephew and Mafia lieutenant Christopher spots Marty Scorsese going into a swank club and, as ebullient fan, yells that he loved *Kundun,* the reference flatters the alert viewer both for its relative obscurity (this is a Scorsese film that few people actually saw) and for its snobby in-jokiness (even the spectator who hasn't seen *Kundun* will know that it is the least likely Scorsese film for a dumb hood like Christopher to have seen). Indeed, HBO has developed a genre all its own with a set of behind-the-scenes fictional series about everyday life in culture industries in the media capitals of Hollywood and New York: *The Larry Sanders Show, Curb Your Enthusiasm, Entourage,* and, most recently, *Unscripted* and the short-lived *The Comeback.*

It may be, however, that these "insider" shows are too caught up in self-reflexivity to work with a broader audience: at this point in time, neither *Entourage* nor *Unscripted* appears to be generating the buzz that was hoped for them. More generally, it may be the case that HBO is coming to discover that hip urban sophistication can be pushed only so far before it becomes itself an unhip cliché. As I am writing these lines, this week's *Variety* sports a cover story on problems the prestige cablers seem to be experiencing as they search to devise new content. If, for instance, Sunday night had been the prime moment in which HBO could introduce and build fidelity for its edgy, envelope-pushing series, there is now the concern that network television has taken back Sunday night with the success of its risqué and quirky genre-mixing suburban soap, *Desperate Housewives.* Hollywood may be an industry of savvy of business acumen, but it also easily works according to a crisis mentality in which one hit show can seem to portend an entire trend. Sporting the title "Trading Places: Cablers Get Addicted to Reality Shows as Networks Steal Their Buzz Machine," the *Variety* article notes that numerous cablers centered on original programming are scurrying to give up on experiment and move into areas that the networks have already exploited, such as reality TV.[25]

As original programming moves into an uncertain future—but in which there is certainty that there are always new revenue streams to be found somewhere—we encounter another reason perhaps why interpretation of individual shows seems beside the point. In the larger economy of media circulation, culture is a mere pretext to reach consumers, and it matters only as long as it continues to do so.

NOTES

1. Peter Grant, "Comcast's Big Bet on Content," *Wall Street Journal,* September 24, 2004, B1.

2. To summarize the show briefly, it deals with the Soprano family and its patriarch, Tony Soprano (James Gandolfini), who is a mid-level Mafiosi. In therapy

after panic attacks, Tony is beset by the problems of his "line of work" as well as domestic issues (an up-and-down relationship with his wife Carmella [Edie Falco] and the growing pains of his two children).

3. Jennifer Baker, "The Unhappiness of Tony Soprano: An Ancient Analysis," in Richard Greene and Peter Vernezze, eds., *The Sopranos and Philosophy: I Kill, Therefore I Am* (Chicago: Open Court, 2004), 28–29.

4. Andrew Ross, *No Respect: Intellectuals and Popular Culture* (New York: Routledge, 1989).

5. On the mutually rewarding interplay between 1950s critics and Golden Age teleplay production, see William Boddy, *Fifties Television: The Industry and Its Critics* (Urbana: University of Illinois Press, 1990).

6. Perry Miller, *Errand into the Wilderness* (Cambridge, MA: Harvard University Press, 1956).

7. Paul de Man, "Semiology and Rhetoric," *Diacritics* 3, no. 3 (1973): 27–33.

8. Ted Johnson, "Great Expectations (Interview with Chris Albrecht)," *Variety*, August 25–31, 2003, A2.

9. Susan Sontag, "Against Interpretation" (1964) in *Against Interpretation, and Other Essays* (New York: Laurel Editions, 1969), 23.

10. David Simon, *Tony Soprano's America: The Criminal Side of the American Dream* (Boulder, CO: Westview Press, 2004), x.

11. David Bernstein, "Philosophy Hitches a Ride with 'The Sopranos': Small Publisher Finds Route to Big Numbers," *New York Times*, April 13, 2004, E3.

12. Meredith Armdur, "Has Success Spoiled HBO?" *Variety*, August 25–31, 2003, A14.

13. The standard account of HBO is George Mair's anecdotal volume, *Inside HBO: The Billion Dollar War Between HBO, Hollywood, and The Home Video Revolution* (New York: Dodd, Mead & Company, 1988).

14. For sex programming as an intermediate step in the elaboration of modern cable, see Karen Backstein, "Soft Love: The Romantic Vision of Sex on the Showtime Network," *Television & New Media* 2, no. 4 (November 2001): 303–17.

15. Much of the information in this paragraph comes from Denise Martin and John Dempsey, "Mobsters Mint Moolah: 'Sopranos' offnet pricetag like to hit record high," *Variety* (on-line), January 27, 2005, http://www/variety.com/.

16. In an essay on the distribution of *The Sopranos* in the United Kingdom—which is in part about the fact that the show has *not* been a hit there—Joanne Lacey notes that in its first two seasons, the program screened on Thursdays. This is a night when many urbanites begin their long weekend of nights out on the town and don't want to be at home in front of the telly. As some of Lacey's interviewees suggested, to not be out on Thursday night is to brand one's self a loser. See Lacey, "One for the Boys: *The Sopranos* and Its Male, British Audience," in David Lavery, ed., *This Thing of Ours: Investigating The Sopranos* (New York: Columbia University Press, 2002), 104 and passim.

17. Richard Florida, *The Rise of the Creative Class, and How It's Transforming Work, Leisure, Community and Everyday Life* (New York: Basic Books, 2002).

18. Allen Rucker, *The Sopranos Family Cookbook as Compiled by Artie Bucco* (New York: Warner Books, Inc., 2002). Factual information on the cookbook in this paragraph comes from e-mail communication from Rucker, to whom I express my appreciation.

19. Mark Wolff, "Site Seeing Internet Lets Fans Call up In-Depth Connection with TV Favorites," *Rocky Mountain News*, September 23, 2002, accessed on-line.

20. Michael Curtin, "Media Capitals: Cultural Geographies of Global TV," in Lynn Spigel and Jan Olssen, eds., *Television After TV: Essays on a Medium in Transition* (Durham: Duke University Press, 2004), 281.

21. John Caldwell, *Televisuality: Style, Crisis, and Authority in American Television* (New Brunswick, NJ: Rutgers University Press, 1995).

22. Jeffrey Sconce, "What If?: Charting Television's New Textual Boundaries," in Spigel and Olssen, eds., *Television after TV*, 98.

23. Caryn James, " 'Sopranos': Blood, Bullets, and Proust," *New York Times*, March 2, 2001, E1.

24. On "cultural citizenship's" re-writing of politics as lifestyle, see Toby Miller's *Cultural Citizenship: Cosmopolitanism, Consumerism, and Television in a Neoliberal Age* (Philadelphia: Temple University Press, 2007).

25. Brian Lowry, "Trading Places: Cablers Get Addicted to Reality Shows as Networks Steal Their Buzz Machine," *Variety*, January 31–February 6, 2005, 1, 70.

Chapter 13

Bank Tellers and Flag Wavers
Cable News in the United States

Toby Miller

Forty-five years ago, John Fitzgerald Kennedy's Federal Communications Commissioner, Newton Minow, called U.S. TV a "vast wasteland."[1] He was urging broadcasters to embark on enlightened Cold-War leadership, to prove the United States was not the mindless consumer world the Soviets claimed. The networks would thereby live up to their legislative responsibilities to act in the public interest by informing and entertaining. Twenty years later, Ronald Reagan's Federal Communications Commissioner, Mark Fowler, argued that "television is just a toaster with pictures" and hence in no need of regulation, beyond ensuring its safety as an electrical appliance.[2] Both expressions gave their vocalists instant and undimmed celebrity. Minow was named "top newsmaker" of 1961 in an Associated Press survey, and he was on TV and radio more than any other Kennedy official. The phrase "vast wasteland" has even, irony of ironies, provided raw material for the wasteland's parthenogenesis: it has been the answer to questions posed on numerous TV game shows, from *Jeopardy!* to *Who Wants to Be a Millionaire?*[3] The "toaster with pictures" is less-celebrated, but it has been more efficacious as a slogan for deregulation across successive Administrations.

Where Minow stood for public culture's restraining (and ultimately conserving) function for capitalism, Fowler represented capitalism's brooding arrogance, its neoliberal lust to redefine use value via exchange value. Minow decries Fowler's vision, arguing that television "is not an ordinary business" because of its "public responsibilities."[4] But Fowler's phrase has won the day, at least to this point. Minow's lives on as a recalcitrant moral irritant, rather than a central policy technology.

Four decades on from Minow, and two from Fowler, it seems to me that U.S. cable TV is two things:

- a bank—because it has been taken over by conglomerates with little concern for or experience of media ownership's role in democracy, and is increasingly characterized by a model of governance at a distance that informs and entertains U.S. viewers by addressing them in terms of their own financial risk and security; and
- a flag—these corporations ground their otherwise rootless quest for profit in a violent, destructive nationalism that addresses U.S. citizens in terms of their physical risk and security and moral superiority over others.

My case study will be recent coverage of U.S. militarism.

The influence of economic discourse on news has a special power in the United States. Business advisors dominate discussions on dedicated finance cable stations like CNBC and Bloomberg, and they are granted something akin to the status of seers when they appear on MSNBC or CNN. Former Federal Reserve Chair (and husband of NBC correspondent Andrea Mitchell) Alan Greenspan was filmed for decades getting in and out of cars each day as if he were *en route* to a meeting to decide the fate of nations, each sclerotic upturned eyebrow subject to hyper-interpretation by a bevy of needy followers. The focus is always on stock markets in Asia, Europe, and New York; reports on company earnings, profits, and shares; and portfolio management. There is a sense of markets stalking everyday security and politics, ready to punish all anxieties and any political activities that might restrain capital. The veneration, surveillance, and reportage of the markets is ever-ready to point to infractions of this anthropomorphized, yet oddly subject-free sphere, as a means of constructing moral panics around the conduct of whoever raises its ire.

TV-as-bank is dedicated to knowing and furthering the discourse of money and its methods of representing everyday life, substituting for politics and history. Economic and labor news has become corporate news, and politics is measured in terms of its reception by business. In the case of the assaults of September 11, 2001, TV produced an immediate calculation of the likely impact of the attacks on Wall Street, an already-emergent recession, and the global economy. There was no critical discussion of labor issues and minimal participation by academic economists. CNBC even ran pitiful quasi-public-service announcements for the market as an indomitable force that would trump the effects of the violence.[5]

This selfish appeal to monetary gain is both ramified and legitimized by a corresponding flag-embossed chauvinism that identifies with an extreme style of American exceptionalism and the dire dogmas of *e pluribus unum*: loyalty tests, anti-cosmopolitanism, xenophobia, and critiques of the rest of the world. In November 2001, CNN released a thirty-second montage that identified "the U. S., war, [George] Bush, and CNN in a harmonious unity of patriotism and goodness" by intercutting the World Trade Center, firefighters, police officers, U.S. President Bush, and the U.S. flag.[6] During the invasion of Iraq eighteen months later, both MSNBC and Fox adopted the Pentagon's grotesque *cliché* "Operation Iraqi Freedom" as the title of their coverage. MSNBC accompanied stories of U.S. troops with a banner reading "God bless America. Our hearts go with you."[7] The U.S. flag was a constant backdrop in coverage, correspondents identified with the army units they traveled with, and jingoism was universal.[8] The proliferation of U.S. flag pins on reporters, along with the repeated crass use of such othering devices as "we," would not be permitted by major global newsgatherers elsewhere, whether regionally or nationally based or funded.[9] But Erik Sorenson, the president of MSNBC, chortled that "one can be unabashedly patriotic and be a good journalist at the same time."[10] When the University of Missouri's TV station, in keeping with best practice in functioning democracies, decided that presenters should not wear nationalistic symbols on air, the state legislature immediately threatened massive cuts to the school's budget.[11] The bank struck back, as did the flag, revealing the profound influence of the state on American media.

My approach here is in many ways a killjoy argument in large circles of U.S. television and cultural studies, where it is something of a *donnée* that the mainstream media are not responsible for—well, anything. This position is a virtual *nostrum* in some research into, for instance, fans of TV drama or wrestling, who are thought to construct connections with celebrities and actants in ways that mimic friendship, make sense of human interaction, and ignite cultural politics. This critique commonly attacks opponents of television for failing to allot the people's machine its due as a populist apparatus that subverts patriarchy, capitalism, and other forms of oppression. Commercial TV is held to have progressive effects because its programs are decoded by viewers in keeping with their own social situations. All this is supposedly evident to scholars from their perusal of audience conventions, web pages, discussion groups, quizzes, and rankings, or from watching television with their children. But can fans be said to resist labor exploitation, patriarchy, racism, and U.S. neo-imperialism, or in some specifiable way make a difference to politics beyond their own selves, when they interpret texts unusually, dress up in public as men from outer space, or chat about their romantic frustrations? And why have such practices become so popular in the First World at a moment when media policy fetishizes consumption, deregulation, and self-governance? As Theodor Adorno said, while there are problems with "cultured snobbism," populist TV disempowers audience knowledge in the key areas of public life affected by politics.[12]

The strand of U.S. TV and cultural studies that I am questioning is a very specific uptake of venerable U.K.-based critiques of cultural pessimism, political economy, and current-affairs-oriented broadcasting. These critiques originated in reactions against a heavily regulated, duopolistic broadcasting system—in 1970s Britain—in which the BBC represented a high-culture snobbery that many leftists associated with an oppressive class structure. Hence the desire for a playful, commercial, non-citizen address as a counter. Change the angle a few degrees to the United States. When cultural studies accounts of TV made their Atlantic crossing, there was no public-broadcasting behemoth in need of critique—there was just an amoeba. And there were lots of not-very-leftist professors and students seemingly aching to hear that U.S. audiences learning about parts of the world that their country bombs, invades, owns, misrepresents, or otherwise exploits was less important, and less political, than those audiences' interpretations of actually existing local soap operas, wrestling bouts, or science-fiction series. They even had allies amongst reactionary political scientists, who extolled the virtues of market-driven minimization of news, pared down to the essentials for the survival and entertainment of audiences.[13] But anything beyond a cursory examination of how TV news magazines select and investigate their topics gives the lie to such neoliberal fantasizing.

Since its inception, TV has principally been regarded as a means of profiting and legitimizing its controllers and entertaining and civilizing its viewers. Pierre Bourdieu refers to these rather graceless antinomies as "populist spontaneism and demagogic capitulation to popular tastes" versus "paternalistic-pedagogic television."[14] "Populist spontaneism" dovetailed nicely with certain other effusively utopic pretensions from the early cultural studies era. Both the free-cable, free-video social movements of the 1960s and 1970s, and the neoclassical, deregulatory intellectual movements of the 1970s and 1980s, saw a people's technology allegedly emerging from the wasteland of broadcast television. Porta-pak equipment, localism, and unrestrained markets would supposedly provide an alternative to the numbing nationwide commercialism of network television. The social-movement vision saw this occurring overnight. The technocratic vision imagined it in the "long run." One began with folksy culturalism, the other with technophilic futurism. Each claimed it in the name of diversity, and they even merged in the depoliticized "Californian ideology" of community media, which quickly embraced market forms.[15] Left-liberals and neoliberals adopted a similar attitude to cable. Neither formation started with economic reality. Together, they established the preconditions for unsettling a cozy, patriarchal, and quite competent television system that had combined, as TV should, what was good for you and what made you feel good, all on the one set of stations—i.e., a comprehensive service. In place of the universalism of the old networks, where sport, weather, news, lifestyle, and drama programming had a comfortable and appropriate *frottage*, a system of highly centralized but profoundly targeted consumer networks was developed that fetishized lifestyle consumption over a blend of purchase and politics, of fun and foreign policy. It was almost as oligopolistic as the previous environment, with 90 percent of the top cable stations owned by the conglomerates that own the networks and cable systems.[16] The laughable claims from the 1990s that 24-hour news channels compromised the capacity of governments to obfuscate look as silly now as they did to media historians at the time.[17]

CNN and Friends

My particular focus in this chapter is CNN, though I also pay attention to Fox News, MSNBC, and CNBC. I shall maintain that while cable news appears to embody the best of the rhetoric of both Minow and Fowler—it addresses issues of public concern on a round-the-clock basis, and does so in a deregulated environment that presumptively follows public tastes—the result is pitiful. Part of this conclusion comes from my belief that Yanqui journalism in general is subnormal, and so I engage its occupational ideology of flag-waving and othering. And part follows from the deregulatory logics of Fowler and his followers—hence the engagement with TV as a bank. My case studies will be of news and current affairs at times of crisis, of life at the limit, where citizens need the best information possible.

Seventy percent of the U.S. public obtained "information" about the 2003 invasion of Iraq from television.[18] Whilst all media increased their audiences during the crisis, the largest growth was achieved by cable. By the time of the 2004 U.S. presidential election, of the 78 percent of U.S. citizens who followed the campaign on TV, 47 percent did so via cable news (the respective figures for the 2000 contest were 70 percent using television, of whom 36 percent relied on cable).[19] Did they get diverse opinions? Studies of the two major cable news channels, Fox and CNN, reveal that despite the former's claim that it is less liberal, after September 11, 2001, each delivered a pro-Bush position on foreign policy, adhering closely to the line of the Pentagon.[20] The distinctiveness lies in their political economy. CNN and Fox market themselves differently—the former to urban, highly educated viewers, the latter to rural and suburban, relatively uneducated viewers. One functions like a broadsheet, the other like a tabloid, with CNN punditry coming mostly from outsiders, and Fox punditry as much from presenters as guests. CNN costs more to produce and attracts fewer routine viewers (but many more occasional ones). It has brought in higher advertising revenue because of the composition of its audience, and because its fawning and trite business coverage addresses and valorizes high-profile investors and corporations in ways that Fox's down-market populism does not. CNN has 23 satellites, 42 *bureaux,* and 150 foreign correspondents, but you'd never know it from watching their parochial domestic station, with its blinking, winking, walking-dead presenters, dedicated to eastern-seaboard storms, missing white children, and entertainment news. Fox News Channel, which employs few journalists and staffs only four foreign *bureaux,* has the most pundits on its payroll of any U.S. network—over fifty in 2003.[21]

Strong similarities nevertheless bind the networks—specifically, their diurnal practice of rolling out the flag, associating truth and light with the United States executive government. A study conducted at the beginning of 2002 disclosed that CNN had covered 157 events featuring Bush operatives, and just 7 that centered on Democrat politicians.[22] More than half of U.S. TV-studio guests talking about the impending action in Iraq in 2003 were U.S. military or governmental personnel.[23] Television news effectively diminished the available discourse on the impending struggle to one of technical efficiency or state propaganda. Research undertaken through the life of the Iraq invasion reveals that U.S. broadcast and cable news virtually excluded anti-war or internationalist points of view: 64 percent of all pundits were pro-war, while 71 percent of U.S. "experts" favored the war. Anti-war voices were 10 percent of all sources, but just 6 percent of non-Iraqi sources, and 3 percent of U.S. speakers. Viewers were more than six times as likely to see a pro-war than an anti-war source, and amongst U.S. guests, the ratio increased to 25 to 1.[24] When the vast majority of outside experts represent official opinion, how is this different from a state-controlled media?[25]

In justifying this state of affairs, then-CNN anchor Aaron Brown complained that "'there was no center to cover'" in opposition to the Administration because Democrats had not opposed invasion, while Fox News accompanied anti-war protests in Manhattan with a ticker-news crawl taunting the demonstrators.[26] When *The Simpsons* mocked this (on the Fox broadcast network) via a ticker that read "Do Democrats cause cancer? Find out at foxnews.com," the latter immediately threatened the creator with legal action.[27] At the same time, Viacom, CNN, Fox, and Comedy Central were refusing to feature paid billboards and commercials against the invasion.[28] And having repeatedly instructed Phil Donahue to feature more right-wing people on his program, MSNBC fired its liberal talk-show host immediately prior to the 2003 invasion, even though he had the network's top-rated program, because his "anti-war agenda" would look bad when "our competitors are waving the flag . . . a difficult face for NBC in a time of war." This of a program that showcased more pro-war than anti-war guests.[29]

MSNBC also hired as a talk-show host the improbably hypocritical Republican, Joe Scarborough. As a congressman, he had appeared on Fox during the 1999 bombing of Serbia and described the assault as "an unmitigated disaster," specifying "the people in Belgrade we've killed . . . the refugees that we've killed . . . the people in nursing homes . . . the people in hospitals." Four years later, as the focal point of *MSNBC Reports,* he attacked "leftist stooges for anti-American causes" whose beliefs "could hurt American troop morale" by criticizing military actions. He ranted that such people must be held accountable for their views—unlike Scarborough himself.[30]

Meanwhile, because MSNBC's Ashleigh Banfield occasionally reported Arab perspectives during the 2003 conflict, Michael Savage, then a talk-show host on the same network prior to being removed for telling a caller he hoped the person would contract HIV, called Banfield a "slut," a "porn star," and an "accessory to the murder of Jewish children" on air. NBC executives rewarded this conduct by naming him their "showman."[31] Banfield told a Kansas State University audience during the Iraq invasion that "horrors were completely left out of this war. So was this journalism? . . . I was ostracized just for going on television and saying, 'Here's what the leaders of Hezbollah, a radical Moslem group, are telling me about what is needed to bring peace to Israel.'"[32] She was immediately demoted and disciplined by NBC for criticizing journalistic standards.

Such conduct is hardly surprising given that the network's parent, General Electric, is among the world's biggest arms suppliers, a conflict of interest rarely discussed openly. A study of NBC and MSNBC's use of metaphor during the 2003 build-up discloses a fixation on the strong and highly structured metaphors of sport and business ("Countdown," "Showdown," "Timetable," "Selling the Plan," and "Target" were banners and segment headings) that erase indeterminacy and subordinate a critique of ends to a focus on means.[33]

For its part, CNN has not always been so much in lockstep with reactionary politics. Consider the Republican reaction when CNN's Peter Arnett disclosed during the 1991 Gulf War that the Yanquis had bombed a factory that made baby formula, not a chemical-weapons facility, as had been claimed. He was immediately described by the administration's news maven Marlin Fitzwater as a "conduit for Iraqi disinformation," and branded "the Joseph Goebbels of Saddam Hussein" by Representative Laurence Coughlin. Thirty-four Congresspeople petitioned CNN in protest of the coverage, and Arnett became a target for Pentagon retribution in the years to come. The contrast between CNN and the other networks in 1991 was stark—CBS notoriously offered to edit its footage in order to conclude segments positively prior to commercials.[34] A decade later, CNN had learned its lesson. Its gleeful coverage of the invasion of Iraq was typified by one superannuated military officer who rejoiced with "Slam, bam, bye-bye Saddam" as missiles struck Baghdad.[35]

CNN also broke every precept of responsible journalism by sending armed private-security forces to accompany its journalists during the invasion, leading to the notorious incident when bodyguards shot at Iraqi troops in Tikrit. This drew censure from *Reporters sans Frontières*, the advocacy organization for press freedom and safety, as being "contrary to all the rules of the profession."[36] At such moments as the one in Tikrit, a form of uncompromising nationalism supplements the ordinary business of U.S. journalism in a neoliberal age. The bank and the flag go mutual guarantor.

Noted CNN foreign correspondent Christiane Amanpour told CNBC after the invasion that

I think the press was muzzled, and I think the press self-muzzled. . . . I'm sorry to say, but certainly television and, perhaps, to a certain extent, my station was intimidated by the administration and its foot soldiers at Fox News. And it did, in fact, put a climate of fear and self-censorship, in my view, in terms of the kind of broadcast work we did.

She was immediately derided by Fox's Irena Briganti as "a spokeswoman for Al Qaeda."[37]

To the rest of the world, CNN staffers had become pale parodies of their erstwhile selves. Pakistan's *Friday Times* offered this guide on how to "look like a CNN correspondent":

10. Pretend to be in grave danger while reporting from the roof of the Marriott, Islamabad.
9. Never learn how to pronounce "Pakistan."
8. Get a U.S. marine escort to help you do your groceries.
7. Bond with the locals by hanging out at Muddy's Café.
6. Carry big black cameras with CNN stickers pasted all over them.
5. Always wear a safari jacket (esp. when in big cities).
4. Wear a CNN t-shirt.
3. Wear a CNN hat.
2. Wear CNN underwear.
1. Hunt for the biggest lunatics to put on air.[38]

This makes CNN's rejection of Ted Turner as a war correspondent because of his inexperience entirely laughable.[39]

In *Q-News: The Muslim Magazine* editor Fuad Nahdi's words, dispatching "young, inexperienced and excitable" journalists who are functionally illiterate and historically ignorant means that the U.S. media depends on "clippings and weekend visits" overseas—filings that are of dubious professional integrity.[40] No wonder CNN's Jerusalem Bureau chief, Walter Rodgers, insensitively proclaims that "[f]or a journalist, Israel is the best country in the world to work in . . . [o]n the Palestinian side, as is the case in the rest of the Arab world, there is always that deep divide between Islam and the West."[41] CNN, of course, reached its Middle Eastern *nadir,* and lost viewers to Al Jazeera and others, when one of its reporters stated that some nomads would be thunderstruck by seeing "camels of steel" (cars) for the first time.[42] Not to mention the notorious exchange between two reporters on air during the Afghanistan invasion, where one suggested that an assault on an arms depot may have been part of the civil war, and the other offered, "Oh, are they having one?"[43] Reactionaries celebrate such changes, regarding them as responses to market demand from Yanquis and the flexible supply of new technologies, while ignoring the special responsibility for U.S. citizens to know about the domination and destruction wrought in their name.[44]

When added to the speeded-up routines of 24-hour news channels, that particular grotesquerie of flag-waving TV—embedding journalists with military forces—led to disgraces like the day when nine separate announcements were made that Umm Qasr had fallen to the invaders, which was untrue, and a Fox news producer saying that "Even if we never get a story out of an embed, you need someone there to watch the missiles fly and the planes taking off. It's great television."[45] The Project for Excellence in Journalism's analysis of ABC, CBS, NBC, CNN, and Fox found that in the opening stanza of the Iraq invasion, 50 percent of reports from the 1,000 journalists working embedded with the invaders depicted combat. Zero percent depicted injuries. As the war progressed, the most we saw were deeply sanitized images of the wounded from afar, in keeping with the 50 contractual terms required of reporters in return for their "beds."[46] No wonder Bernard Shaw, the former CNN anchor, saw these journalists as "'hostages of the military.'"[47]

Such tendencies were not merely a reaction to September 11. Their causes went deeper. After the 2000 election, CNN's Judy Woodruff had told Bush's first chief-of-staff, Andy Card, on-air that "we look forward to working with you."[48] This remark would be unsayable by a journalist in a functioning democracy, where official sources are starting-points for work, not results—and the idea of "working with" a government is seen as "Soviet-style."[49] But inside the United States, there is a long heritage of reliance on official sources, dating from the evolution of journalistic codes and norms as tools for monopolistic owners to distract attention from their market domination by focusing on non-partisan journalism and cloaking themselves in professionalism.[50] The White House, the State Department, and the Pentagon are referred to pleasurably as "the Golden Triangle," a sure sign of the deluded faith in official sources that dogs contemporary Yanqui journalism's "stenographic reporting."[51] This is the occupational ideology that drapes the bank in Stars and Stripes. It reaches its apogee in the treatment of others.

Casualties

Thirty-eight percent of domestic CNN's coverage of the 2003 bombardment of Iraq centered on technology, while 62 percent focused on military activity without history or politics—as a matter of technical specifications and instrumental rationality. Al Jazeera dedicated only a third of its stories to war footage. Unlike CNN, it emphasized human distress rather than electronic effectiveness, vernacular reportage rather than patriotic euphemism.[52]

This extraordinary disconnection between the fantasy world of U.S. cable TV and international news organizations came into stark relief during a notorious exchange over the U.S. occupation. In April 2004, CNN's Daryn Kagan interviewed Al Jazeera's editor-in-chief, Ahmed Al-Sheik. What might have been an opportunity to learn about the horrendous casualties in the Fallujah uprisings, or to share professional perspectives on methods and angles of coverage, turned into a bizarrely unreflective indictment of Al Jazeera for bothering to report the deaths of Iraqi noncombatants at the hands of the invaders. Kagan complained that "the story" was "bigger than just the numbers of people who have been killed or the fact that they might have been killed by the U.S. military."[53]

At least this incident represented interaction with Al Jazeera, in stark contrast to CNN's refusing to appear on a Nordic TV panel with Al Jazeera representatives.[54] And some analysts suggest that CNN's dependence on Al Jazeera for direct images and reportage from the Afghanistan conflict helped to make for a semblance of balance between technocratic celebrations and humanitarian discussions of death.[55] That was not needed by the likes of Fox News, whose operatives described the Taliban as "rats," "terror goons," and "psycho Arabs" during the 2001 conflict.[56]

The U.S. networks' censorship of footage of Afghan civilian casualties in October 2001 was as predictable as it was appalling.[57] The thousands of civilian Afghan deaths reported by South Asian, Southeast Asian, Western European, and Middle Eastern news services went essentially unrecorded here, because they could not be "verified" by U.S. journalists/officials.[58] Fox News Managing Editor Brit Hume said that civilian casualties may not belong on television, as they are "historically, by definition, a part of war." CNN instructed presenters to refer to September 11 each time Afghan suffering was mentioned, and Walter Isaacson, the network's president, worried aloud that it was "perverse to focus too much on the casualties or hardship."[59] This "perversity" applied only to domestic CNN viewers— those *in* the real world were judged sturdy enough to learn *about* the real world.[60]

Cable had previewed such callous ignorance in 2001. Consider the following from Lawrence Eagleburger, a former secretary of state, who was called in by CNN to comment after the attacks on the United States: "There is only one way to begin to deal with people like this, and that is you have to kill some of them even if they are not immediately directly involved." Meanwhile, Republican house intellectual Anne Coulter called on the government to identify the nations where terrorists lived, "invade their countries, kill their leaders and convert them to Christianity."[61] Katie Couric, NBC's apolitical headliner, was caricatured by Coulter as "Eva Braun" and "Joseph Goebbels." Coulter was also the author of the notorious rebuke on TV to a disabled Vietnam veteran, saying, "People like you caused us to lose that war," and proceeding to propose that the Right "physically intimidate liberals, by making them realize that they can be killed too."[62] Coulter's reward for such hyperbolic ignorance was frequent appearances on CNN, MSNBC, HBO, and a raft of other cable programs, in addition to admiring profiles in the *New York Times, Newsday,* the *Wall Street Journal,* and the *New York Observer.*[63] Bill O'Reilly of Murdoch's Fox News referred to Islamic fundamentalism as "the enemy of the U.S." (without defining his terms via either the adjective or the noun (or engaging Hindu and Christian fundamentalism and the reigns of terror they have perpetrated).[64] O'Reilly described college students being required to read about the Koran as equivalent to assigning *Mein Kampf* during World War II,[65] and called for the government to "bomb the Afghan infrastructure to rubble—the airport, the power plants, their water facilities and the roads. . . . We should not target civilians, but if they don't rise up against this criminal government, they starve, period."[66] During the invasion of Iraq, he called U.N. Secretary General Kofi Annan "a villain."[67]

Conclusion

This kind of chauvinistic, anti-democratic, theo-idiotic hysteria has an inevitable impact in the rest of the world. A study by the International Federation of Journalists in October 2001 found blanket global coverage of the September 11 attacks, with very favorable discussion of the United

States and its travails—even in nations that had suffered terribly from U.S. aggression. But it is no surprise that:

- the giant advertising firm McCann-Erickson's evaluation of 37 states saw a huge increase in cynicism about the U.S. media's manipulation of the events;[68]
- the Pew Research Center for the People & the Press' 2002 study of 42 countries found a dramatic fall from favor for the United States since that time;
- Pew's 2003 follow-up encountered even lower opinions of the U.S. nation, population, and policies worldwide than the year before, with specifically diminished support for anti-terrorism, and faith in the U.N. essentially demolished by U.S. unilateralism and distrust of the White House; and
- after Bush's 2004 reelection, a poll across 21 sovereign-states found 47 percent of people opposed to the United States in general—not just its government, now, but its population, too, while 58 percent thought his mandate made life riskier. Longitudinal studies indicated that the big change between the contemporary moment and the last time the United States was led by a similarly ideological president, Ronald Reagan, was that while the U.S. government was loathed across the globe at that time for its imperialism, the U.S. population was exempted from responsibility. Now it isn't excused the violence undertaken in its name.[69]

In terms of empirical knowledge of the events of September 11, 2001, and the subsequent invasions of Afghanistan and Iraq, the U.S. public knew less about the world and its own government than seems possible. Comprehensive studies by the Program on International Policy Attitudes and Knowledge Networks found that a minority of the population realized that clear majorities all over the world opposed the 2003 invasion, and a significant minority thought the war was supported globally.[70] These people also believed weapons of mass destruction were in Iraq, and that there were indisputable ties between Iraq and September 11. These delusions held firm a year after the invasion, and were directly correlated with viewers' support for the Republican Party—and their consumption of commercial TV news. When the Iraq Survey Group's final report to Congress just prior to the 2004 presidential elections confirmed what the Left, the International Atomic Energy Agency, and the United Nations had painstakingly explained eighteen months earlier—that there were no significant weapons of mass destruction in Iraq until the United States arrived—the report attracted such minimal Yanqui press attention that the Republican half of the population continued to believe the lies it had been told.[71] The truth was known only to those who watched or listened to public broadcasting or news from elsewhere. The lies were believed by those watching U.S. cable news. The storytelling function of journalism no longer applied —U.S. cable news employees had become akin to tellers in a bank, ensuring returns to their conglomerate owners while covering this goal in a prophylactic *façade* of flags. Goodbye discourse and the active audience, hullo ideology and the passive recipient.

To return to the quotations from the FCC commissioners with which I began, we must re-regulate the media to ensure we have an informed citizenry. We need to swim with the minnows, not hunt with the fowlers. That doesn't mean *Buffy the Vampire Slayer* is unimportant and fandom apolitical. They just might not be *quite* as important or political as knowing that hundreds of thousands are being killed and millions more having their lives transformed in your name. If that means a little less focus on the pleasures of interpretation, then U.S. TV studies will have done something worthwhile. So let's tell the tellers to step out from behind their work-stations and put their flags to bed.

NOTES

1. Newton N. Minow, "The Broadcasters are Public Trustees," in Allen Kirschener and Linda Kirschener, eds., *Radio & Television: Readings in the Mass Media* (New York: Odyssey Press, 1971), 207–17.

2. Jeff Chester, "Strict Scrutiny: Why Journalists Should Be Concerned about New Federal Industry Media Deregulation Proposals," *Harvard International Journal of Press/Politics* 7, no. 2 (2002): 106.

3. Newton N. Minow and Fred H. Cate, "Revisiting the Vast Wasteland," *Federal Communications Law Journal* 55 (2003): 408.

4. Ibid., 415.

5. Eric Alterman, *What Liberal Media? The Truth about Bias and the News* (New York: Basic Books, 2003), 125.

6. Douglas Kellner, *From 9/11 to Terror War: The Dangers of the Bush Legacy* (Lanham: Rowman & Littlefield, 2003), 105.

7. Quoted in Jacqueline E. Sharkey, "The Television War," *American Journalism Review* (2003), accessed on http://www.ajr.org.

8. Ibid.; see also David Folkenflik, "Fox News Defends its 'Patriotic' Coverage," *Baltimore Sun,* April 2, 2003.

9. Justin Lewis et al., *Too Close for Comfort? The Role of Embedded Reporting During the 2003 Iraq War: Summary Report* (London: British Broadcasting Corporation, 2004), 25.

10. Quoted in Stuart Allan and Barbie Zelizer, "Rules of Engagement," in Stuart Allan and Barbie Zelizer, eds., *Reporting War: Journalism in Wartime* (London: Routledge, 2004): 7.

11. Sandra L. Borden, "Communitarian Journalism and Flag Displays After September 11: An Ethical Critique," *Journal of Communication Inquiry* 29, no. 1 (2005): 31.

12. Theodor W. Adorno, *The Culture Industry: Selected Essays on Mass Culture,* trans. Gordon Finlayson, Nicholas Walker, Anson G. Rabinach, Wes Blomster, and Thomas Y. Levin; ed. J. M. Bernstein (London: Routledge, 1996), 89.

13. Matthew A. Baum, "Sex, Lies, and War: How Soft News Brings Foreign Policy to the Inattentive Public," *American Political Science Review* 96 (2002): 91–110; John Zaller, "A New Standard of News Quality: Burglar Alarms for the Monitorial Citizen," *Political Communication* 20, no. 2 (2003): 109–30.

14. Pierre Bourdieu, *On Television,* trans. Priscilla Parkhurst Ferguson (New York: New Press, 1998), 48.

15. Megan Mullen, "The Rise and Fall of Cable Narrowcasting," *Convergence* 8, no. 1 (2002): 62–83; Richard Barbrook and Andy Cameron, "The Californian Ideology," *Science as Culture* 6 (1996): 44–72.

16. Michael J. Copps, "I Dissent Because Today the FCC Empowers America's New Media Elite with Unacceptable Levels of Influence . . . ," CommonDreams.org (accessed June 2, 2003).

17. Piers Robinson, "Researching U.S. Media-State Relations and Twenty-First Century Wars," in Allan and Zelizer, eds., *Reporting War,* 99.

18. Mark Fitzgerald, "Study Shows Readership at Crossroads," *Editor & Publisher,* May 6, 2003; see also Sharkey, "The Television War."

19. Lavine and Readership Institute, *Beyond Impact: Engaging Younger, Lighter Readers: A Joint Venture of NAA, ASNE, and Media Management Center,* 2003; Pew Research Center for the People & the Press, *The Internet and Campaign 2004* (2005).

20. Pew Charitable Trusts, *Return to Normalcy? How the Media Have Covered the War on Terrorism* (January, 2002).

21. Paul Farhi, "Everybody Wins," *American Journalism Review* (2003); Alterman, *What Liberal Media?,* 136–37; Daya Kishan Thussu, "Live TV and Bloodless Death: War, Infotainment and 24/7 News," in Daya Kishan Thussu and Des Freedman, eds., *War and the Media: Reporting Conflict 24/7* (London: Sage Publications, 2003), 118; Alina Tugend, "Pundits for Hire," *American Journalism Review* (May 2003).

22. Alterman, *What Liberal Media?*, 206

23. FAIR, "Some Critical Media Voices Face Censorship" (2003).

24. Steve Rendall and Tara Broughel, "Amplifying Officials, Squelching Dissent," *EXTRA! Update*, May/June 2003.

25. Peter Johnson, "Media Question Authority Over War Protests," *USA Today*, February 24, 2003.

26. Amy Goodman et al., *Democracy Now!*, April 4, 2003; David Folkenlik, "Fox News Defends its 'Patriotic' Coverage," *Baltimore Sun*, April 2, 2003.

27. Ciar Byrne, "Simpsons Parody Upset Fox News, Says Groening," *Guardian*, October 29, 2003.

28. Michael Hastings, "Billboard Ban," *Newsweek*, February 26, 2003.

29. Quoted in FAIR, "Media Should Follow Up on Civilian Deaths," April 4, 2003, and Ralph Nader, "MSNBC Sabotages Donahue," CommonDreams.org, March 3, 2003; also see Rick Ellis, "The Surrender of MSNBC," AllYourTV.com, February 25, 2003.

30. Quoted in Rendall and Broughel, "Amplifying Officials."

31. Quoted in David Lieberman, "NBC Hopes Big Investment in News Coverage Pays Off," *USA Today*, March 24, 2003.

32. Quoted in Danny Schechter, "The Media, the War and Our Right to Know," Alternet.org (May 1, 2003).

33. Jack Lule, "War and its Metaphors: News Language and the Prelude to War in Iraq, 2003," *Journalism Studies* 5, no. 2 (2004): 179–90.

34. Peter Arnett, "You are the Goebbels of Saddam's Regime," *The Guardian*, February 14, 2003; Fitzwater quoted in Chris Jones, "Peter Arnett: Under Fire," BBC News Online, April 4, 2003; Patrick Martin, "The Firing of Peter Arnett: Right-Wing Straitjacket Tightens on the U.S. Media," *Guardian*, April 1, 2003; Will Hutton, *A Declaration of Interdependence: Why America Should Join the World* (New York: W. W. Norton, 2003), 156

35. Quoted in Richard Goldstein, "The Shock and Awe Show," *Village Voice*, March 26–April 1, 2003.

36. *Reporters sans Frontières*, www.rsf.org (2003).

37. Quoted in Antonia Zerbisias, "The Press Self-Muzzled its Coverage of Iraq War," *Toronto Star*, September 16, 2003, and Allan and Zelizer, eds., *Reporting War*, 9.

38. Quoted in Danny Schechter and Aliza Dichter, "The Role of CNN," in Danny Schechter, ed., *Media Wars: News at a Time of Terror* (Lanham: Rowman & Littlefield, 2003): 49.

39. Ken Auletta and Ted Turner, "Journalists and Generals," New Yorker.com, March 24, 2003.

40. Fuad Nahdi, "Doublespeak: Islam and the Media," openDemocracy.net, April 3, 2003.

41. Quoted in Dina Ibrahim, "Individual Perceptions of International Corre-

spondents in the Middle East," *Gazette: The International Journal for Communication Studies* 65, no. 1 (2003): 87–101, 96.

42. Neil MacFarquhar, "Arabic Stations Compete for Attention," *New York Times*, March 25, 2003.

43. Quoted in Schechter, ed., *Media Wars*, 6.

44. John Maxwell Hamilton and Eric Jenner, "Redefining Foreign Correspondence," *Journalism: Theory, Practice and Criticism* 5, no. 3 (2004): 301–21.

45. Chris Tryhorn, "When Are Facts Facts? Not in a War," *Guardian*, March 25, 2003.

46. Sharkey, "The Television War"; Oliver Boyd-Barrett, "Understanding: The Second Casualty," in Allan and Zelizer, eds., *Reporting War*, 30–31.

47. Quoted in Andrew Bushell and Brent Cunningham, "Being There," *Columbia Journalism Review*, March/April 2003.

48. Judy Woodruff quoted in Norman Solomon, "Media War Without End," *Z Magazine* December 2001.

49. Michael Massing, "Press Watch," *The Nation,* October 15, 2001.

50. Jessica Clark and Robert McChesney, "Nattering Networks: How Mass Media Fails Democracy," *LiP*, September 24, 2001; Robert McChesney, "The Problem of Journalism: A Political Economic Contribution to an Explanation of the Crisis in Contemporary U.S. Journalism," *Journalism Studies* 4, no. 3 (2003): 299–329; Edward S. Herman, *The Myth of the Liberal Media: An Edward Herman Reader* (New York: Peter Lang, 1999), 83, 87, 158.

51. Maryann Cusimano Love, "Global Media and Foreign Policy," in Mark J. Rozell, ed., *Media Power, Media Politics* (Lanham: Rowman & Littlefield, 2003), 246; Susan D. Moeller, "A Moral Imagination: The Media's Response to the War on Terrorism," in Allan and Zelizer, eds., *Reporting War*, 71.

52. Amy E. Jasperson and El-Kikhia O. Mansour, "CNN and al Jazeera's Media Coverage of America's War in Afghanistan," in Pippa Norris, Montague Kern, and Marion Just, eds., *Framing Terrorism: The News Media, the Government, and the Public* (New York: Routledge, 2003), 119, 126–27; 113–32.

53. Quoted in FAIR, "CNN to Al Jazeera: Why Report Civilian Deaths?" April 6, 2004.

54. Elizabeth Eide, "Warfare and Dual Vision in Media Discourse," in Stig A. Nohrstedt and Rune Ottosen, eds., *U.S. and the Others: Global Media Images on "The War on Terror"* (Goteborg: Nordicom, 2004), 263–84, 280.

55. Jasperson and Al-Kikhia, "CNN and al Jazeera's Media Coverage of America's War in Afghanistan," 120, 125.

56. Quoted in Thussu, "Live TV and Bloodless Death," 127.

57. Michael Hudson et al., "Covering the War on Terrorism," *Transnational Broadcasting Studies* 8 (2002).

58. Marc W. Herold, "Who Will Count the Dead?," *Media File* 21, no. 1 (2001).

59. Quoted in Kellner, *From 9/11 to Terror War*, 107, 66

60. Bruce A. Williams, "The New Media Environment, Internet Chatrooms, and Public Discourse after 9/11," in Thussu and Freedman, eds., *War and the Media*, 177.

61. National Review Online (October 13, 2001).

62. Quoted in Alterman, *What Liberal Media?*, 3–4.

63. Ibid., 5.

64. Bill O'Reilly, Fox News, June 15, 2002.

65. Alterman, *What Liberal Media?*, 37.

66. Bill O'Reilly, Fox News, September 17, 2001.

67. Bill O'Reilly, Fox News, April 24, 2003.

68. Claire Cozens, "Viewers Greet September 11 Coverage with Cynicism," *Guardian*, October 26, 2001.

69. "Global Poll Slams Bush Leadership," January 18, 2005; *BBC News Online*; Pew Research Center for the People & the Press, Global Opinion: The Spread of Anti-Americanism, 2005.

70. Program on International Policy Attitudes and Knowledge Networks, *Misperceptions, the Media and the Iraq War* (2003); and *U.S. Public Beliefs on Iraq and the Presidential Elections,* 2004; Steven Kull et al., "Misperceptions, the Media and the Iraq War" *Political Science Quarterly* 118, no. 4 (2003–2004): 569–98.

71. Program on International Policy Attitudes and Knowledge Networks, *U.S. Public Beliefs on Iraq and the Presidential Elections.*

Chapter 14

Dualcasting

Bravo's Gay Programming and the
Quest for Women Audiences

Katherine Sender

In the summer of 2003, gays were big news in the United States and Canada: the U.S. Supreme Court overturned sodomy laws in all states, the Canadian government decided to award marriage licenses to same-sex couples, and Gene Robinson was confirmed as the bishop of New Hampshire, making him the first openly gay and partnered Episcopalian bishop in the Anglican church. The television show that catalyzed the national imagination was Bravo cable channel's *Queer Eye for the Straight Guy*, a makeover show in which five gay men worked with the raw material of a stylistically and socially incompetent heterosexual in order to "build a better straight man."[1] A measure of the show's success was its spoof spin-offs, including three episodes of *Straight Plan for the Gay Man*, and a special episode of *South Park*, "South Park is Gay!"—all of which aired on Comedy Central. Bravo also ran an original gay-themed dating series, *Boy Meets Boy*, which led into *Queer Eye*. For a channel formerly known for its signature show *Inside the Actors Studio*, this assertively gay-themed programming seemed a happy moment of serendipitous timing. Yet the broader context of the fragmentation of mass television audiences across increasing numbers of channels and Bravo's own history of programming gay content help to make sense of why the channel was a major player in 2003's "summer of gay love."[2] This chapter considers the deployment of gay-themed programming on Bravo as an example of a new approach to attract a fragmented and volatile audience, hitherto loyal to the Big Three networks, to niche cable channels. Did Bravo's executives position the channel as the de facto gay channel in the gap left by MTV and Showtime's stalled dedicated gay cable channel, Outlet (see Freitas, in this volume), or was gay-themed programming like *Queer Eye for the Straight Guy* and *Boy Meets Boy* part of a broader strategy to appeal to audiences, gay and straight alike?[3]

Bravo's gay-themed shows of 2003 arrived at a unique moment in U.S. television history that saw a confluence of two cultural trends, one industrial, the other representational. From a television industry standpoint, as the number of channels available to the average television household increased from ten in 1980 to ninety in 2003, the Big Three broadcast networks experienced a significant erosion of their audience, seeing their shared prime-time rating drop by half in the same period.[4] As Joseph Turow observes, in the early 1980s advertisers began to consider cable as a means to reach the dissipating audience,[5] the Big Three networks consequently suffered a decline in ad revenue from 98 percent to 46 percent of total advertising bought on television between 1980 and 2003.[6] The smaller broadcast and cable channels enjoyed a growing share of advertising revenue. Fox, UPN, and the WB won almost 11 percent of advertising income by 2003, and the ad-supported cable networks combined took 36 percent of television ad revenues.[7] The smaller channels not only offered an alternative and cheaper means to reach audiences, but also were focused on a highly targeted one: from their inception, cable channels such as MTV, Lifetime, and BET offered fare designed to attract young, female, and African-American viewers, respectively. Cable executives scheduled programs designed to signal to viewers within a particular demographic and lifestyle niche that there was a special relationship between the channel and that niche, as well as to signal to advertisers that the channel had efficiently separated the desirable group from those viewers outside the target market.[8] But cable channels must navigate a narrow line between signaling a niche appeal and retaining large enough audiences. In order to be included in the Nielsen Cable Activity Report, the cable equivalent to the broadcast television ratings, a cable channel has to be available in at least 3.3 percent of U.S. television households and to generate a minimum 0.1 rating in those households (approximately 100,000 households). In the increasing competition for audiences in the 1990s, cable channels "couldn't afford to see themselves as so targeted as to fall below Nielsen's radar. On the other hand, they were aware that with the proliferation of offerings they needed advertisers and viewers to go to their format because it had a distinct personality."[9] Like its cable competitors, Bravo had to develop programming that signaled a niche appeal, but could still garner large enough audiences to gain advertiser attention.

Seismic shifts in audience activity were paralleled by profound trans-formations in television representations. Since the mid-1990s, gay and les-bian images have proliferated on broadcast network, basic cable, and pre-mium cable channels. Gay and lesbian (though rarely bisexual and almost never transgender) characters are now not only tolerated but often wel-comed on U.S. televisions: NBC's *Will & Grace*, for example, drew an aver-age weekly audience of 16.8 million viewers in the 2003–2004 season.[10] The reality television genre has been particularly hospitable to gay charac-ters: MTV's *The Real World* has included at least one gay, lesbian, or bisex-ual participant in almost every season since its debut in 1992, and, with the exception of dating shows (but sometimes here, too), gay members are a regular part of the "diversity" of reality competitor line-ups. Indeed, Larry Gross argues that gay participants are not incidental but fundamental to the realism of reality television: "Whereas, as recently as the early 1990s, the inclusion of a gay character would typically be the focus of some dra-matic 'problem' to be resolved, today, particularly for programs that aim at coveted younger viewers, it seems that the presence of gay people is a nec-essary guarantor of realism."[11]

The relative ubiquity of gay and lesbian characters on reality-TV shows reflects the expectations of a genre that demands diversity and conflict among participants, as well as a long history of gay activist agitation to-wards media visibility. Especially since the Stonewall Riots of 1969, gay ac-tivists have put much emphasis on media representations of gays, pressur-ing producers to show the world and isolated gays that other gay people exist, and campaigning against the most egregious stereotypes of homo-sexuals as pathological, criminal, and pathetic. The AIDS epidemic chal-lenged many people to come out to their families and colleagues, increas-ing the number of people who were aware that they knew gay people and encouraging more openly gay and lesbian media executives to lobby for more and better representations. President Clinton's attempt to lift the ban on gays in the military, though failed, nevertheless made homosexu-ality a topic for national debate. Watershed shows such as *Ellen, Will & Grace, Dawson's Creek, Queer as Folk, Sex in the City,* and *Six Feet Under* brought new, likeable, and increasingly complex gay and lesbian characters to prime-time broadcast and cable television.

Perhaps most significant in the growing presence of gay-themed pro-gramming was the development of the gay market, which was in forma-tion by the late 1970s and rapidly consolidated in the 1990s.[12] Gay and les-bian consumers went from a marginalized and largely stigmatized group to a desirable marketing niche in this period, with two distinct effects for television programming. First, gay marketing taught media producers of all kinds that there is a potentially sizable gay and lesbian audience for their programs, as well as emphasizing (usually by exaggeration) the affluence and loyalty of that audience. Second, the construction of the ideal gay consumer as not only wealthy but trend-setting was so successful that advertisers wanted to be associated with the gay market in order to appeal to *heterosexual* consumers. By including gay and lesbian characters in shows, programmers and advertisers could reach two distinct audiences: gays and lesbians in search of people who look (sort of) like them and heterosexuals attracted to the hip cachet of gay taste. Television executives hoped to attract a sizable combined audience with lots of money and cultural capital, an audience apparently primed for advertisers' messages.

These two trends—the dissipation of audiences and the proliferation of gay images—are separable in theory only. In practice, the rise of increasingly differentiated and consolidated target markets and the tailoring of program content for niche media and are entirely interdependent. Turow describes two perspectives on the fragmentation of audiences across media channels. One version argues that the technological and industrial changes that facilitated hundreds of television channels led to an inevitable process of segmentation of the audience, as viewers went in search of varied fare. The other perspective argues that media fragmentation was a response to, not a cause of, audience fragmentation. The civil rights movement, the anti–Vietnam War movement, the women's and gay rights movements highlighted identity affiliation and politics in unprecedented ways. According to an *Advertising Age* editorial, in the 1970s America seemed "split asunder into innumerable special interests—gray power, gay power, red power, black power, Sunbelt and frostbelt, environmentalists and industrialists . . . all more aware of their claims on society."[13]

The truth, however, lies somewhere between these versions; marketers did indeed respond to identity movements, but as audiences showed they were willing to be organized into newly distinct segments, media producers made increasingly concerted efforts to both consolidate and further split those segments. Cable television proved especially agile in this process of segmentation: because revenues come from both cable distributors that pay the cable network for content and from advertisers buying space on shows, cable networks can afford to target smaller niche audiences than the mass demanded by the much more expensive broadcast networks. Cable also has a tradition of cheap programming, showing reruns in syndication and developing inexpensive original programming, such as reality shows and documentaries, that do not necessitate paying high salaries to writers and actors. Cable TV is thus ideally placed to appeal to narrow segments of the overall audience in small but select groups as a target for advertisers.

Cable has also been able to afford to take more risks with gay programming than the broadcast networks can, because of its different sources of revenue and types of regulation. Cable can be less concerned by advertiser backlash, because cable channels have an additional source of sponsorship from distribution and subscription. Cable channels are not subject to the same FCC regulations on content as the networks, allowing them to take greater risks with controversial content. Premium cable channels like HBO and Showtime have pushed the envelope on gay themes, modeling a television environment more friendly to gay characters and viewers. Bravo's gay-themed programming of 2003 thus reflects not a brave attempt by a cable renegade to bring gays to basic cable, but a confluence of existing industry, marketing, and representational trends that made the channel ideally placed to develop gay television.

Indeed, *Queer Eye* and *Boy Meets Boy* were not Bravo's first shows to feature gay content, but continued the channel's history of gay-themed programming. Bravo executives had commissioned a number of short-run, reality-format, gay- and lesbian-themed shows in the early 2000s: *Fire Island* (2000), *Gay Riviera* (2001), and *Gay Weddings* (2003) had built small audiences before *Queer Eye* and *Boy Meets Boy* appeared in 2003. Bravo had also profiled openly gay and lesbian celebrities, such as k.d. lang (1996), and celebrities popular with gay and lesbian fans, such as Cher, whose much-advertised retirement concert was rerun on Bravo as a lead-in to the premier episode of *Queer Eye*. Bravo continued this run of gay-themed reality shows with a short-lived spin-off, *Queer Eye for the Straight Girl*, in 2005.

Bravo's gay-themed programming also reflects the channel's changing affiliation within the television industry: NBC's parent company, General Electric, purchased Bravo from Cablevision Systems Corporation in November 2002, wanting to "improve the network's cable presence and give [NBC] another outlet for programming."[14] *Queer Eye*'s executive producer David Collins admitted being nervous that NBC would find the show, already in development, too controversial, and cancel it: "We thought for sure it was all over . . . We thought 'OK, that was fun. We got to make a pilot, and it's going to stay on the shelf.'"[15] Instead, NBC executives were enthusiastic about *Queer Eye,* seeing it as one of a number of new shows that would help to change Bravo's reputation from its "artsy," "highbrow" tradition to something "edgier" and more youth-oriented.[16] By doing so, they hoped to shift the dominant demographic from older viewers (in early 2003 half the audience was over fifty) towards a younger population.[17] As one industry insider said, Bravo had been perceived to have "a 'Masterpiece Theatre' kind of audience, and the perception [among Bravo and NBC executives] is that there's more to offer."[18]

The debut of *Queer Eye* and its lead-in, *Boy Meets Boy,* two weeks later, created a flurry of press activity, much of which addressed the prospective audience Bravo was after. As Bravo President Jeff Gaspin said, "Does this mean that Bravo is becoming a gay network? Absolutely not. . . . Not that there's anything wrong with that." He continued, "On the surface [the program block] might seem designed for gay audiences, but it's really not. . . . When we discussed our advertising plans for how we are going to promote it, the first group of people we are going to promote it to are women [aged] 18 to 49."[19] Elsewhere, Gaspin said, "We have had success with gay audiences in the past . . . but the primary audiences for these shows will be women. We don't sell a gay audience to advertisers."[20] With one exception, all the articles about the show in the trade and popular press duly emphasized that women (presumably heterosexual), not gays or lesbians, were the primary audience for the shows. Only one article talked about the show's potential to market products, like the Fab Five's chariot, a General Motors Yukon Denali SUV, to gay consumers: "GM officials say the biggest draw for them for *Queer Eye* is that it features top-shelf brands, such as Thomasville Furniture and Ralph Lauren, and [General Motors], with its 'professional grade' image, fits nicely into that mold."[21] This was a rare acknowledgement of an advertiser's wish to be associated with gay consumers' "professional grade"—read "affluent"—reputation, and of the value of that association for product placement on a show that profiles gay taste.

Gaspin insisted that Bravo has a "dual target" for their gay programs; gay audiences were considered a "secondary priority to female viewers," and three-quarters of the advertising budget for both *Queer Eye* and *Boy Meets Boy* was allocated to attracting women audiences.[22] One television commercial for *Queer Eye,* for example, asked, "Ladies, is your man an embarrassment? . . . Is his place a pig sty? Meet five gay men out to make over the world one straight guy at a time." A marketing representative from Bravo saw the additional spending as justified: women "wouldn't gravitate as naturally to 'Queer' or 'Boy' as gay viewers would." She continued, "Women, as a broader target, are not as easy to convince to see these programs. . . . Communication in the gay community is such that [gays] would find their way to these shows quickly."[23] Bravo executives also acknowledged that gay viewers were harder to target through conventional advertising venues, so they adopted other methods to raise the profile of the shows, including distributing 44,000 whistles at gay pride parades in New York and Los Angeles with *"Queer Eye for the Straight Guy"* on one side and "Bad style really blows!" on the other.

Gay audiences, then, were not Bravo's primary target for their gay-themed shows, for a number of reasons. First, sexual identification is not included in mainstream ratings data, meaning that it is hard to sell a gay audience to advertisers. As a senior vice president for marketing and advertising services at NBC said, "Gay men are not measured by Nielsen [Media Research] . . . Women 18 to 49 is a more saleable demo[graphic]."[24] Second, at an estimated 5 to 6 percent of the adult population, gays represent a much smaller potential audience than women ages 18 to 49 do. Third, just as women 18 to 49 will inevitably have a diverse range of viewing habits and tastes (a diversity that is the lifeblood of niche cable programming), gay and lesbian audiences also have highly variant preferences: simply including gay characters or participants on a show is no guarantee of winning a large sector of the gay audience. A glance at gay-niche print media bears this out: few gay and lesbian magazines have managed to achieve circulation rates higher than 100,000, even though the GLBT-identified population numbers in the millions.

Boy Meets Boy is an especially interesting example of where gay-themed programming, even when produced by gay personnel, might not necessarily appeal to gay audiences. The show's premise—a gay *Bachelor* dating show with the "twist" that some of the suitors were heterosexual—was especially controversial, despite its producers' claims that it was a "sociological experiment" designed to teach the heterosexual participants (and their audience counterparts) about the stress and pain of hiding one's "true" sexuality.[25] Executive producer Doug Ross speculated that "adding straight guys to the dating pool . . . will bring straight viewers to a show that might otherwise have attracted only a gay audience."[26] At the same time, however, the deceptive strategy—revealed to the show's leading man only midway through the series—risked alienating gay viewers who, like the gay "bachelor" James, felt manipulated by what seemed like a particularly homophobic twist. Bravo focused its efforts on reaching women, then, because women constituted a larger, measurable, and therefore more desirable audience. Further, because gay viewers were hard to reach through conventional advertising, because they would watch the shows anyway (Bravo executives hoped), and because they might prove a more critical or fickle audience, there was less incentive to actively court them.

Finally, Gaspin's refutation of the suggestion that Bravo was becoming a de facto gay channel reflects a conventional industry distancing from too close an association with gay consumers, and a related claim that marketing should be a matter of "business, not politics."[27] Emphasizing Bravo's courting of the female audience makes obvious business sense, but also avoids thornier questions about the politics of developing a gay niche channel.

Understanding why gay and lesbian viewers were only a secondary target for Bravo's gay-themed offerings is an easier task than understanding why women are a primary target. Why produce gay-themed programming if you are not primarily interested in reaching gay audiences? This strategy seems to go against a common-sense assumption that people watch shows that portray people somewhat like them—that there is a direct, if aspirational, identity connection between audiences and characters. When Bravo produced shows that featured gay men, it went against this assumption and instead tapped into long-standing associations both between gay men and sophisticated consumption and between gay men and heterosexual women in order to appeal to a sizable female audience.

In different ways, *Queer Eye* and *Boy Meets Boy* endorse the adage that gay men are women's best friends. In *Queer Eye,* the Fab Five are the on-screen women's proxies, making over their mates on the show and transforming them into better romantic and domestic partners. The Fab Five also help female viewers train their menfolk, either by example, when women audiences can get their husbands and boyfriends to watch, or by passing on tips female viewers can then use to reform their men. The Fab Five, with their tart critiques and camp rejoinders, manage to achieve more than any amount of womanly nagging can do. *Queer Eye* makes explicit what has been a common assumption for decades: that gay men are uniquely positioned to guide those around them through the intricacies of domestic and style matters. As Lisa Henderson writes, the show's "stereotypical possibilities exist because gay men have had historical access to the style trades when others were denied them, because that is where they could be safely sequestered as inverts among women."[28] Because of this historical professional association between gay men and women, gay makeover experts are ideally positioned to be the conduit for a feminized, and female-audience-friendly, training of heterosexual men.

Making a space for the women audience in *Boy Meets Boy* is a harder task. How can heterosexual female viewers insert themselves into an improvised script that involves men courting other men? The appeal to straight women in *Boy Meets Boy* comes from the strategic deployment of carefully regulated manifestations of gay masculinity, the inclusion of a number of "closeted" heterosexual men among the suitors, and the presence of James's female best friend, Andra. As Joshua Gamson notes, all the participants in the show are normatively masculine—there are no sissy boys here. Gamson quotes the eventual winner, Wes, who says, "There have always been these stereotypes of gay men not being athletic, gay men not being masculine . . . and this show blows that out of the water."[29] All the contestants are fit and muscular, well groomed, stylish, and charming. The discovery that some of them are in fact heterosexual only adds to the appeal: it may be said that all the best men are gay, but if you can't tell the gay ones from the straight ones, some of those gay-ish straight men might be available as romantic partners for women.

Andra is the straight female audience's on-screen proxy, a reversal of the classic model of the gay man as the straight woman protagonist's sidekick. Discussing *My Best Friend's Wedding*, James Allan describes the role of George, played by Rupert Everett: he "gives Jules [Julia Roberts] support and advice and is there for her when her romantic machinations fail. But he has no storyline of his own, nor does he have any romantic or sexual life that the audience knows about."[30] Similarly, Andra's role is to facilitate James's selection process: she does crowd control, taking a group of guys out while James gets to know the others; she investigates the guys to assess their sincerity, values, and suitability; and she offers advice to James before the weekly elimination ceremony. Crucially, however, Andra is not James's sexual competitor. After the reveal of the "twist," when we learn that in fact some of the suitors are heterosexual, it is Andra, more than James, who expresses shock, betrayal, and outrage on her best friend's behalf. She does not show delight, at least publicly, at the opening up of romantic possibility that the twist affords her—no hint of "Yummy! Some for me!"

It is not surprising, then, that Bravo's emphasis has been on gay men, and not lesbians. Because of gay men's historical association with taste and fashion, their status as straight women's best friends, and their availability as objects of desire (however unattainable), gay men are a more attractive draw for heterosexual women audiences than lesbians would be. Further, as I elaborate elsewhere, lesbians' lower average household income and a historical association between lesbian-feminism and anti-consumerism makes lesbians less desirable or recuperable within a model of ideal consumption (Showtime's lesbian drama *The L-Word* notwithstanding).[31]

With *Queer Eye* and *Boy Meets Boy* Bravo thus skillfully harnessed the reputation of gay men as experts in conventionally feminine image professions and the tradition of special friendships between gay men and straight women to the newly popular genre of reality television. Bravo reworked two types of reality-TV programs—dating and makeover shows—that were already popular with women; for example, women make up 71 percent of both *Bachelor* and *Bachelorette* audiences, and 67 percent of *Extreme Makeover* audiences.[32] The channel did so with a twist (twists themselves being a staple of reality television): many of the principal participants were gay. The generic conventions that both *Queer Eye* and *Boy Meets Boy* deployed put less familiar protagonists into very familiar formats, involving the least necessary stretch for heterosexual audiences to understand, identify with, and enjoy the programming: the genre is familiar, even if the sexuality of the participants may be less so.

Bravo's strategy of offering women audiences entry points into the gay-inflected worlds of *Queer Eye* and *Boy Meets Boy* proved successful in boosting the channel's ratings and shifting the demographic profile of its audience. Between the first quarter of 2003, before Bravo's "summer of gay love," and the same period in 2004, the channel increased its total viewership by 75 percent, and went from number 30 in cable ratings for the Tuesday evening slot to number "one, two, or three—depending upon what FX or MTV had on against *Queer Eye*," according to Jeff Gaspin.[33] The show averaged 1.77 million viewers in its first year, "small by broadcast standards but an increase of 634 percent over the pre–*Queer Eye* average in that time slot."[34] In addition to boosting ratings in that prime-time segment, there was also a "halo effect": as more viewers tuned in to see the Fab Five, they were successfully courted for shows such as Bravo's staple, *Inside the Actors Studio,* and reruns of *The West Wing,* newly purchased from sibling network channel NBC. Bravo was also successful in attracting more affluent and younger viewers, increasing the median annual income of its audience from $61,429 to $65,952, and lowering its median age from 50.8 years to a more advertiser-friendly 45.3 years.[35] And advertisers were indeed impressed. According to *Advertising Age,* Bravo increased its upfront commitments from advertisers for the 2004 fall season by 100 percent over the previous year.[36] Interestingly, however, there is no ratings information in the popular or trade press suggesting how successful these shows were with women.

Bravo's executives may have been less concerned to reach gay audiences than to build audience ratings and to shift its demographics from an older audience to a younger, female one, but gay-themed programming was nevertheless central to this strategy. Because gay characters and participants on television shows in general (as opposed to reality TV in particular) are still somewhat unusual and contentious, gay-themed shows get more media attention than similar shows do. Christian fundamentalists play their predictable part in stimulating controversy: invited to comment on *Boy Meets Boy,* a spokeswoman from the Traditional Values Coalition asked, "What's next, 'Boy Meets Sheep'?"[37] Bravo tapped into a strategy used before by advertisers looking to increase the profile of low-budget campaigns: by including gay content, with the notoriety this inevitably brings, a campaign will get extra, free publicity. Indeed, some companies get a great deal of press attention for ads that are never or rarely broadcast. *New York Times* advertising columnist Stuart Elliott explained to me that a 1998 Virgin Cola television commercial that showed two men kissing during a commitment ceremony "never ran. Nobody ran it. A lot of times they'll put a gay theme in an ad because they know it'll be controversial, and they know it won't be accepted and then they get a lot of publicity, . . . you know: 'Virgin Cola redefines cutting edge.' "[38] Similarly, Bravo capitalized on the still-edgy reputation of homosexuality to gain a great deal of press attention for the channel.[39] When *Adweek* asked president Jeff Gaspin how advertisers reacted to *Queer Eye for the Straight Guy,* he said, "It was the title more than anything that scared them. The title was a statement, and one of our goals this year for Bravo was to get people talking about it."[40] Once the show was successful, its gay themes were not "an issue at all" for advertisers, according to Gaspin.

Bravo also tapped into particular characteristics associated with gay men to appeal to audiences, to advertisers, and to product-placement sponsors alike. The formation of the gay market since the 1970s consolidated the ideal image of the gay consumer as a trend-setter, an image Bravo deployed to get a younger, hipper audience. *Queer Eye* most aptly puts this reputation to work, where the Fab Five draw upon stereotypes that gay men "naturally" have impeccable manners, enviable cooking skills, and great taste in clothes, hair, and interior design. Not only is this image appealing to heterosexual women in search of advice about how to improve their menfolk, but it is also appealing to advertisers, who want to associate their products with gay male consumers' reputation for copious, high-end shopping. Because gay men are assumed to have not just great taste, but also abundant disposable income, too, Bravo simultaneously tapped into a class-specific appeal associated with affluence. Channel executives wanted to keep their "highbrow reputation," while also appealing to younger audiences with more "popular" fare. But because reality television is cheap to produce, uses non-actors, and tends to be preoccupied with domestic and feminized concerns, reality shows risk bringing a lowbrow reputation to the channel. Using gays, with their upscale associations, helps deflect the trashy shadow of reality television. One article makes explicit the device of using gay men as a prophylactic against the taint of cheapness that comes with the reality genre: "*Queer Eye* . . . plays into the reality-makeover trend, but the Bravo twist is to do the show with gay men styling straight ones. 'We don't want to become low rent with the programming we do,' Mr. Gaspin says. 'We don't want to become common.'"[41] If reality TV is a "low-rent" genre, gays come to the rescue with their high cultural capital and abundant incomes. Like the British queen, the Fab Five don't carry money; many products are acquired in the process of a day's frenetic shopping, but prices are never discussed, affordability is never considered, and cash never changes hands. Practically, this is because the goods are donated as part of companies' product placement strategies; getting a product on the show increased some companies' sales by more than 300 percent.[42] But the effacement of the financial transaction as part of the makeover has the added effect of implying such abundant wealth that no one needs to ask awkward, embarrassing questions about money. Bravo thus tapped into the association between gay men and affluent, high-class style in order to attract audiences to cheap reality programming without risking the "low-rent" association that reality makeover shows have.[43]

Gay content has therefore not been incidental but crucial to Bravo's success. Of all basic cable channels, Bravo has most aggressively pursued the strategy of using gay-themed programs to appeal to a range of audiences, not just GLBT-identified ones. By linking the channel's content with characteristics consolidated by gay marketing—that gay men are trend-setting, affluent, female-friendly, and newsworthy—Bravo shed some of its staid reputation to become "more relevant and current" to the affluent, younger audience it sought.[44] At the same time, however, Jeff Gaspin repeatedly resisted the charge that Bravo was becoming "the gay channel." In one article, for example, he clarified: "I don't think people think of *Queer Eye* as a gay show. I think people think it's a show with five gay leads."[45] In another, he discussed future programming plans:

> I do think it's important that we have a mix of programming, and if we did another gay-themed show, then the accusations that were made against the channel six, seven months ago—that Bravo was a gay network—would resurface. At the time, I was trying to manage that. At the same time, I didn't want to back off—it would counteract everything we were trying to do. By no means do we want to abandon the gay audience that's coming to Bravo . . . but I do want to service a broader audience.[46]

This is a very different model of niche programming than that of Logo, Viacom's new channel targeted at GLBT viewers. In contrast to Logo's strategy of gay narrowcasting, which requires that a large proportion of a relatively small target market be attracted to shows and watch advertising, Bravo is "dualcasting": targeting two specific audiences, gays and women aged 18 to 49, with the same shows. Given the struggles PrideVision, Canada's dedicated gay cable network, has faced in amassing a large enough subscriber base to become profitable, such dualcasting tactics might prove necessary to make gay television financially viable.

Using gay content to dualcast to two distinct audiences not only has been essential to shift Bravo's audience demographics and increase the channel's profile, but is also part of a larger strategy of diversification by NBC. As audiences dispersed across a range of television channels and other media, the Big Three networks could no longer simply carve advertising revenues up among them. Most media companies have responded to the profound shifts in consumption patterns by acquiring an increasingly diversified portfolio, purchasing competitors so that even as viewers choose ever more segmented media, all slices ultimately come from the same company pie. NBC is far from unique in this strategy. On the contrary, Viacom, for example, owns CBS and UPN broadcast networks as well as a broad range of cable channels, including MTV, VH1, and CMT music channels, Nickelodeon, Spike TV, Showtime, Sundance, and Logo. General Electric, traditionally wary of overextending itself in the volatile world of media and entertainment, held back from this approach and focused its efforts on building the profitable NBC network and its news-oriented cable channels CNBC and MSNBC. Emboldened by its success with Bravo, purchased in 2002, GE acquired 80 percent of Vivendi's remaining holdings in September 2003, forming the new NBC Universal company that now included the cable channels Sci Fi, Telemundo, Trio (a pop-culture channel replaced early in 2006 with mystery-themed Sleuth), and USA Network, as well as Universal's movie and television studios and theme parks. Tellingly, Bravo president Jeff Gaspin was promoted to president of NBC Universal Cable Entertainment. Such a diversification strategy allows NBC Universal to offer viewers a broader range of gay-themed fare, with the very popular but safe sitcom *Will & Grace* airing on NBC to the more experimental and cheaper gay-themed reality shows like *Queer Eye* and *Boy Meets Boy* shown on NBC's baby sister Bravo. Like other media conglomerates, NBC need not fear the dispersal of audiences to cable when those cable channels are owned by NBC's parent company; indeed, NBC encouraged viewers to seek out *Queer Eye* on Bravo by airing a half-hour version of one episode in the summer of 2003.

Once Bravo has capitalized on the publicity and halo effect that its gay-themed shows have brought the channel, it remains to be seen if executives will remain committed to gay content. Bravo's post–*Boy Meets Boy* reality show, *Manhunt: The Search for America's Most Gorgeous Male Model,* drew an interested audience from—once again—both gay male and heterosexual women audiences. After *Manhunt*'s first episode, *Entertainment Weekly* declared Bravo "the gayest television network of all time. Approximately half an hour into this *America's Next Top Model* knockoff . . . the 16 hottie contestants go tandem skydiving. In Calvin Klein underwear. To the tune of 'It's Raining Men.' "[47] But the *Queer Eye* spin-off, *Queer Eye for the Straight Girl,* survived only a few episodes, sunk by uncharismatic hosts and poor ratings. *Straight Girl* aside, though, dualcasting has clearly proven successful for Bravo, as well as Bravo's parent company, NBC Universal. And if dualcasting is the way to garner large enough audiences and sufficient publicity to be seen as successful by cable's modest standards, GLBT audiences may not care that they are not the primary focus of programmers' attention. Yet however much they offer, dualcasting approaches might leave gay viewers wondering if they are just Bravo's new best friend, as we saw with the explosion of one-season characters on a slew of shows after the success of *Will & Grace* in the late 1990s (for example, Malcolm on *Beggars and Choosers,* Ford on *Oh, Grow Up,* and George on *The Profiler,* all unmemorable shows from the 1999 fall season). The channel offers no guarantees that once gay programming has successfully boosted ratings and shifted audience demographics, gay characters and topics won't simply be sidelined, especially as the success of *Queer Eye* has waned in the fall of 2004, with ratings for the show dropping by 40 percent compared with the previous year.[48] Bravo's gay programming strategy may be an example, then, of the fragility of gay representations in a commercial television marketplace, in which sizable audiences are the nonnegotiable bottom line for programmers in a medium so expensive to produce. Once the novelty has worn off for heterosexual viewers—and perhaps for many gay ones too—Bravo executives have no necessary loyalty to develop new gay topics for their shows. It may be that an ongoing commitment to gay-themed programming will need to come from a dedicated gay cable channel such as Logo. It remains to be seen, however, whether Logo will offer sufficiently diverse programming to appeal to a large enough audience to be profitable, and thus whether it can afford to provide a sustained television environment committed primarily to GLBT audiences.

NOTES

1. "Bravo to Offer Man-on-Man Fashion Action," Zap2it.com, 2003 (accessed February 10 2003).

2. Stuart Elliott, "Stuart Elliott in America," *Campaign,* August 15, 2003.

3. This chapter is based on analysis of more than forty popular, gay, and trade-press articles about Bravo's gay-themed shows and segmentation strategies, and on an analysis of all 2003–2004 episodes of *Queer Eye for the Straight Guy* and *Boy Meets Boy,* as well as some episodes of *Gay Weddings.* The author thanks Paul Falzone for his research assistance.

4. Ed Papazian, *TV Dimensions 2004* (New York: Media Dynamics, 2004), 43.

5. Joseph Turow, *Breaking Up America: Advertisers and the New Media World* (Chicago: University of Chicago Press, 1997).

6. Papazian, *TV Dimensions 2004,* p. 38.

7. Ibid.

8. Turow, *Breaking Up America,* 55.

9. Ibid., 103.

10. Bernard Weinraub and Jim Rutenberg, "Gay-Themed TV Gaining a Wider Audience," *New York Times,* July 29, 2003.

11. Larry Gross, "The Past and the Future of Gay, Lesbian, Bisexual, and Transgender Studies," *Journal of Communication* 55 (2005): 520.

12. Katherine Sender, *Business, Not Politics: The Making of the Gay Market* (New York: Columbia University Press, 2004).

13. *Advertising Age* editorial, 1981, quoted in Turow, *Breaking Up America,* 40.

14. "GE's NBC Wins Clearance to Buy Bravo Cable Network," Associated Press, November 20, 2002.

15. Michael Giltz, "Queer Eye Confidential," *Advocate*, September 2, 2003, 43.

16. Daisy Whitney, "Bravo Stretches, Adds Viewers and Advertisers," *Advertising Age*, June 9, 2003.

17. Lisa de Moraes, "A Hipper, Edgier Bravo Orders 'Queer Eye,'" *Washington Post*, February 11, 2003.

18. Whitney, "Bravo Stretches."

19. Quoted in Andrew Wallenstein, "Bravo Targets Two Demographics with Gay Shows," *BPI Entertainment News Wire*, July 29, 2003.

20. *Cable's Bravo Is Carving out Gay Turf with Two New Shows*, Infobeat Entertainment, 2003 (accessed July 15, 2003).

21. Christie Schweinsberg, "Perfectly Queer," *Ward's Business Dealer*, May 1, 2004.

22. Wallenstein, "Bravo Targets Two Demographics."

23. Ibid.

24. Ibid.

25. Lisa de Moraes, "A Dating Game with No Straight Answers," *Washington Post*, May 28, 2003, C7.

26. Ibid.

27. Sender, *Business, Not Politics*.

28. Lisa Henderson, "Sexuality, Cultural Production and Foucault/Conjunctures," paper presented at the *Sexuality After Foucault Conference*, Manchester University, UK, November 29, 2003, 5.

29. Joshua Gamson, "The Intersection of Gay Street and Straight Street: Shopping, Social Class, and the New Gay Visibility," *Social Thought and Research* (2005).

30. James Allan, "Fast Friends and Queer Couples: Relationships between Gay Men and Straight Women in North American Popular Culture, 1959–2000" (Ph.D. Dissertation, University of Massachusetts, Amherst, 2003).

31. Sender, *Business, Not Politics*, chapter 6.

32. "U.S. Ratings: American Network Scorecard," *World Screen News*, October 27, 2004, http://www.worldscreen.com/ (accessed October 27, 2004).

33. James Hibberd, "Bravo's Gaspin Talks Content; Net President Discusses What's Ahead," *Television Week*, December 1, 2003; Bill Keveney, "Bravo Has Eye for 'Straight Girl,'" *USA Today*, April 7, 2004.

34. Keveney, "Bravo Has Eye for 'Straight Girl.'"

35. Abby Ellin, "Bravo Learns to Make Noise and Have Fun," *New York Times*, July 5, 2004; Hibberd, "Bravo's Gaspin Talks Content."

36. Claire Atkinson, "Cable vs. Broadcast Fight Thaws Out, a Bit," *Advertising Age*, June 14, 2004.

37. De Moraes, "A Dating Game."

38. Sender, *Business, Not Politics*, 122.

39. As a rough measure, a Lexis-Nexis search of major papers that included "gay" within five words of "Bravo" in 2002 yielded only 8 hits, compared with 144 hits in 2003.

40. "Q&A: Jeff Gaspin on the Spot," *Adweek*, December 22, 2003.

41. Whitney, "Bravo Stretches."

42. Ellen Florian, "Queer Eye Makes over the Economy!" *Fortune*, February 9, 2004.

43. It is not the participants' sexuality alone that brings this highbrow reputation to reality shows, but the association between queerness and a particularly affluent image of the gay male consumer. Working class queers' participation in an earlier incarnation of reality TV, "trashy" daytime talk shows, affirmed, rather than elevated, these shows' lowbrow status, as Joshua Gamson skillfully demonstrates in *Freaks Talk Back: Tabloid Talk Shows and Sexual Nonconformity* (Chicago: University of Chicago Press, 1998).

44. Hibberd, "Bravo's Gaspin Talks Content"; Keveney, "Bravo Has Eye for 'Straight Girl.'"

45. Hibberd, "Bravo's Gaspin Talks Content."

46. Ibid.

47. Entertainment Weekly, "Manhunt: The Search for America's Most Gorgeous Male Model," *Entertainment Weekly*, October 15, 2004.

48. Andrew Wallenstein, "Ratings Not So Fab as 'Queer Eye' Fad Fades," *Hollywood Reporter*, September 10, 2004.

Chapter 15

"I'm Rich, Bitch!!!"

The Comedy of Chappelle's Show

Christine Acham

In the Winter Quarter of 2003, I taught my first African-American television course. I introduced my students to the rarely screened *The Richard Pryor Show*, which debuted on NBC in 1977. The thematically innovative and racially provocative show lasted just six episodes. While there are several factors that led to the downfall of *The Richard Pryor Show*, one of the key causes was the inability of Pryor to freely express himself on network television. He could not invest himself in the medium on a long-term basis because of the limitations imposed on him by NBC. In newspaper and magazine articles, Pryor complained about network censors who he believed interfered with the creative process. After watching episodes of Pryor's show, several of my students asked if I had seen *Chappelle's Show*. They could not stop talking about Clayton Bigsby, a blind, black, white supremacist, and I was duly informed that I was missing the cutting edge of new black televised comedy.

I tuned in soon thereafter and witnessed what has now arguably become one of the most talked about shows on cable television, especially in the arena of comedy. In the very first episode, Dave Chappelle introduced a skit that was a parody of the PBS investigative show *Frontline*. The fictional *Frontline* explores the life of Clayton Bigsby, the leading voice in the white supremacist movement for the past fifteen years. In order to increase the audience for his ideas, the typically reclusive Bigsby has granted *Frontline* an interview. Upon arrival at Bigsby's backwoods home, the reporter discovers that Bigsby, a man who engages in the most racist diatribes, is black. He is also blind and was apparently the only black child to attend the local blind school. In order to make his life easier, the teachers told him that he was white. Shocked at his discovery, the reporter attempts to understand Bigsby.

> *Reporter*: What would you say is the overall message of your books?
>
> *Bigsby*: Sir, my message is simple. Niggers, Jews, homosexuals, Mexicans, A-rabs and all kinds of different Chinks stink, and I hate them.
>
> *Reporter*: I notice you referred to African Americans. What exactly is your problem?
>
> *Bigsby*: How much time you've got, buddy? Where will I start? Well, first of all, they're lazy, good for nothing tricksters, crack-smoking swindlers, big butt havin', wide nose breathing all the white man's air. They eat up all the chicken. They think they're the best dancers, and they stink. Did I mention that before?
>
> *Reporter*: Yes, I believe you did, sir.
>
> *Bigsby*: As a matter of fact, my friend Jasper told me one of them coons came by his house to pick his sister up for a date. He said, look here, nigger, that there is my girl. Anyone who has sex with my sister, it's going to be me.

Bigsby refuses to listen to the reporter who attempts to inform him of his racial background. By the end of the skit, the reporter follows him to a white supremacist rally. At the audience's urging, Bigsby removes his KKK hood revealing his true black identity. During the *Frontline* wrap-up, the journalist reports that Bigsby has come to terms with the fact that he is black and has filed for divorce from his wife of nineteen years. When asked for a reason, Bigsby replies that it is because she is a "nigger lover."

Chappelle explains that the skit is based on his grandfather in reverse. His grandfather, who was also blind, was born of a white mother in a white hospital and looked white. However, he grew up in Washington, D.C., in black neighborhoods as a black man. Chappelle discusses an incident in which his grandfather was riding a bus in D.C. on the day after Martin Luther King was shot. He heard some black men harassing a white man, asking him how he dared ride that bus in D.C. on such a historic day. It took his grandfather some time to realize that the men were talking to him.[1]

While the skit is outrageous on many levels, at its core, it pokes fun at the unyielding culture of white supremacy. At the same time, however, discussing southern white hate groups—racism at its most overt—overlooks contemporary covert and insidious racism that has as powerful an effect on minority communities. Chappelle could use the platforms of cable and comedy in a more politically incisive fashion that looks at the ways in which racism impacts a contemporary audience on an everyday basis. Moreover, while it would be naïve to believe that the language in the skit has disappeared in contemporary America, does Chappelle's comedy also bring back hated language of decades past into common discourse through cable television?

The first black friend Chappelle showed the sketch to hated it and told Chappelle that he had set back black people. Yet this skit has become one of the most popular and most discussed. These mixed responses to Chappelle's comedy are not unusual. In my classroom some of the African-American students believe that the show at times perpetuates negative aspects of black culture while others celebrate Chappelle's sharp wit and success. This is a bind that most black comics become tied into once their comedy enters mainstream American popular culture. What might appear humorous within all-black spaces suddenly takes on different meanings when viewed by an audience of multiple races. These are some of the issues that would surround Chappelle's comedy throughout its tenure on Comedy Central and eventually questions that the comic himself would ask with his decision to leave Comedy Central and a lucrative $50 million contract.

Using *Chappelle's Show* as an example, this article explores several key issues about the intersections of network television, cable, and blackness. How have network creative or political decisions impacted the black artist? Does cable television facilitate or deter the black performer, and in what ways? Finally, what have Chappelle's use of cable and Comedy Central specifically envisioned about black people?

Dave Chappelle: "America's No. 1 Source for Offensive Comedy"

In 2004, Dave Chappelle officially crossed over to the mainstream: he made the cover of *TV Guide* for August 8–14, 2004, an issue that named him the "Funniest Man on TV." "I'm Rick James, Bitch!"—a line from his most celebrated skit in the show's second season—became a part of the American lexicon. But to those familiar with comedy circles and perhaps more specifically black comedy circles, Chappelle was a well-known name long before the success of *Chappelle's Show*. Like Pryor before him,

Chappelle had a large underground following of comedy fans before his mainstream breakthrough on cable television.

In order to understand the comedy of *Chappelle's Show* and its presence on Comedy Central, it is first relevant to trace the roots of Chappelle's performance. Born in 1973 of two college educators, Dave Chappelle grew up in a middle-class family. After his parents' divorce, Chappelle moved between Ohio and D.C. As he explains, "I'm not the guy with the sad childhood. Everybody in the family was educated, when you look at old photo albums, you don't see much liquor in the pictures. I had the benefit of both of my parents, which a lot of my peers didn't have. So I felt real lucky."[2]

When he was fourteen, his mother, a college professor and minister, chaperoned him as he performed at comedy clubs around Washington, D.C. He presented jokes based on Jesse Jackson's bid for the presidency and the sitcom character Alf. Often questioned about why his mother let him perform at such a young age, Chappelle explained, "D.C. was the murder capital of the year. Selling crack was the No. 1 after-school job. In contrast to all the bad things I could have been doing, how benign does telling jokes sound?"[3]

Chappelle went to New York to break into the comedy club circuit, and, when performing at the Boston Comedy Club in Greenwich Village, he met his eventual writing partner, Neal Brennan. The fact that Brennan is white is an issue about which the media consistently questions him. "All I'm saying is, I want to exist as I am, me and my partner. My partner's white, it's not a racial thing—we write these racial jokes in racial harmony."[4]

Chappelle signed his first television deal when he was a teenager. He then made eleven pilots of—as Chappelle describes—"different versions of Urkel."[5] Chappelle also attempted to make a pilot with the Fox network which suggested that he add more white characters to give the show a more mainstream appeal. As Chappelle comments about this suggestion,

> It was racist. Look, I don't think these people [Fox executives] sit around their house and call—call black people niggers and all this kind of thing, but the idea that unless I have white people around me on my show, that it's unwatchable or doesn't have a universal appeal is racist. You know, they don't—they don't make them put black people on "Friends," or they don't make them put black people on "Seinfeld." But all of a sudden I get in the room, and it's like, "Well, where's all the white people?"[6]

After the incident with Fox, Chappelle walked away from network television and focused on his stand-up and simultaneous film career. Chappelle has had roles in over fourteen films, including the *Nutty Professor* (1996); *Con Air* (1997); *Half Baked* (1998), which he co-wrote with Neal Brennan; *You've Got Mail* (1998); *Blue Streak* (1999); and *Undercover Brother* (2002).

Chappelle's failure on network television, especially considering his enormous success on cable, is evidence of the inability of mainstream television to provide the creative atmosphere necessary for such a comic to succeed. According to Chappelle,

> At a certain point, I abandoned that whole philosophy [and said], "Well, what is it that I want to do, and is there a way I could do something that's more in line with my sensibility?" And cable was kind of the remedy. It was the only thing I never tried.[7]

As he further explains,

> On network TV you don't have much freedom because the audience is so large and you have to keep everything nice and generic. This is a comedy network, and they're at a place corporately where they're willing to take these kinds of chances. [This show] is kind of a celebration of my freedom.[8]

While cable has become accessible to an increasing number of American households, broadcast television, available to all television owners, continues to produce primarily white homogenous programming aimed at the rarely changing in-demand demographic of 18–39-year-old consumers. Black-oriented television is relegated to the secondary networks of the WB, UPN (now the CW). While the shows on these networks rarely vary much from preconceived notions of ghetto blackness, those that do rarely survive. Therefore programming that shifts outside of this narrow representational scope of blackness has to seek alternate spaces for expression. For example, black dramatic programming such as *Soul Food* (2000–2004) was produced through Showtime. Additionally, black-themed, made-for-television films treating social and historical issues have appeared on several cable stations, while provocative African-American comedian Chris Rock aired his specials and weekly series on HBO. HBO also produced Chappelle's comedy special *Killin' Them Softly* in 2000; Showtime produced his *For What It's Worth* in 2004.

Cable television remains the most viable arena for the expression of

traditional black humor, overtly political or otherwise.[9] While it is significant that these spaces exist for African Americans and other underrepresented groups, it is still highly problematic that opinions, images, and
representations outside of race, gender, and sex norms are still unable to
find a space on mainstream television. To be sure, the fact that Chappelle
landed on Comedy Central is the key to his transition to a mainstream
cultural figure. A smaller network fighting to define its position within a
multi-channel cable system was able to target a niche market, and it was
willing to take a chance on a show such as Chappelle's. However, another
consequence of the move to cable is that Chappelle's audience is more
limited in scope.

"We're All Gonna Die, Watch Comedy Central"

Comedy Central has come a long way from its somewhat unfocused beginnings and heavy rotation of reruns. Through adept and savvy programming deals, the network has become a key competitor in the coveted 18–39
male demographic. Born of a merger between MTV's HA! The TV Comedy Network and HBO's The Comedy Channel, Comedy Central was
launched on April 1, 1991; by the end of the year, the network boasted over
22 million subscribers and won three cable ACE awards.[10]

Using the programming freedom and lowered censorship standards allowed by cable, the network brainstormed creative ways to attract viewers with original programming in a variety of genres. In February of 1992,
the network began its own stand-up show hosted by Richard Lewis entitled *The A-List*. In 1993, the network then aired the first all gay and lesbian comedy special "Out There," which would become a yearly event.[11] Al
Franken covered both the Democratic and Republican National Conventions and produced an election night show entitled *Indecision '92: Election
Night*. Comedians such as Bill Maher, Dennis Miller, Chris Rock, and Al
Franken would continue political coverage at major Democratic and Republican National Conventions as well as on election nights during the
upcoming years.[12] Also in 1993, Comedy Central introduced weekly programming such as *Politically Incorrect with Bill Maher*, and in 1996, Craig
Kilborn was hired to host *The Daily Show*, a satirical network news program. The network's most controversial hit—*South Park*—premiered on
August 13, 1997, and has gained much notoriety for its representations of

characters such as the black school cook Chef, Big Gay Al, Saddam Hussein, and Satan.

In 1999, Jon Stewart took over the desk of *The Daily Show* as anchor, writer and co-executive producer. By July, Stewart's show was given the task of covering the now traditional *Indecision* nights and provided the first live telecast for the network entitled *Indecision 2000: Election Night Choose and Lose.* The program has become a prominent stop for the who's who of American politics, such as Hillary Rodham Clinton, Bill Clinton, John Edwards, John Kerry, Pat Buchanan, John McCain, Ralph Nader, and Jesse Jackson. In 2004, Comedy Central was available to over 85 million subscribers, in 94 percent of homes with cable and 79 percent of all U.S. households.[13]

Senior Vice President of Original Programming Lauren Corrao admits that when she inherited the network in 2002, Comedy Central had the reputation of "the white-frat-boy-network" and that one of the goals was to "diversify the portfolio."[14] Thus, during Black History Month 2002, the network produced its first documentary series, a weekly in five parts entitled *The Heroes of Black Comedy.* Narrated by Don Cheadle, the series discussed such comedy greats as Richard Pryor, Chris Rock, Redd Foxx, and The Original Kings of Comedy, as well as the new hip-hop comics. Wanda Sykes' weekly show, *Wanda Does It,* also premiered in 2004.

Corrao, who met Chappelle when they were both working at the Fox network, knew of his talent and wanted his show to succeed on Comedy Central. It was she who recommended the live studio audiences and wrapping the stand-up routines around the show's sketches. According to television producer Peter Tolan, who had worked with Chappelle during the failed Fox experiment, the networks "always tried to make him [Chappelle] into a homogenous, vanilla, black man."[15] "She [Corrao] has put Dave in a position to shine. He was notoriously wasted in network television."[16] Indeed, having appeared on several Comedy Central shows before the debut of *Chappelle's Show,* including *Crank Yankers, Comic Justice,* and *Second Annual Comics Come Home,* Chappelle felt at home on Comedy Central. In several interviews he indicated that the network gave him a lot of freedom on his show, a fact that was supported by network honchos. Comedy Central's Vice President for Original Programming Lou Wallach stated, "He is a major talent that everyone wants to be in business with. We're never going to water down his point of view."[17] Bill Hilary, executive vice president and general manager of Comedy Central, added,

[*Chappelle's Show* is] brave, smart and has really original writing and great ideas. It also reinvented the sketch variety format for a new generation. We had a tremendous response for the first season, but no one could have predicted the runaway success of the second season, when its audiences grew by almost 50 percent, and it became the highest-rated show on the network since *South Park.* [18]

Comedy Central also sponsored "Dave Chappelle is Blackzilla," a national comedy tour.

Killin' Them Softly: *The Comedy of Dave Chappelle*

Looking at Chappelle's comedy special, *Killin' Them Softly* (2000), one can see the laid back, yet critically sharp style that Chappelle eventually brought to his weekly show. Produced by HBO, the show was taped at the historic Lincoln Theatre in Washington, D.C., and marked Chappelle's return to the neighborhood he left in the 1980s for New York and eventually Los Angeles. Chappelle's roots are clearly in stand-up comedy, and like the black comedians before him who worked the comedy circuits, he sees the importance of heading back to small arenas that allow him to work more closely with the audience. As he explains,

> I like the genesis of the material to come from a casual conversation with an audience of people. Of course, you're going to have your favorites or you're going to have stuff that you know is going to work, but it's an evolution. And I do put a lot of faith in the audience.[19]

While much of Chappelle's comedy revolves around observational humor, targeting popular culture, cultural differences, and relationships between the sexes, *Killin' Them Softly* begins with his addressing his roots within the specific D.C. community. Discussing the crack era of the 1980s and the practically all-black landscape of the inner city, Chappelle points out that D.C. looks different now since there are so many more white people walking around. As he remembers, in the 1980s, "White people be looking at DC from Virginia with binoculars. 'That looks dangerous. Not yet.'" His comedy special deals with racism in the police department and states that whites ignored police brutality until they read about it in *Newsweek*. He suggests that many blacks were assaulted and killed by the police who would "sprinkle some crack on" them in order to make the arrest or killing look legitimate.

Most importantly, however, Chappelle also broadens the discussion of race beyond the black and white dichotomy. At one point he states that he would not want to be the first black president, but that if he were, he would be sure to have a Mexican vice president for insurance: "You can shoot me if you want, but you're just gonna open the borders up. You might as well leave me and Vice President Santiago to our own devices." *Killin' Them Softly* set the stage for the comic style and topicality that Chappelle brings to his weekly cable show.

Although he is often compared to Chris Rock, it is clear that the fact that these two men are black, more than anything else, elicits the need for this connection within journalistic circles. For, coming from different backgrounds and having distinct life experiences, the two men have unique approaches to their craft. While Rock often displays thinly veiled anger and sharp political critique as he paces furiously across the stage, Chappelle appears more relaxed and observational and thereby allows the members of the audience to reach their conclusions at their own pace. Although these differences are marked, having two famous black people of any profession means that reporters will connect them, and do so in ways that give one priority over another. Is Denzel Washington the next Sidney Poitier? Is Chris Rock the next Richard Pryor? Is Dave Chappelle the next Chris Rock? While there is a certain "next-ness" in popular culture in general, it appears historically difficult to let African-American performers have their own particular identities, as well as identities that are not based on supplanting their predecessors.

For all their differences, Chappelle and Rock certainly do discuss racial problems within American society. Insofar as their jokes contest the status quo, they work as overtly political statements. However, as with any comic, the source of humor most often comes from real life experiences —in this case, the racism both men confront living in the United States. To be a black comedian, unless one revels entirely in the crude or the integrationist, necessitates a level of political awareness. This is the similarity shared between Rock and Chappelle and many who existed before them, such as Redd Foxx, Moms Mabley, Paul Mooney, and Richard Pryor, who not only had political awareness but also the desire to express it. It is therefore understandable that each of these outspoken comics, although stylistically different, had a difficult time gaining acceptance on mainstream network television. Due to mainstream television's historical appeal to the white middle-class audience, performers who have spoken out about racism or other social injustices have been silenced by network television.

It is significant to note that the most famous African-American performer and comedian to gain entrance to and adoration on mainstream television is Bill Cosby, whose integrationist style does little to ruffle American feathers.[20] Indeed, Cosby's acceptance by white mainstream society has designated him a spokesperson for the black community in the eyes of white America. Whether he is a suitable one is another matter. Cosby's inflammatory comments made in Washington, D.C., at the fiftieth anniversary celebration of the *Brown v. Board of Education* ruling and at other public events suggested that poor black people were responsible for their status in life. The issue is not just that his words ignored the contemporary and historical circumstances of African Americans in the United States, but that they were repeatedly played on the mainstream television networks and discussed on numerous talk shows and news programs. Cosby's status as all-American, albeit sitcom, father gives him the power to confirm negative beliefs about black people held both within and outside of the community. Other comics like Rock, Chappelle, and Margaret Cho, who do not fit the integrationist mold and offer differing views on race in U.S. society, do not get regular time on network television and must locate an alternate space for expression. With each of these individuals, cable television has been that site.

Chappelle's Show: *"It's the Comedy Equivalent of Crack; One Try and You're Hooked"*

Chappelle's Show premiered on Comedy Central on January 22, 2003. Boasting 2.5 million viewers, it was the highest rated series premiere for the network since 2001's *That's My Bush! Chappelle's Show* was also the first to increase the viewing audience from its *South Park* lead-in.[21] The numbers are indicative not only of the strength of Comedy Central's marketing but also of the large number of Chappelle followers from his nightclub comedy performances, Comic Relief appearances, and HBO and Showtime comedy specials and films, especially the cult, stoner favorite *Half Baked.* So a number of questions arise. Some specifically concern the medium of cable: What has been the impact of cable on Chappelle's comedy? And what is the significance of his departure from Comedy Central?

Other questions are more broad: What does Chappelle bring to comedy as a whole and to African-American comedy in particular? How does the show participate in contemporary black public politics?

Addressing such questions can be challenging because Chappelle's comedy is difficult to categorize. While many comics find their strength with one particular type of humor, Chappelle's stand-up routines and sketches vacillate widely. He does, however, repeatedly address certain themes in his role as racial observer. Within the sketches where he assumes this role, Chappelle's discussion of different ways in which he sees race operating within American society moves beyond the traditional limitations of black and white. Plays upon America's ingrained stereotypical beliefs about a variety of racial groups, he makes us think more generally about the absurd process of making race-based assumptions.

This type of comedy is most clearly evidenced in a season-two sketch in which Chappelle sets out to disprove the adage that white people cannot dance. Instead he proposes that "They like certain musical instruments . . . electric guitar." With white musician John Mayer, Chappelle visits various sites such as an office boardroom and a fine restaurant. In each case, once Mayer begins to play, the white people lose control and engage in Woodstock-inspired dancing and mosh-pit-like thrashing. Chappelle, however, understands that he needs a control group to support his experiment. He thus heads to a local barbershop in Harlem in which everyone is either black or Latino. When Mayer begins to play the electric guitar, he gets a series of hostile stares from the customers. One yells, "Hey Yo! Shut the f—— up!" Understanding the black customers' disgust at the electric guitar, Chappelle unveils the figure of Quest Love, of the group The Roots, who begins to play drums—to the enjoyment of the black clientele who form ciphers and rap. He then introduces Sanchez, a man who plays electric piano to appeal to the Latinos, who then dance while Chappelle yells unintelligible Spanish-sounding words over a microphone. With the experiment completed, Chappelle draws some conclusions about racial groups and their preferences for music. "We learned that white people can dance if you play what they like, electric guitar. Of course, we the blacks can't resist drums and Latins love congas and electric pianos with Spanish gibberish over it." But this is not the end of the matter. While filming his wrap-up outside the barbershop, he is confronted by two police officers, one white and one black, who demand to see his film permit. He calls on John Mayer for help, and Mayer begins to play a song from the white rock group Guns 'n' Roses. Much to Chappelle's disdain, the black police officer starts to sing along. As he explains to Chappelle, "I'm from the suburbs man. I can't help it."

One of the common topics in the study of race in media studies concerns the problematic nature of stereotyping. While this is an important issue, especially since stereotypical representations often become the only representations of particular racial groups in film and television, what is often overlooked in such discussion is that there is some truth in stereotypes and that these truths have to do with exposure to certain racial and ethnic cultural practices. Thus, in the sketch Chappelle exposes the role of environment in one's cultural preferences through the black cop in the final scene who, due to his life in the suburbs, is a fan of the white rock group Guns 'n' Roses. Chappelle's very broad approach to race, the critical twists in the sketch, and the seemingly obvious and equal-opportunity race-baiting make it difficult to take offense to his characterizations. On the contrary, it permits a diverse audience to have a guilt-free experience of engaging in traditionally stereotypical humor. But does this then allow the audience more freedom to engage in and express such stereotypical beliefs? This is a question that Chappelle would eventually ask himself.

Another significant characteristic of *Chappelle's Show* is its ability to make critical statements or indictments about racism in American society. In the fourth episode of the first season, Chappelle discusses an appearance he made on *Donahue*, in which he was asked to join a panel to discuss the topic of "Angry White Men." As Chappelle explains, he was truly disappointed in himself as he was unable to respond to other panelists who said things that made him want to "choke" them. For example, at one point a panelist, John Leboutillier of Newsmax.com, says that affirmative action has made white people angry and that when they talk amongst themselves about a black employee, they say that "He is an affirmative action hire." On his show, Chappelle is able to respond as he had wanted and says, "Hey that's a lot better than saying 'hey that nigga's homeless.'" Back on the *Donahue* show, another white man in the audience says that he feels like affirmative action is forced onto him and he does not liked to be forced. Chappelle responds on his own show: "Forced? [Pause.] Like slavery forced? Remember that thing where you forced us to work? What do you think black people was like 'No problem boss. I'd love to.'" Taking another moment in which he felt disempowered on television, Chappelle brought it back to his show for a rematch.

Alluding to common, although problematic, accounts of U.S. history,

Chappelle takes on some of the myths of the New Right, especially as told by Ronald Reagan, and challenges the idea that affirmative action was responsible for displacing qualified white men with unqualified African-American and Latino men. Indeed, while Reagan and others on the New Right ignored the fact that most jobs lost during the 1970s and 1980s were due to a stagnant economy, de-industrialization, liberalized trade policies, and other factors, those who accused people of color of taking jobs from deserving white men obscured the fact that middle-class white families—that now relied on two incomes to maintain their falling standards of living—were, in fact, a primary economic beneficiary of affirmative action policies as middle-class white women began to enter career fields once unavailable to them. If those who may have initially tuned into Chappelle's show did so for his bathroom and stoner humor, they are soon exposed to sharp social and political commentary.

Inasmuch as Chappelle calls attention to racial injustices, he also critiques the consumer culture ingrained into segments of black society in a sketch on reparations, while at the same time acknowledging the impact of poverty and American racism. The sketch begins as a news report that announces that one trillion dollars have been given to African Americans as reparations for slavery. Moving to the stock market, the news story includes a reporter who notes that Sprint stock has just gone up as $2 million in delinquent cell phone bills have been paid, that gold and diamonds are at their highest, that eight thousand record labels have been started, and that three million Cadillac Escalades have been sold. The reporter suggests that the recession is over because black people have spent all of their money, willing to just give it all back to whites. In this way, Chappelle critiques what he sees as the priorities of certain segments of black society. At the same time, though, he refuses to let mainstream society off the hook. Unlike Bill Cosby, who does not treat the position of black people in context, Chappelle includes reports that the crime rate dropped, thereby suggesting that if blacks and other minorities did not live in poverty, then perhaps there would be less of an emphasis on consumer goods since access to them would not be as remote a possibility. He also suggest that there would be less crime, as this often evolves from an environment of poverty and need.

Clearly, Chappelle does not hesitate to incorporate what he considers problems within African-American society into his humor. Public sites such as television provide powerful arenas to address problems within black society and create conversations. While it is important to encourage open dialogue about such issues, the fact that the critical thrust of *Chappelle's Show* coexists with the writers' love of "bathroom humor" complicates the situation. In most skits this humor factors in at some level, as Chappelle and Brennan agree, whenever they need a punch line—take, for instance, "crack, masturbation, poop or titties."[22] Under the circumstances, one has to wonder whether such references actually make some of the more critical assertions palatable to the Comedy Central audience or whether the humor simply deflates the impact. Does this audience, which the producers admit were primarily made up of "white frat-boys" to begin with, read the more critical moments? Or are those moments, and the audience, lost in the bathroom humor? If not, are viewers able to interpret assertions about racism and, more importantly, understand the gist of Chappelle's critique of issues within African-American society and their relation to the whole? Or does the critique itself reinforce the stereotypical beliefs about African Americans that are already in existence?

Some of Chappelle's sketches trigger this ambivalence and concern that some in the black community, including myself, have about the show's representation of issues of race and blackness in particular. In one episode Chappelle complains that reality shows such as *The Real World* often put one black person into the situation and then makes that person look like he or she is crazy. Therefore, to counteract MTV, Chappelle created a sketch in which one white person, Chad, is housed with five black roommates, Lisa, Zondra, Tron, Tyree, and Faze 2, who are the craziest black people you would ever meet. I admit that I found several parts of the sketch very funny, and I acknowledge that comedic sketches use exaggeration. However, the skit revels in every contemporary black caricature and is in this way quite troubling.

Chad's roommates play craps and smoke pot. Tyree also masturbates in the bedroom when Chad's girlfriend Katie visits. When given a job at a juice bar, the black women paint their nails, while Tron (Chappelle) adds alcohol and cigarette butts to the drinks. Chad is later incapacitated by a chokehold, and his father, while visiting, is stabbed by Tyree.

Of course, it can be argued on the other hand that Chad is played as stereotypically white. But in terms of the sketch, what that simply means is that he is responsible, wants to do a good job at work, has solid relationships with his father and girlfriend, and looks for the best in his crazy roommates. Does a skit like this reinforce a particular image of blackness?

Perhaps, for instance, everything that Clayton Bigsby and Bill Cosby suggest are characteristics of black people? Even Chappelle realizes that these sketches, while funny, can be taken out of context and can confirm preconceived notions about black people. As he recalls,

> People are coming up to me like, white people come up to me like, man, that sketch you did about them niggers, that was hilarious. Take it easy. You know, I was joking around. You start to realize these sketches in the wrong hands are dangerous. You know?[23]

In this case, while setting out to critique MTV for its positioning of the black people in their casts, Chappelle and Brennan confirm the worst fears about black people to a mainstream viewing audience.

It is clear that *Chappelle's Show* achieved a certain cultural currency. Outside of the magazine covers and newspaper articles, Chappelle signed an unheard-of $50 million deal with Comedy Central to complete seasons three and four. Chappelle was also slated to receive a cut of the DVD sales from these seasons.[24] In a press release Chappelle stated, "The success of the show has been phenomenal and overwhelming . . . now I'm actually rich, bitch!"[25] In critical circles, the show also gained further recognition; it was nominated for three Emmy Awards in 2004, one for Outstanding Variety, Music or Comic series, as well as gaining nominations for directing and writing.

"They Got Me in Touch with My Inner Coon"

In the spring of 2005, rumors emanated from the production stages of Comedy Central that Chappelle had walked off the set and was taking drugs. The third season was delayed, and then magazines, newspapers and Internet sources reported that Chappelle had checked himself into a mental health facility in South Africa. Soon Chappelle was forced to address the rumors, and in an interview with *Time* on-line he explained that he was taking a break from the show and was staying with friends in South Africa. Chappelle told the reporter that he needed to take a step back and deal with how his financial success was impacting the relationship with non-family members around him. He also wanted to make sure that he would stay true to his vision. Chappelle eventually explained his actions on public sites such *The Oprah Winfrey Show* and Bravo's *Inside the Actor's Studio*. With the pressure to make a successful third season for Comedy Central, Chappelle began to question some of the sketches that he was writing. As he told Oprah,

> I was doing sketches that were funny but socially irresponsible. I felt like I was deliberately being encouraged and I was overwhelmed so it's like you get flooded with things and you don't pay attention to things like your ethics.[26]

Chappelle cites one moment when he was shooting an episode in which he was dressed as a black pixie. As the skit opens, Chappelle is on a plane, and the female flight attendant gives him the choices for his in-flight meal, fish or chicken. Immediately a small black "pixie" appears on top of the seat in front of him. Played by Chappelle, the pixie is actually a traditional old black servant in black face, a bellboy wearing a red coat and hat with gold trimmings and buttons, white gloves, and black pants and holding a cane. The pixie is the most stereotypical image of blackness whose role is to encourage the black person to live out his most stereotypical self. The pixie immediately begins laughing, encouraging Chappelle to order the chicken while tap dancing on the seat to banjo music.

Overall the skit follows one of the key innovations of the *Chappelle Show* mentioned previously, the opening of the dichotomy of race from black and white while making people consider the ludicrous nature of stereotyping itself. In terms of the cross-racial stereotypical jokes—there are Latino, Asian, and white pixies. However, in recalling when he was taping the part of the skit with the black pixie, Chappelle said,

> I'm on the set, and we're finally taping the sketch, somebody on the set that was white laughed in such a way—I know the difference of people laughing with me and people laughing at me—and it was the first time I had ever gotten a laugh that I was uncomfortable with. Not just uncomfortable, but like, should I fire this person.[27]

What this experience reinforced for Chappelle is that in televised comedy, there are differences within the audience that have broader ramifications. What might be understood within one community as having one particular meaning may take on a different meaning in a different environment.[28] As he stated,

There are a lot of people who understand exactly what I am doing. Then there is another group of people, who are just fans . . . that is along for a different kind of celebrity worship, and they are going to get something completely different—that concerned me. I don't want black people to be disappointed in me for putting that out there.[29]

This in the end is one key issue that pulled Chappelle away from Comedy Central.

Considering his fears about the direction of the show, it is important to see what writing partner Neal Brennan, whom Dave had criticized for his lack of support during his difficulties, and the producers of Comedy Central did with the show when the "Lost Episodes" were aired and the DVD was released in August of 2006. Most significant is the way in which the pixie episode was handled. It was hosted by Darnell Holloway and Charlie Murphy, two of the most popular figures to come out of the show, who state that they questioned whether or not they should have actually aired the episode. Considering Chappelle's reaction and vocal statements, it seems as if it would have been respectful to have left the episode alone.

However, the presentation of the skit is interesting. After the sketch is screened, the comics open it up to a "Town Hall" meeting on the set and ask the audience for their opinions on the show. One person states that Chappelle does good job of bringing race to the surface so that it is something comfortable to talk about. Another says that it is derogatory to blacks and Latinos and not to white people. Another claims that it makes people feel too comfortable with racist jokes. However, it is the comment that is chosen to close the episode that is the most critical and appears to address Chappelle directly. A woman states, "I don't think that it is the responsibility of the show to educate everyone in the world . . . even if it is being a responsible comedy show . . . so you have to stick to what your true goal which is making people laugh." This clearly addresses Chappelle's comments on *Oprah* and his decision to leave Comedy Central because of his concern over his lack of "social responsibility" to black people. It also suggests that it is the opinion of Comedy Central that comedians should just stick to what they do best, which is to make people laugh.

One cannot deny the importance of having new black voices on the television screen, especially ones that continue to question the status quo and diversify the medium. Chappelle's decision to leave Comedy Central and his feeling of disempowerment, especially when $50 million was at stake, is an almost unbelievable statement about the power of a black artist to impact the industry. Whereas cable had given Chappelle an opportunity to present his brand of humor, due to Comedy Central's desire to compete within the cable industry as well as for the broadcast networks' audience, it also attempted on some level to control the ideas of a black artist. Screening the pixie episode showed little respect for the artist's discovery of the problematic nature of the subjective interpretive community of television.

When mainstream U.S. society still looks to older, out-of-touch-with-contemporary-black-America, integrationist, broadcast television stars such as Bill Cosby for its views on black society, the relegation of alternate voices to cable is extremely problematic. Nonetheless, while Chappelle sought to make sure that he was actually providing an alternate voice and not a replication of the status quo, Comedy Central seemingly had the last word with the structure of the "Lost Episodes." But then since Chappelle is no longer a part of the channel, he cannot be used in ways that are detrimental to the black community. His absence is the last word.

NOTES

1. Charlie Rose, "An Interview with Dave Chappelle," Transcript: 042801cb.111, Section: News; International. Accessed through Lexis-Nexis.

2. Bruce Fretts, "What!? Dave Chappelle!?," *TV Guide*, August 8–14, 25, 2004.

3. Lola Ogunnaike, "Nothing Is Out of Bounds for Dave Chappelle's Show," *New York Times*, February 18, 2004, Arts/Cultural Desk, Late Edition Final.

4. Ellen Gray, "Chappelle Speaks Out About Slights, Stress and Price of Success," *Philadelphia Daily News*, August 5, 2004, Entertainment News.

5. Fretts, "What!?" 25.

6. Bob Simon, "Dave Chappelle!; Dave Chappelle's Life and Career," *60 Minutes*, CBS News Transcripts, October 20, 2004. Accessed through Lexis-Nexis.

7. Dave Walker, "Standup and Deliver," *New Orleans Times-Picayune*, September 4, 2004, Living Section.

8. Sommer Mathis, "Experience, Humor Make Chappelle Stand Out," *UCLA Daily Bruin*, February 13, 2003, Interview, U-Wire.

9. Traditional black humor is derived from comedy performed in all black spaces. This began from the days of slavery and continued on through segregation into such outlets as the clubs of the Chitlin' Circuit. For more information see Mel Watkins, *On The Real Side: Laughing, Lying and Signifying* (New York: Touchstone Books, 1994).

10. Comedy Central: Press Central, http://www.comedycentral.com/press/network/milestones.jtml.

11. *Mystery Science Theater 3000, Dr Katz: Professional Therapist* (1995), *Win Ben Stein's Money* (1997), *The Man Show* (1998), *That's My Bush* (2001), *Crank Yankers* (2002), *Comic Groove* (2002), *Reno 911* (2003), *I'm With Busey* (2003), and *Drawn Together* (2004) are all original programs aired on Comedy Central.

12. Information and dates for segment on comedy central history retrieved from Comedy Central's Press Central, http://www.comedycentral.com/. Coverage of these major political events is a tradition that continues on Comedy Central today.

13. Comedy Central: Press Central, http://www.comedycentral.com/press/network/milestones.jtml.

14. Allison Romano, "Laugh Riot; Corrao keeps Comedy Central Red Hot," *Broadcasting and Cable,* April 19, 2004, 10.

15. Fretts, "What!?" 25.

16. Romano, "Laugh Riot," 10.

17. Hillary Atkin, "Chappelle's Show; Variety, Music or Comedy Series," *Television Week,* May 31, 2004, 36.

18. Ibid., 36.

19. Rob Owen, "A Blue Streak Stand-Up Reenergizes Outspoken Dave Chappelle," *Pittsburgh Post-Gazette,* September 3, 2004, Arts & Entertainment, Sooner Edition.

20. Herman Gray, *Watching Race: Television and the Struggle for Blackness* (Minneapolis: University of Minnesota Press, 1995).

21. Comedy Central Press Release, "'Chappelle's Show' Sets Ratings Records," New York, January 23, 2003.

22. Paramount Pictures, *Chappelle's Show: Season 1 Uncensored,* Audio Commentary on DVD.

23. Charlie Rose, "An Interview with Dave Chappelle," Transcript: 042801cb.111, Section: News; International.

24. Phil Rosenthal, "Chappelle Lights Up, Takes Off," *Chicago Sun-Times,* September 1, 2004, Features; Television.

25. Comedy Central Press Release, "Dave Chappelle Inks New Pact With Comedy Central Keeping the Host of the Emmy-Nominated 'Chappelle's Show' at the All-Comedy Channel for Two More Years," New York, August 2, 2004.

26. *The Oprah Winfrey Show,* February 3, 2005.

27. Ibid.

28. For a further exploration of the ways in which audiences read, interpret and use texts, see Stuart Hall, "Encoding, Decoding," in Simon During, ed., *The Cultural Studies Reader* (New York: Routledge, 1993), 90–103.

29. *The Oprah Winfrey Show,* February 3, 2005.

Chapter 16

Worldwide Wrestling Entertainment's Global Reach

Latino Fans and Wrestlers

Ellen Seiter

In 1999, community leaders in a working class, ethnically diverse neighborhood of a large Southern California city invited me to design a computer lab and an after-school program for children aged 8–12. This was the year that the federal government initiated its Community Technology Center grants program, and optimism about the educational benefits of Internet access ran high. As I conducted research for my book, *The Internet Playground*, I found that conflicts immediately arose between children's desire to use the Internet to pursue their fan interests (largely derived from television) and the more narrowly educational goals of most technology center directors and volunteers.[1] The program was planned to follow a child-centered pedagogy in which the students would generate their own study topics. This meant that popular culture websites, easily accessible through the lab's T-1 line, were the first choice for the majority of students. First and foremost, the students sought out websites that related to their television viewing—to TV shows, such as *Pokemon, Digimon, Yu-Gi-Oh!*, and *The Powerpuff Girls*, or TV channels, such as nickelodeon .com and kidswb.com. Other fads came and went during the four years in which I taught the program: NeoPets; free games at bonus.com; music videos by N'Sync, Backstreet Boys, Aaliyah, Destiny's Child, Britney Spears, Christina Aguilera, and Jennifer Lopez. Yet nothing matched the class's ardent, sustained interest in Worldwide Wrestling Entertainment (WWE).[2] Most of the students showed at least some interest in professional wrestling, but the biggest fans of all were Latino boys, who identified both with the United States and with Mexico, were bicultural and bilingual, and had strong family ties on both sides of the border.

While my students pursued their interest in the WWE through the "new" media channel of high-speed Internet access, the WWE also manifested itself in material form throughout the everyday world of the children I taught. Unlike high-ticket consumer goods such as PlayStation machines, WWE merchandise was available in the neighborhood, as close as the corner convenience store. WWE trading cards and magazines were coveted items, but the boys rarely had enough spending money to purchase them, and local shopkeepers eyed the boys suspiciously, seeing them as potential shoplifters when they browsed these goods. Much of this beloved media was originated for tiers of television programming—expanded cable, pay-per-view, and home videotapes and DVDs—out of reach of many of their families. Under these circumstances, access to the WWE website provided a highly desirable substitute. The boys were most pleased with their ability to print out the prized portraits and action photos of their favorite wrestlers to be displayed on their binders and at home. Free Internet access and our classroom printer provided for these boys beloved media that, for largely economic reasons, they were unable to access by other means.

My lengthy contact with children in this neighborhood alerted me to the brilliance of the WWE's marketing strategy on the World Wide Web and the effectiveness of its ties among broadcast, cable, and pay-per-view events and on-line content and product merchandising. This strategy—and its success with the Latino boys in my class—are based on securing as broad a global market as possible and on finding ways to attract youth with storylines and characters whose overt themes are the injustices of ethnic stereotyping and racial conflict. Such themes, especially as embodied in the career of the WWE's biggest star, The Rock, resonated with the boys' own identity struggles as blue-collar Mexicanos. I begin this essay with an analysis of the aspects of the WWE that most appealed to these children, in terms of forms of delivery (television and the Internet) and heavily, strategically promoted star wrestlers. I conclude by showing how these aspects of WWE content have been part and parcel of its strategies of expansion into new media and new markets, reaching these young fans—members of a marginalized group in the domestic television market—as well as fans in other regions who are targeted by WWE's global marketing and presented with its very particular version of American patriotism, edgy entertainment, and exaggerated performances of ethnic and gender stereotypes.

WWE on Television

Vince McMahon, Chairman of the Board of WWE, is unique among cable programming executives. He emerged from a career as a third-generation wrestler and wrestling promoter to take over a publicly traded firm while still functioning as on-screen boss, evil ringmaster, and occasional wrestler. His wife, Linda McMahon, holds the position of CEO (their adult children, Shane and Stephanie, are vice-presidents of, respectively, the Global Media and Creative Writing divisions). Under Vince and Linda McMahon's control, the WWE has come to hold a virtually unassailable monopoly over an entire entertainment genre. As financial analyst Robert Routh explained, "It would be too capital intensive for anyone to recreate what WWE has."[3] Eventually, the WWE could conceivably launch its own cable channel, rather than continue as a seller of content to media giants. And why shouldn't it? According to another industry observer, the company remains in a remarkably advantageous position: "WWE has practically no debt, a big pile of cash, a stranglehold on a niche market and an annual dividend yield of about 4.5 percent."[4]

Despite the WWE's command of not only a TV genre but virtually an entire industry, its growth and scope might not have been easily predicted. Professional wrestling has appeared on television since the medium's earliest years: the broadcast network NBC aired wrestling in a two-hour block on Wednesday evenings during the 1948–1951 seasons. Subsequently, most televised wrestling originated from the local and regional markets. For example, Atlanta-based WTCG (later WTBS) began to broadcast Georgia Championship Wrestling matches (GCW) in 1972, long before media mogul Ted Turner turned the struggling UHF station into the "superstation" TBS, uplinked by satellite to cable markets nationwide. But wrestling never seemed so pervasive across the television spectrum as it came to be in the 1990s. Nor had wrestling's popularity ever seemed so volatile, as audiences at times grew rapidly, then just as dramatically declined. For some time, World Championship Wrestling (WCW) dominated the market, especially after it was acquired in 1988 by Turner, who scattered wrestling shows across TBS's schedule. The centerpiece of the line-up became *WCW Monday Night Nitro* (1995–2001); its runaway success helped to make the WCW such an attractive property that it was acquired by Time Warner along with Turner's media assets in 1996. *Nitro*'s ratings peaked spectacularly in 1997, at the height of a fierce rivalry with the World Wrestling Federation (WWF), the original name for the WWE. Then, *Nitro*'s viewership plummeted to a small fraction of its former size.

The WCW lost its battle with the WWF on a number of fronts. The WWF had slipped into TV markets with *WWF Superstars* on the cable channel USA Network (1984–1996), the syndicated *WWF Wrestling Challenge* (1986–1994), and *WWF Saturday Night's Main Event* on NBC (1985–1992). But its big splash began with the launch of *WWF RAW* in 1993.[5] *RAW* originally aired on the USA Network, moving in 2000 to TNN, which would be rebranded by its owner, Viacom, as Spike TV in 2003.[6] *RAW* was scheduled on Monday evenings, to compete directly for *Nitro*'s viewers, and skyrocketed to the tops of cable-ratings charts. In 1999, seeking to capture viewers who may not subscribe to cable or satellite services —and gambling that some of its viewers would simply never get enough —the WWF added a Thursday primetime series, *SmackDown!*, on the Viacom-owned broadcast network UPN. In addition to the smash-hits *RAW* and *SmackDown!*, the WWE/WWF spun out related titles including *Sunday Night Heat* (1998–), the syndicated recapper *WWE The Bottom Line* (2002–), and the behind-the-scenes *WWE Confidential* (2002–2004).

The WCW's loss in the ratings game led to its demise. In 2001, McMahon acquired his rival from AOL TimeWarner. The WWF (soon to become the WWE) came then to nearly monopolize professional wrestling. But without the raucously exploited WCW-WWF rivalry, audiences seemed to lose interest.[7] Viewership for *SmackDown!* dropped from a pre-merger 4.4 million at its 2000 peak to a post-merger 3.2 million in 2002, with summer viewership down 25 percent despite the fact that, unlike most of TV's other entertainment genres, there are no reruns, even of popular programs *SmackDown!* and *RAW*. Among the key demographic—men ages 18–34, who are coveted by advertisers and perceived as difficult to reach with other TV genres—ratings for *SmackDown!* fell 41 percent from 2000 to 2003, and *RAW*'s ratings dropped 28 percent over the same period.[8]

Still, the ratings slide would not unhinge the WWE's grasp on the market, in part because the enterprise operates according to a formula that depends on its saturation of multiple TV markets. This formula is virtually unique in today's television world. Instead of supplementing revenue with ads sold during re-runs, the WWE's extensive library of programming is monetized through home video/DVD sales, pay-per-view events, and, beginning in 2005, Video on Demand. The television programs are shot live at events where audiences pay for tickets; thus, WWE collects double, in ticket sales and television rights. Fixed studio facilities are unnecessary. Rather, the programs are shot in stadiums and sports arenas across the United States and Canada, as well as in Germany, the United Kingdom, Japan, and Australia. In 2004, home video revenues were nearly $6 million in a single quarter in which a popular title of highlights of the career of wrestler Chris Benoit sold 65,000 units—an impressive figure in such a short period of time for a nontheatrical title. In addition, pay-per-view events, such as *WrestleMania, Summer SmackDown!, No Mercy, Unforgiven*, and *Valentine's Day Massacre*, are all sold quickly as home videos. In 2005, the WWE moved RAW from Spike to the USA Network, whose parent company, NBCU, would also begin screening its programs on its Spanish-language broadcast network Telemundo.

Listening to the boys in my class discuss wrestling, I realized how precious and how rare access to cable television was in their neighborhood. Their viewing of WWE programming was mostly confined to the series *SmackDown!*, since it was aired by UPN, a free-to-air broadcast network. Their idea of highly prized viewing was cable, or better yet, pay-per-view events. Consider this conversation between an Anglo, middle-class college student visiting the class, whom I'll call Hunter, and the students Américo and Andre:

> *Américo*: WWE is where almost all the best wrestlers are. And today at 9 to 11 o'clock there's *SmackDown!*
> *Hunter*: Can you watch *SmackDown!* tonight on regular TV?
> *Américo and Andre*: Yeah.
> *Hunter*: So you guys have cable at home?
> *Andre*: No.
> *Hunter*: Neither do I.
> *Américo*: I have [cable] but I don't have pay-per-view.
> *Hunter*: So you can't watch the good matches unless you have cable huh?
> *Andre*: Yeah.
> *Américo*: And pay-per-view.
> *Hunter*: So even if you have cable you still have to pay to watch some of these matches.
> *Andre*: Yeah.
> *Hunter*: So do your parents buy that for you sometimes?
> *Américo*: No, sometimes I go to my friends' house and watch it there.
> *Hunter*: So some of your friend's parents pay for pay-per-view?
> *Américo*: They have the box.

This exchange expresses some of the intensity of the boys' desire to have access to pay-per-view. It also reveals Hunter's lack of understanding of the entire value system the boys' share in relation to coveted but unaffordable cable content. For Hunter it is a sign of education to eschew watching television or even having much access to it. He is unfamiliar with the whole system of pay-per-view events or the technology required. Hunter also assumes that wrestling is something only children watch ("do your parents buy that for you?"), when in fact, the boys' fathers would be as likely to spend money on wrestling matches or WWE goods for themselves, since wrestling is a cross-generational family activity in their households.

WWE on the Web

McMahon revitalized the business of professional wrestling by dropping the pretense that it is a sport, dubbing its productions "sports entertainment," and highlighting its soap-opera features through backstage dramas, fixed matches, and conflicts between managers and wrestlers. In doing so, McMahon created a perfect serial narrative to gratify website regulars, who encounter celebrity publicity supplemented by fictional material about the status of its stars' careers. A hybrid of traditional publicity and celebrity biographical material, WWE's overwrought narratives concern the wrestlers, their rivalries, and their struggles with management and change on a weekly and sometimes daily basis. The WWE website provided the boys with what they wanted and kept them coming back to the site by updating the content continually.

It also functioned unusually well from a technological standpoint. Most other websites performed only intermittently at our lab. In the early years of this millennium, required plug-ins still caused frequent technical glitches. Often websites did not work well on Macintosh computers or froze the machines. The WWE, as part of its global ambition and targeting of a fan base in the armed services, goes to great effort to provide different downloading options for different machines, access bandwidths, and software operating systems. The WWE site had its audio-streaming abilities up and functioning seamlessly in 2000, before music downloads were a commonplace of commercial web design. Video downloads are available and updated regularly, and instructions for downloading plug-ins to play them are easy to understand. These video streams—exceedingly popular with the boys—are offered at variable frame rates designed to suit slower CPUs and connections. The WWE is still one of the most sensitive sites to the fact that web surfers may not all have the latest computers or fastest connections, and this has served it well in its international expansion. In 2005, the WWE began offering expanded trivia games to publicize its 7,000 hours of wrestling matches to be made available through a video-on-demand service called WWE 24/7. This expansion of gaming capacity is tied explicitly to WWE's efforts to help cable systems sell high-speed Internet and digital video services. According to Tom Barreca, Senior Vice President, WWE Enterprises, "We also believe that nothing sells the benefits of high-speed better than video, animation and tremendously compelling content and interactivity. Downstream, we believe this killer promotional WWE 24/7 content should lead to other free-standing broadband and iTV projects and services that our fans can play and use through next-generation set-top boxes and other devices."[9] The WWE currently claims 7 million unique visitors to its website per month, with over a million players of its interactive games.

There is another important factor that contributes to the perfect fit between wrestling and the web. Given the disparaged nature of wrestling as a popular culture genre, the Internet proved to be an exceptionally useful space for fan activity. As folklorist Terry McNeill Saunders writes:

> the Internet has reinforced the outsider/insider aspect of being a wrestling fan. No one in your family may understand why you like it, but there are hundreds of people to talk about it with you every day. . . . This need to defend oneself engenders a perverse pride in wrestling fans. This pride stems from persevering in the face of constant ridicule from spouses, family, friends, and co-workers.[10]

The WWE's Latino Heat

In the United States, the WWE has focused on the Latino audience, among whom it claims *SmackDown!* is the highest-rated cable show in the United States. Increased publicity for wrestlers with ties to Mexico, such as Eddie Guerrero, Chavo Guerrero, and Rey Mysterio, is one sign of this effort. In addition, the WWE is exploiting interest in the *lucha libre* tradition through its promotion of the wrestler Rey Mysterio. Initially, the phrase *lucha libre* was never used in profiles of Mysterio found on the WWE web-

Fig. 16.1. Toy-sized wrestling ring bears an image of the popular Rey Misterio, Jr.

sites and in *SmackDown! Magazine,* but the emphasis on his mask, inces-
sant repetition of the term "high-flying," and references to his small size
placed him clearly in the Mexican tradition. Since 2004, increasing WWE
publicity has been devoted to the proud traditions of Mexican wrestling
and to the connections of its "Latin" wrestlers to their "Spanish" heritage.
Eddie Guerrero and Mysterio are now featured on *SmackDown! Magazine*
covers two or three times a year.

According to the *SmackDown!* website, Mysterio lives in San Diego
County (in National City, which is close to the border between California
and Mexico) but has moved back and forth between the United States and
Mexico. He speaks fluent Spanish and enjoys a huge Mexican (and Japanese) following. Mysterio was trained by his uncle in Mexico City. He is
small (5'6", 165 pounds), but has the capacity to beat much larger opponents through his wrestling style:

> Blinding speed and springboard leaps allow Mysterio to run circles around
> most opponents with ease. . . . Acrobatic attacks and stunning counter

moves spring off the ring ropes like an Olympic diver. . . . Fires off a relentless offense of high-risk maneuvers. . . . Atomic-charged wrestling artist that
leaves them limp with his signature Huracanrana move. . . . Has been called
the best pound-for-pound wrestler in the business. *SmackDown!*[11]

Each aspect of the Mysterio character translates into merchandise and a
stylistic association with southern California urban youth and Chicano
styles. Mysterio wears a silver cross (a replica is available for sale through
the website), displays a large tattoo across his abdomen that reads "Mexican," enjoys hip hop, and calls his signature move the "619," which is the
area code for the southern half of San Diego county that extends to the
Mexican border. He began his career in Tijuana, initially wore a mask in
the United States, and was unmasked in the ring in 1998, but has returned
to wearing a mask.

In Mexican *lucha libre,* the wrestlers, called *luchadores,* perform with
tremendous acrobatic skill, using moves that involve leaps into the ring
and tumbling stunts and are both more athletic and more like gymnastics
than the moves that are performed in the WWE. The *luchadores* themselves
are much smaller and lighter than most WWE wrestlers, who appear to be
heavy steroid users and typically weigh over 250 pounds. The worst thing
that can happen to a *luchadore,* unlike the bloody pummeling of the WWE,
is unmasking. *Lucha libre* is associated with working-class urban audiences and its themes of overcoming corrupt, authoritarian, and oppressive
forces. Just as American professional wrestlers are conventionally characterized as either "babyfaces" (good guys) or "heels" (bad guys), *luchadores*
come in two types: *tecnicos* (the equivalent of babyfaces) and *rudos* (heels).

Some critics have seen in the WWE themes of social injustice that resonate with the *lucha libre* tradition. Media scholar Henry Jenkins argues
that WWE "stories hinge upon fantasies of upward mobility, yet ambition
is just as often regarded in negative terms, as ultimately corrupting. . . .
Virtue, in the WWF moral universe, is often defined by a willingness to
temper ambition through personal loyalties, through affiliation with others, while vice comes from putting self-interest ahead of everything else."[12]
Wrestling has traditionally been a cultural forum for slurs, stereotypes, entrenched ethnic rivalries, and bald racial prejudice. According to folklorist
Saunders,

In the 1960s and 1970s heel wrestlers thought nothing of insulting entire
ethnic groups of fans to "draw heat," that is to make people very angry at

them. One wrestler was famous for a routine where he speaks gibberish that sounds like a foreign language to wrestlers who are (supposedly) non-English speakers and their fans in the audience. In New York, he would often grab the microphone from the announcer, and say that he wanted to speak a few words to the Puerto Rican fans. Then he would begin speaking with an inflection and accent that sounded vaguely like Spanish, enraging the Spanish-speaking fans in the audience, which was his intent all along. At the time, he was a "hell manager," and Pedro Morales was the babyface champion, beloved by all wrestling fans and the pride of the Latino community. He would say at the end of a show that next week he was going to bring his wrestler in and show all the "Mexican garbage truck drivers" that their champion was no good.[13]

In the WWE, racial insults have been reinvigorated. Storylines involving Rey Mysterio typically contain narrative arcs that take up themes of insult and respect. As Jenkins argues, "The plots of wrestling cut close to the bone, inciting racial and class antagonisms that rarely surface this overtly elsewhere."[14] Mysterio figured prominently in a live WWE *SmackDown!* event held in San Diego in 2003. The scripted conflict that was the centerpiece of the show was between Brock Lesnar—a heel represented as cheating for his title in collusion with the bad-guy *SmackDown!* manager—and Rey Mysterio.

Lesnar taunted the audience, in a manner reminiscent of Saunders's description of "drawing heat" from Latino audiences. Over and over again, he referred to the match as being held in Mexico, and when corrected by his manager, Lesnar claimed that he could not tell the difference, given the look of the audience. As the crowd booed, he turned on them with the threat that if he did not get any respect from the audience, he would get all of them deported. The audience, at least two-thirds Latino, and about a third composed of children, held handmade signs reading "619" against a background of the Mexican flag as well as others reading "Viva La Raza" and "Latino Heat." Then Mysterio appeared to tremendous cheering from the crowd and challenged Lesnar to a match, saying he could not let him disrespect his people.

Eddie Guerrero, whose signature wrestling move was called "Latino Heat," was developed by the WWE as a heel with an overt anti-Mexican image: his slogan, set to music with a flashy entrance video, was "Lie, cheat, steal." When he appeared at live events, he was driven into the arena in a classic "lowrider" convertible with hydraulic lifts, filled with gorgeous women. Much of Guerrero's backstory concerned his prowess as a "Ladies' Man," and he was typically pictured in publicity photos surrounded by four or five women. With the WWE's greater interest in the Latino viewership, however, his image underwent considerable softening prior to his death in 2005, at age thirty-eight, from heart failure.

A January 2004 *SmackDown! Magazine* cover story suggested that Eddie Guerrero was a highly talented wrestler (he was usually portrayed as a cheater in all matches) who was grossly underrated. His lack of career success was attributed to prejudice by management thwarting his career advancement:

> There are many reasons one wins a world championship, and not all of them have to do with ability. Given the chance, Eddie would have several WWE championships by now.
>
> Rumor has it that someone very high in the WWE pecking order has something against Latino Heat and has enlisted a number of Superstars to keep Eddie down. Based upon the past few months and all that has happened to Guerrero it seems more than likely that there is some type of conspiracy to keep Eddie form achieving the ultimate prize.[15]

This is typical of the versatility of wrestling's stereotypes and of the new WWE twist of soap-opera plots involving conflicts between wrestlers and management. Guerrero won the championship shortly after this article appeared, in a match against Brock Lesnar, which once again pitted a relatively small (Guerrero was 5'8") Mexican-American against the hulking Aryan-archetype Lesnar who, at 6'3", weighs in at close to 300 pounds and is always costumed in a wrestling suit that bears the American flag.[16]

The Rock was, for a long time, the WWE's biggest star and probably held the greatest international appeal. He was also one of the very favorite wrestlers of the boys in the after-school computer class. His implacable demeanor, enormous size, and insistence on getting respect offered fantasies of power that spoke to the boys' denigrated position and feelings of powerlessness at school. They especially enjoyed the power of The Rock vis-à-vis leather-bound Aryan types such as the Undertaker, A-Train, and Brock Lesnar. While other wrestlers exerted their own fascination as figures of contempt or hatred, The Rock, whose real name is Dwayne Johnson, was always treated with great reverence; his photos were regularly downloaded and posted on children's walls at home or made into special posters or used to adorn their binders. For the boys in my class, following The Rock's career involved the memorization of matches, opponents, and signature winning moves. Consider, for example, this story written by Américo Reta and José Garcia, entitled "Is The Rock the Best Wrestler?"

> We tried to find out if The Rock or Stone Cold Steve Austin is the best wres-
> tler in the WWE, The Rock is called "The People's Champion." He is from
> Florida and is the current champion. There is a debate over whether he is
> really a better wrestler than Stone Cold.
>   The Rock lost to Stone Cold in the past but defeated him in a rematch.
> Stone Cold has more experience and has wrestled longer than The Rock.
> In both matches between the two, Stone Cold wrestled with a serious knee
> injury. Despite the injury, he beat the Rock, but was unable to overcome
> his injury and lost in the rematch. The Rock lost the championship on
> 10/22/00. He lost in a match against Kurt Angle. The Rock lost his champi-
> onship belt because Rakishi put his butt on The Rock.[17]

The Rock was singled out as the figure of continual fascination for his expressions of ethnic pride combined with his hypermasculinity. His imposing physique and classically handsome face are linked in his publicity materials to the constant theme of ethnic pride. In his autobiography, for example, the contents of which are endlessly reiterated on fan websites, The Rock continually refers to his racial identity, his respect for his African-American father and Samoan grandfather, both of whom wrestled professionally, and his hatred of racism—especially racism coming from poor whites. Moreover, Johnson's success in feature film production attests to some of his unique qualities as a performer. Hollywood seems to promote—and fans seem to accept—The Rock as a figure of pan-ethnic Latino identity, despite the fact that Johnson's only ties to Latino culture are his marriage to a Cuban American. My students clearly identified strongly with The Rock as a man of color, one who had no fear of standing up to white racism and who scrupulously maintained that honor and respect were the most important elements of his identity. Every student—wrestling fans and not—had viewed all of The Rock's movies repeatedly.

Websites devoted to The Rock celebrate the highlights of his career, and the boys in my class memorized these stories. Throughout these stories, he has emphasized strong connections to family members; absolute contempt for racism of any kind; and a frank discussion of the ways that his own life has been shaped by his racial and ethnic heritage. As The Rock tells the story of his father, Rocky Johnson:

> When my dad broke in, all of the top black wrestlers—not to knock them or anything—but they were all jive-talking caricatures. They'd come out to cut their promos and you'd swear they'd just stepped off the set of *Shaft* or *Superfly*. They'd eat watermelon on camera and do all sorts of degrading things, because that's what was expected of them. My father wouldn't do that. He was the first black wrestler to insist on being very intelligent in front of the camera. When he cut his promos, there would be no jive in his voice. . . . And when he stepped through the ropes, there was no bullshit funk strut or anything like that. . . . He was a fearless man, a man who welcomed the chance to break down barriers.[18]

In describing his courtship of and marriage to his wife, Dany, he notes the obstacles stemming from her parents' prejudices: "Her parents were Cuban immigrants. . . . [T]hey wanted their children to assimilate, to adapt and succeed. What they did not want was their daughter dating me, mainly because I was a person of color. So was Dany, of course, but that didn't seem to matter. I was half black, and that made me an unsuitable suitor."[19] These are stories that speak very directly about the experience of racism and the complexities of negotiating identity as a man of color.

While The Rock always appeared as a hero to the children in my class, his career has included transitions from a babyface to a heel. When The Rock switched to playing a heel, a role developed through a central conflict with Stone Cold Steve Austin, his popularity soared. The Rock would assail white crowds at WWF events with comments like: "There are twenty thousand pieces of trailer-park trash here tonight!" The Rock offers fans a figure of ethnic hypermasculinity, athleticism, and overt hostility against Anglos who insult him. At the same time, his star image emphasizes close ties to his extended family: his book is dedicated to his mother, "the strongest person I know."

WWE Fights the Ratings War, at Home and Abroad

After wrestling's heyday in the 1990s, the WWE struggled against flat TV ratings and U.S. ticket sales both by aggressively expanding its pay-per-view and international events and by diversifying its media holdings. In

2002, WWE spun off a film division, hiring Joel Simon, one of the producers of *X-Men*, to head the operation whose goal was to find vehicles for WWE stars who might follow in The Rock's footsteps to Hollywood stardom. Previously, the WWF, as it was still known, earned production credits on *The Mummy Returns* (2001) and *The Scorpion King* (2002), due largely to the fact that it owned the rights to The Rock's name. WWE Films was credited more prominently on The Rock's next two movies, *The Rundown* (2003), a Universal Pictures/Columbia Pictures co-production, and the *Walking Tall* remake (2004), from Metro-Goldwyn Mayer. Both were moderate successes that prompted WWE Films to develop feature film projects for Stone Cold Steve Austin, John Cena, and Kane. Yet the success of the film division weakens the cable franchise by drawing the biggest stars away from live events and therefore from television. In 2005, Dwayne Johnson repurchased the rights to the name "The Rock" and greatly reduced his appearances on WWE shows.

According to financial analysts, the lack of a big star has slowed growth of the company, which had its initial stock offering in 1999. In many ways, the WWE served as a paradigm for other television reality shows in its exploitation of cheap on-air talent, who are in many respects interchangeable and disposable. Yet only big stars such as The Rock lead viewers who are not routinely interested in professional wrestling to tune in or consider attending live events. Thus the lowly status of wrestlers and the eager pool of unknowns desiring a chance in the ring keep salaries low, but the lack of a higher-priced star leads to a downgrading of the company's financial interests overall.[20] It is doubtful that the WWE roster includes many who can duplicate The Rock's success as a pan-ethnic hero, attractive to male and female fans, and to fans across a range of race and ethnic backgrounds. Even Austin, with his violent, working-class macho image (his career stalled due to injury as well as alcohol-related problems and criminal prosecutions for domestic violence), may be less successful in today's more global WWE market and more heavily Latino U.S. audience base.

In the attempt to shore up failing TV ratings, McMahon has increased the sensationalistic elements of the programs, exploiting not only racial and ethnic tensions but also titillating audiences with sexually provocative storylines that sometimes feature same-sex relationships between wrestlers. The promotion of female wrestlers, catfights, and WWE Divas has accelerated since the dip in ratings that began in 1999. A 2002 script called for WWE Divas Dawn Marie and Torrie to meet in the ring following an (offscreen) sexual tryst at a hotel; Torrie would seek to exact revenge on Dawn Marie, who had blackmailed her into the liaison. Their ensuing catfight was part of the wrestling "card" for that week's televised event. Although the show followed such a "coerced lesbian sex" plotline, Vince McMahon was worried enough to call for the omission of the word "lesbian" from publicity materials.[21] As executive producer Kevin Dunn explained the strategy:

> We watch our ratings minute by minute. We know our ratings go down when a match ends. We know viewers, men especially, are flipping around, back and forth, and when they see something they like, they stay with it. And our viewers are really interested in women. Dawn's interaction with Torrie tonight will be the second-highest rated event of the evening, after the four-way tag team match.[22]

For the summer 2005 season of *RAW,* the WWE began a talent search for the new *RAW* Diva, introducing eight finalists and a competition to run from June 27 to August 15, when the winner was announced. Ratings for the first episode of the Diva search were the highest for *RAW* in well over a year.

McMahon (and his writers, wrestlers, and fans) has frequently acknowledged that the WWE's brand of "sports entertainment" blends elements of several other genres, media, and skills: it is action-adventure, soap opera, and comedy; "variety show" and "performance art"; stuntwork, acrobatics, and acting.[23] Together, these elements serve a formula that tends to zero in on areas of social conflict—especially on the fault-lines of social conservatism, on the one hand, and political correctness, on the other. Thus, for example, as ratings were slipping in 2002, a heel tag team, Chuck and Billy, became engaged and a "gay commitment ceremony" was scheduled for *SmackDown!*:

> The plot took a wholly unforeseen twist when Chuck got down on one knee and asked Billy to be his lifepartner on an episode of *SmackDown!!* "It was fun as hell. People weren't booing us anymore. They were straight up loving us. We did *Good Morning America.* We did *Howard Stern.* Everybody wanted in on it." The world was enraptured when Billy and Chuck stepped into the ring for their much awaited commitment ceremony. To the dismay of many, it was revealed that Billy and Chuck weren't gay after all. The whole thing was a publicity stunt.[24]

Wrestling addresses the dark side of the American obsession with competition and achievement. It is sports for those who know they will never be included in the NBA or the NFL. As Terry McNeill comments on the outsider status of much of the wrestling audience:

> wrestling seems to reflect back to them their own difficulties and problems
> . . . and give voice to the cultures and classes that are frequently ignored and
> neglected. Wrestling turns on its head the very notions of fair play and good
> sportsmanship that American culture says it reveres—whether it be on the
> baseball field or on the battlefield.[25]

When calls for censorship of the WWE are launched, the outrage expressed is not only based on the bare skin, violence, and vulgarity of its cable programs—what *GQ* referred to as its "porniness."[26] The objections also stem from recognition of something essentially unpatriotic about its jaded depiction of competition. When the Parents Television Council (PTC) launched a campaign against the then WWF on the basis of overt violence and sexuality, they also accused the organization of creating disdain for authority figures, patriotism, and religion. In response, McMahon launched a storyline featuring a team of wrestlers known as the Right to Censor, whose antics, according to Kathleen S. Lowney, satirized wrestling's critics. The Right to Censor "was portrayed as obsessed with cleaning up the WWF," appearing to endorse the PTC's campaign against wrestling's violence and sexual themes, even as they acted as the worst of heels themselves and "would cheat with impunity."[27]

Conclusion

Despite the fickleness of audiences, slipping ratings, a decreased number of television hours per week, and the rapidly cyclical booms and busts of pop culture trends, McMahon is a survivor. He has successfully bought out every competing league in the last decade, evaded criminal prosecution for providing steroids to wrestlers, and led the company, now the largest programmer of pay-per-view events in the world, toward intense global outreach. Recently, the WWE hired an Olympics expert to increase the lucrativeness of WrestleMania by soliciting bids from competing cities, scheduling events years in advance, and planning an extravagant range of supporting local activities.

The WWE used cable, satellite, pay-per-view, the Internet, and home video sales to push aggressively into an international market and cope with domestic declines. By 2002, the WWE boasted that its televised wrestling programs were available in 10 languages in 130 countries. In the same year, the WWE produced 300 live events in the United States, even as the popularity of these events waned. According to one report, ticket sales in 2002 slumped by 17 percent.[28] International events are better-attended, command higher ticket prices (roughly $70 per ticket, as compared to $35 in the United States) and sell twice as much merchandise; audiences are growing especially in the United Kingdom, South Korea, and Australia, while the WWE is also trying to expand its audience in Japan. One sign of its confidence in the global marketing strategy is the fact that it signed a new five-year contract with SKY TV in 2004. After forays into India, Malaysia, and the Middle East starting in 2000, the WWE refocused its efforts to stage live events in countries where Rupert Murdoch's satellite channels dominate sports programming, as well as in Korea and Japan. In the summer of 2005, weekly WWE programming was available on Sky Italia, on Fox 8 in Australia, and on Sky Sports in the United Kingdom.

In India, the WWE renewed its contract with Ten Sports in 2005 for another five years. Ten Sports broadcasts to thirty million households via cable or satellite, almost half of the total TV households in India. According to Chris McDonald, CEO of Taj Television Ltd., owners of Ten Sports,

> We are delighted to extend our very strong relationship with the WWE. Ten Sports' experience with the WWE is that the programs appeal to every demographic sector of our viewer base, including some very surprising ones such as housewives in India. We look forward to building upon our successful partnership with WWE by introducing new and innovative promotional angles around the programs.[29]

The next newly targeted region for staging live events is Central America.

While the WWE has perfected a mix of television programming, live events, and merchandising that is admirable in today's cable marketplace, it faces enduring challenges as a business entity in pursuit of global expansion. Some analysts doubt that its popularity will continue in the international marketplace, after the exoticism of the WWE has worn off. The WWE's success in the future seems to depend on its adeptness in navigating highly charged social and political territory: How will its enactment of ethnic taunts play in a world increasingly troubled by war and terrorism?

How will its depiction of female wrestlers be received in nations where religious fundamentalism is on the rise? How entertaining will the display of
American brute force be in a world where the actions of U.S. military are
increasingly coming under scrutiny for human rights violations? Back in
the United States, will the WWE be able to continue to create characters
and storylines that appeal both to its core audiences—mostly male, Latino, and youthful—and to new potential audiences? Finally, will the
WWE be able to address its domestic and global audiences with the same
content, through the same media, simultaneously?

NOTES

1. Ellen Seiter, *The Internet Playground: Children's Access, Entertainment, and
Mis-Education* (New York: Peter Lang, 2005).

2. When the after-school program began, the WWE was still known as the
World Wrestling Federation (WWF). In 2002, the World Wide Fund for Nature
won a long battle in the British courts, enabling the charity to continue to use the
acronym WWF and forcing the wrestling organization to rename itself. See "Wrestling Body Submits to Name Change," BBC News, May 8, 2002, http://news.bbc.co
.uk/1/hi/uk/1974600.stm (accessed August 11, 2005). This essay uses both acronyms
as needed for clarity or historical accuracy.

3. Katie Benner, "WWE: Get in the Ring!" CNN/Money, May 20, 2005, http://
money.cnn.com/2005/05/20/markets/spotlight/spotlight_wwe/index.htm (accessed
July 10, 2005).

4. Ibid.

5. *WWF Monday Night RAW* (since 2003, *WWE Monday Night RAW*) underwent a series of adjustments to its title, responding to the logic of TV industry
economics and to external events. Changed to *RAW Is War* in 1997, the title is said
to have been shortened in response to the terrorist attacks of September 11, 2001,
to simply *WWF RAW.* For many years, the second hour of the two-hour program
was referred to as *The War Zone* or *The RAW Zone,* so that its advertising time
might be sold separately, taking advantage of higher rates if audiences tended to
tune into one hour more than the other.

6. Not only did professional wrestling come to TNN in 2000; the same year,
TNN changed hands and underwent rebranding. Grand Ole Opry and Opryland-
owning Gaylord Entertainment sold TNN to Viacom. TNN, which originally
stood for The Nashville Network, was known henceforth as The National Network, disassociating the channel from its country-music roots.

7. The WWE's monopoly has not been seriously threatened. Though a few
hopeful independents entered the market, none has met with much success: *WXO*

Wrestling (2000), for example, managed to distribute only three syndicated episodes. Competitors sought unsuccessfully to dislodge the WCW and later the WWE's dominance in the arena of women's professional wrestling with several short-lived syndicated series, such as *GLOW: Gorgeous Ladies of Wrestling* (1986), *POWW: Powerful Women of Wrestling* (1987), and *WOW: Women of Wrestling* (2000–2001).

8. Paige Albiniak, "WWE Feels the Pain: Wrestling Stumbles and Looks for a New Brute Star," *Broadcasting & Cable*, November 17, 2003, 1.

9. Ibid.

10. Terry McNeill Saunders, *Play, Performance and Professional Wrestling: An Examination of a Modern Day Spectacle of Absurdity*, Ph.D. dissertation (University of California, Los Angeles, 1998), 209.

11. World Wrestling Entertainment, Inc.: Rey Mysterioso Profile, smackdown .wwe.com/superstars/mysterio/index.html (accessed December 8, 2003).

12. Henry Jenkins, " 'Never Trust a Snake': WWF Wrestling as a Masculine Melodrama," in Aaron Baker and Todd Boyd, eds., *Out of Bounds: Sports, Media, and the Politics of Identity* (Bloomington: Indiana University Press, 1997), 59.

13. Saunders, *Play, Performance and Professional Wrestling*, 5–6.

14. Jenkins, " 'Never Trust a Snake,' " 66.

15. Aaron Williams, "Eddie Guerrero: SmackDown's Best-kept Secret," *Smack-Down! Magazine* 23, no. 1 (January 2004): 20.

16. Guerrero's nephew Chavo Guerrero has also received a flood of publicity that links the Guerrero family to decades of the history of Latino wrestlers in the United States. See Keith Elliot Greenberg, "Latino Legacy: Eddie and Rey Carry on the Storied Tradition of Morales and Mascaras," *SmackDown! Magazine* 23, no. 12 (December 2004): 18–27.

17. Américo Reta and José Garcia, "Is The Rock the Best Wrestler?" *Normal Heights Kids* 2, no. 2 (Spring 2001).

18. Joseph Layden, *The Rock Says: The Most Electrifying Man in Sports-Entertainment* (New York: Reagan Books, 2000), 10–11.

19. Ibid., 117.

20. Katie Benner, "WWE: Get in the Ring!" CNN/Money, May 20, 2005, http:// money.cnn.com/2005/05/20/markets/spotlight/spotlight_wwe/index.htm (accessed July 10, 2005).

21. Phil Kloer, "As the Smackdown Turns," *Atlanta Journal-Constitution*, December 12, 2002, 1G.

22. Ibid.

23. Ibid.

24. John Christofferson, "World Wrestling Entertainment Takes Beating in Ratings," *Associated Press Wire*, November 29, 2002.

25. Saunders, *Play, Performance and Professional Wrestling*, 209.

26. Andrew Corsello, "I Am The Rock," *Gentleman's Quarterly* 57, no. 10 (December 2003): 206–11.

27. Kathleen S. Lowney, "Wrestling with Criticism: The World Wrestling Federation's Ironic Campaign Against the Parents Television Council," *Symbolic Interaction* 26, no. 3 (Summer 2003): 429.

28. Tara Ebrahimi, "Between The Rock and a Hard Place," *Seattle Times*, August 12, 2002, E1.

29. "World Wresting Entertainment® and Ten Sports Sign New Five-Year Agreement in India," World Wrestling Entertainment, Inc.: 2005 News Releases, February 2, 2005, http://corporate.wwe.com/news/2005/2005_02_02.jsp (accessed August 12, 2005).

About the Contributors

Christine Acham is Associate Professor in the Program for African American and African Studies at the University of California, Davis. She is the author of *Revolution Televised: Prime Time and the Struggle for Black Power.* She co-edited an issue of *Screening Noir: Journal of Black Film, Television and New Media Culture* entitled "Blackploitation Revisited," and is currently on the editorial board of *Film Quarterly.*

Sarah Banet-Weiser is Associate Professor in the Annenberg School for Communication at the University of Southern California. She is the author of *The Most Beautiful Girl in the World: Beauty Pageants and National Identity,* and *Kids Rule!: Nickelodeon and Consumer Citizenship.*

François Bar is Associate Professor of Communication in the Annenberg School for Communication at the University of Southern California. He directs the Annenberg Research Network on International Communication. His research has been published in *Telecommunications Policy, The Information Society, Organization Science, Infrastructure Economics* and *Policy, Communications & Strategies, Réseaux,* the *International Journal of Technology Management,* and elsewhere.

Cynthia Chris is an assistant professor in the Department of Media Culture at the City University of New York's College of Staten Island. She is the author of *Watching Wildlife.*

Anthony Freitas has a Ph.D. in communication from the University of California, San Diego. He works as a media relations consultant for nonprofit organizations in San Francisco.

Cheri Ketchum received her Ph.D. in communication from the University of California, San Diego, in 2005, completing a dissertation entitled *Avoiding Anxiety: How Public Discourses Around Food Construct Lifestyle Realities.* She is currently working as a research analyst with the Culinary Union in Las Vegas.

Amanda D. Lotz is Assistant Professor of Communication Studies at the University of Michigan. She is the author of *Redesigning Women: Television After the Network Era.*

Katynka Z. Martínez is a postdoctoral researcher at the University of Southern California Annenberg Center for Communication and Assistant Professor of Raza Studies at San Francisco State University.

John McMurria received his Ph.D. from the Department of Cinema Studies at New York University in 2004 and is currently Assistant Professor of Communication at DePaul University. He is co-author, with Toby Miller, Nitin Govil, Richard Maxwell, and Ting Wang, of *Global Hollywood 2.* He is currently working on a critical cultural policy history of cable television in the United States.

Toby Miller is Professor of English, Sociology, and Women's Studies at the University of California, Riverside. His most recent books are *Global Hollywood 2* and *A Companion to Cultural Studies.* He is the editor of *Television & New Media.*

Megan Mullen is Associate Professor in the Department of Communication and Director of the Humanities Program at the University of Wisconsin, Parkside, and the author of *The Rise of Cable Programming in the United States: Revolution or Evolution?*

Lisa Parks is Associate Professor in the Department of Film and Media Studies at the University of California, Santa Barbara. She is the author of *Cultures in Orbit: Satellites and the Televisual* and co-editor of *Planet TV: A Global Television Reader.* Her new book project, *Mixed Signals,* explores uses of satellite and wireless technologies in societies in transition.

Dana Polan is the author of six books in film studies, including *In a Lonely Place, Pulp Fiction, Jane Campion,* and *Scenes of Instruction: The Beginnings of the U.S. Study of Film.* He is currently writing a monograph on Julia Child's *The French Chef* television program.

Ellen Seiter is Professor in the School of Cinema-Television, University of Southern California, where she holds the Stephen K. Nenno Chair in Television Studies. In 2006, Seiter produced a documentary entitled *Projecting Culture: Perceptions of Arab and American Film.* She is the author of *The Internet Playground, Television and New Media Audiences,* and *Sold Separately: Children and Parents in Consumer Culture.*

Katherine Sender is an assistant professor at the Annenberg School for Communication, University of Pennsylvania. She is the author of *Business, Not Politics: The Making of the Gay Market* and *Queens for a Day: Queer Eye for the Straight Guy and the Neoliberal Project.* She is also the producer, director, and editor of a number of documentaries, including *Off the Straight and Narrow: Lesbians, Gays, Bisexuals, and Television* and *Further Off the Straight and Narrow: New Gay Visibility on Television.*

Beretta E. Smith-Shomade is Associate Professor in the Department of Media Arts at the University of Arizona. Her teaching and research focus on television and film critical studies with particular attention to representations of race, gender, class, and generation. She is the author of *Shaded Lives: African-American Women and Television* and the forthcoming *Pimpin' Ain't Easy: Selling Black Entertainment Television.*

Jonathan Taplin began his entertainment career in 1969 as tour manager for Bob Dylan and The Band. In 1973 he produced Martin Scorsese's first feature film, *Mean Streets,* and between 1974 and 1996, produced 26 hours of television documentaries and 12 feature films. In 1984 Taplin acted as the investment advisor to the Bass Brothers in their successful attempt to save Walt Disney Studios from a corporate raid. Later, as Merrill Lynch's vice president of media mergers and acquisitions, he helped re-engineer the media landscape in transactions such as the leveraged buyout of Viacom. Taplin was a founder of Intertainer, the pioneer video-on-demand company, and has served as its chairman and CEO since June 1996. He sits on the advisory board of the Democracy Collaborative at the University of Maryland.

Index